Islam's quantum question

Nidhal Guessoum is Professor of Physics and Astronomy at the American University of Sharjah, United Arab Emirates. He has published widely on the mutual compatibility of science and the Islamic tradition. His book *Reconcilier L'Islam et la Science Moderne* was published in France in 2009.

'This book is essential reading for all those who wish to understand the relationship between Islam and science from both historical and contemporary perspectives. From Averroes to al-Ghazzali, and from Iqbal to Nasr, the author provides a well-informed survey and critique of the very different ways in which Islamic philosophers and scientists have contributed to the scientific enterprise. Muslims and non-Muslims alike will find that this fascinating overview fills a gap in the current literature on science and religion. Firmly committed to mainstream science, the author gives short shrift to those who attempt to find scientific truths hidden in different verses of the Qu'ran. Instead Prof. Guessoum sees the theistic framework as providing the basis for the intrinsic rationality and coherence of the universe, a framework within which the scientific enterprise can continue to flourish in a way that is consonant with religious belief.'

> – Denis Alexander, Director, The Faraday Institute for Science and Religion, St. Edmund's College, University of Cambridge

'This is a prophetic and brilliantly written book. Nidhal Guessoum writes as a devout Muslim. But the Islam that can reconcile religious tradition with science cannot be anti-modern and fundamentalist. Drawing from the tradition of Averroes, who combined revelation with reason, the author brilliantly sketches the vision of an integrative Muslim philosophy and theology for our time. *Islam's Quantum Question* is not merely one of the best books on Islam and science yet to appear. It is also a compelling description of the most urgent decision that Muslim intellectuals face today: Will Islam in the twenty-first century define itself in opposition to modernity? Or will it again become a leading voice around the world for integrating faith and reason, science and values?'

> – Philip Clayton, Professor of Religion and Philosophy, Claremont Graduate University, Ingraham Professor, Claremont School of Theology, and author of God and Contemporary Science and The Oxford Handbook of Religion and Science

'*Islam's Quantum Question* is a sensitive and nuanced account of Islam and science by an author who is intimately acquainted with both. Well written and thoroughly researched, it offers a lively and comprehensive introduction to both historical and contemporary issues. For those seeking guidance in a difficult and sometimes contentious field, Guessoum's stimulating book is to be highly recommended.'

> – Peter Harrison, Andreas Idreos Professor of Science and Religion, University of Oxford

Islam's quantum question

reconciling muslim tradition and modern science

NIDHAL GUESSOUM

I.B. TAURIS

LONDON · NEW YORK

Published in 2011 I.B.Tauris and Co Ltd
6 Salem Road, London W2 4BU
175 Fifth Avenue, New York NY 10010
www.ibtauris.com

Distributed in the United States and Canada Exclusively by Palgrave Macmillan,
175 Fifth Avenue, New York NY 10010

ISBN: 978 1 84885 517 5   (HB)
ISBN: 978 1 84885 518 2   (PB)

A full CIP record for this book is available from the British Library
A full CIP record is available from the Library of Congress

Library of Congress Catalog Card Number: available

Printed and bound in India by Replika Press Pvt Ltd

For Amir and Omar
and their generation . . .

contents

conventions and acknowledgements

The Islamic culture, firmly anchored on the fundamental texts and classical works, is intimately connected to the Arabic vocabulary. I have thus often had to use texts written in Arabic, and refer to names and book titles, for which the transliteration to English sometimes posed difficulties. I have tried to simplify things as much as possible, in order to make the reader at ease, even if the simplification often forced me to forego the academic rules of Arabic transliteration. For example, I have not tried to uphold the *med* (lengthening of vowels), which is usually rendered with a hat (â, î, etc.). It was necessary, however, to distinguish between the *hamza* (as in Qur'an), which I have transcribed by a simple apostrophe (') and the Arabic letter *'ain* (as in *'ilm*, knowledge/science), which I have rendered by an inverted apostrophe ('). Ambiguities arose sometimes between some Arabic letters which are close to one another and which do not exist in English, as for *dh* (like *dhikr*, invocation) and *Dh* (like *Dhulm*, injustice), which in the literature is sometimes rendered with a Z. I hope these all will be clear, and the reader will not be too strict in this regard. (We thank the readers who will kindly bring to our notice any slip or inconsistency.)

There was also the question of names, particularly those of the prophets and the ancient scholars. For prophets, I used their English language names (Jesus, Moses etc.), and for the Prophet of Islam I adopted Muhammad instead of other spellings. For the Muslim scholars, I generally used their

Arabic names (al-Ghazzali, not Algazel), except for the special case of Ibn Rushd, for whom I also frequently used Averroes, and for Ibn Sina (Avicenna, a few times), for two simple reasons: (a) to make both the Western and the Muslim readers at ease, as they are probably familiar with those names; and (b) to avoid repetition when, in some sections, I had to refer to Ibn Rushd/Averroes repeatedly.

Finally, for the Qur'anic verses, I consulted several translations (mainly those of Y. Ali, M. Pickthal and M. H. Shakir), and often 'interpolated' between them. It is thus a 'collective' translation of the verses that I have used.

One last remark regarding footnotes and endnotes: I have tried to remain as rigorous as possible in citing all sources and giving references to all ideas and information that I have presented, but I also wanted to make sure that comments did not clutter the main text, so I have put in footnotes for any clarifications that may help the reader follow the discussion, while I sent additional non-essential text and the bibliographical references to the endnotes section.

There are always many people to thank for their assistance or encouragement on a work of this sort; a number of them will not be named, but I would like them to know that they have not been forgotten and are thanked here collectively.

I must present my warmest thanks to the members of my family, starting with my wife for her unlimited support and patience. I have dedicated this book to my two sons, not just for the time and attention I took away from them to put in this work, but more importantly, because I hope this harmonising effort will help them (and their cousins, the close ones and the far ones, i.e. young Muslims) develop an identity, which is both authentic and modern, without dilemmas, contradictions or crises. My brothers and sisters have never stopped encouraging me; I am grateful to them. I must, however, single out one of them, Ahmed, and express my earnest gratitude to him for having – all these years but especially on this manuscript – constantly presented me with his constructive critiques of all my writings, including several chapters of this book. Finally, a word of deep recognition to my parents: my father who will find parts of him in this book story, and my late mother, who would have been proud of my efforts and intellectual integrity.

A number of friends should be mentioned here, even quickly, for the remarks they made on this or that chapter or for the references they pointed me to: Jamal Mimouni, Karim Meziane, Abdelhak Hamiche, Abdelhaq

Hamza, Mohamed Zayani, Sofiane Sahraoui, Boualem Boashash, Karim Abed, Bruno Abdelhaq Guiderdoni and Réda Benkirane.

In my professional world, there are also a number of persons I should express my thanks to, some more nominally than others. Jean Staune stands out with his constant encouragement and support. Others include Paul Wason, Philip Clayton, Denis Alexander, Ehsan Masood and Zia Sardar. At the American University of Sharjah, a few members of the administration and staff supported this project through parts or all of its development.

I hope this work will please all those who always showed me much respect, believed in my intellectual capacities, and encouraged me to make substantial contributions to the Islamic culture and nation.

prologue: Averroes and I

'Among all the thinkers of the land of Islam, none other [than Averroes] will have had a greater influence on humanity's universal culture.'

Alain de Libera, 2004

yesteryear: Averroes – the man, the times and the spirit

By the twelfth century, Cordova had become one of the main cultural centres of the world, rivaling Baghdad and Constantinople. It had existed for many centuries, as its rich urbanism and varied architectural styles showed, but only recently had it started to produce arts and sciences and attract students from Africa and Europe. Its rulers were proud of its landmarks: the medina (old town), the alcazaba (from the Arabic *al-qasaba*, referring to the Citadel), the Great Mezquita (literally mosque, now the Cathedral), the Madinat al-Zahra (the Versailles of Cordova), to mention only the most famous ones. Its inhabitants were happy to live in one of the rare world cities with paved streets, public baths, street lighting and public entertainment, with the regular impromptu recitals of poetry and music. And its religious groups

(Jews, Christians and Muslims) were satisfied with their peaceful, if not fully harmonious, coexistence.

The Cordova of that period was important not just in its prosperity and shining but particularly in its cultural and intellectual life. The 'jewel city of the world', as it was sometimes referred to, boasted of over 70 libraries, many of which carried hundreds of thousands of volumes. And the scholars who lectured at the Caliph's court and in the Great Mosque were to become some of the most illustrious in the history of the Islamic civilization: Ibn Baja (Avempace), Ibn Tufail, al-Idrissi, Ibn Arabi and Ibn Rushd (Averroes), to name only the giants.

The movie *Destiny*, which the world-acclaimed Egyptian director Youssef Chahine made in 1998, has great visual descriptions of the city and its cultural life. (The movie was made partly to commemorate the 800th anniversary of Averroes's passing and partly to show parallels between the liberal – fundamentalist clash of today and that of twelfth-century Cordova.) If anything, it exaggerates the pop-culture aspect of that time, with so much emphasis on music, dancing, food and other pleasures. Still, the movie scenes inside the Great Mosque, in villas, in the open book markets and, of course, in the royal palaces are visual feasts and accurate descriptions of the place and the time. Like today's sophisticated, cosmopolitan cities, Cordova lived in high style. Music and games, especially chess, were introduced as early as the ninth century (by Ziryab, the famous musician who perfected the lute and sang for emirs at the palace). Gardens and fountains abounded in villas and palaces. The commerce of books and clothes thrived.

In the first few decades and centuries that followed the Muslim invasion and control of Spain, many of the original Christian and Jewish families converted to Islam, being either seduced by its simple religious message or allured by the greater social possibilities offered to Muslims. Still, many retained their own faiths, cultural traditions and quarters; most of those who remained Christians, though, adopted the Arabic language and became known as mozarabes, a Latinised form of *musta'rib*, which is Arabic for 'Arabicized'. Many Jews and Christians were called upon to serve in the administration, at least in the technical professions (e.g. physician). In darker times and episodes, however, Christians and Jews had their churches and synagogues confiscated or turned into mosques and other public buildings; in some cases they were exiled or ousted from the land. One such famous case was the great Jewish scholar, doctor and philosopher Moses ibn Maimon, known as Maimonides (1135–1204), whose family had to flee Andalusia from the Almoravid dynasty's intolerance and persecution. (It is unclear whether Maimonides, who was only nine years younger than Ibn Rushd, did – as many accounts state – study at the hands of Averroes.) What is certain, however, is that Maimonides did not carry any hatred towards Islam; indeed, he chose to settle in Egypt, where he became the physician of the court. More importantly, whether from direct learning or through the books, Averroes's influence on Maimonides is evident and well established[1].

It is often claimed that Andalusia was a land of cultural 'symbiosis' (my own university has a course with that title). Yet, although one cannot subject those times to the modern definition of tolerance, it is perhaps more accurate to describe them as a period of precarious mutual acceptance of 'the other', mixed with fruitful intellectual rivalry and influence. Indeed, it has been shown that many books were translated between the cultures, and Jewish poetry, to mention only one area of influence, adopted many Arab metrics.

The dominance and the influence of the Arabo-Islamic culture on Andalusian Christians, in particular, cannot be denied. It is estimated that as many as 4,000 Arabic words entered European languages then in various areas of society, from culinary arts to irrigation and other sciences, and only a hundred or so were transferred in the reverse direction. Historians have also shown how in those medieval centuries, European royal courts were full of ladies dressed in *adorras* (buttoned tunics), *lihafes* (overgarments), *mobatanas* (fur-lined coats) and cordwains (Cordovan leathers).

It should also be noted that the ruling dynasty, the Almohads (1130–1269), despite clinging to a particular theological school, encouraged cultural and intellectual exchanges and supported thinkers and scholars.

the finest mind in Spain

In such a rich cultural climate was born, in 1120, Abul Walid Muhammad Ibn Ahmad Ibn Rushd, who later became known in the West as Averroes. By the twelfth century, the Ibn Rushd family had become a well-known intellectual dynasty of Cordova. Indeed, three successive generations – including Abul Walid – were to hold the supreme judge title and position in Cordova.

Abul Walid's grandfather, who as per tradition, had the same first name, had been (like his grandson would become after him) a scholar and high judge in Cordova and Seville, the main cities of the Andalusian kingdom; his Islamic jurisprudence books are still read and studied, and his *fatwas* (rulings) sometimes used as benchmarks. Indeed, students sometimes confuse the grandfather with the (greater) grandson. Ahmed Ibn Rushd, Abul Walid's father, was also a learned man, scholar and judge.

As a boy, Abul Walid was educated in the three main fields of the times (religion, law and medicine) at the hands of private master teachers, but it was his sharp mind that quickly made him stand out in discussion groups, even among adults. By the age of 12 his teachers (including his father) had realised that the boy had 'the finest mind in Andalusia'; he soon was able to debate teachers and scholars on matters of Islamic law, science and medicine. By the time he was an adult, he had been certified not only as a master in all those fields but had also discovered theology and developed a passion for philosophy, Greek and Islamic culture.

At the age of 27, Ibn Rushd did not mind being appointed as an inspector of schools in Marrakech, the faraway Moroccan city, which was known for high culture and learning and famous for its bookshops and libraries. Indeed, it was during Abul Walid's residency there that the city's most famous mosque, the Kutubiyya (or Booksellers' Mosque) was constructed. And so it was in Marrakech that he started his intellectual productions, which included – with equal strengths – law, medicine, astronomy and philosophy. He started to conduct astronomical observations and soon wrote a commentary on Aristotle's *On Heavens* as well as, later, a critique of Ptolemy's model, proposing to replace the epicyclical and excentrical motion of a planet by a helical one. (One of the great astronomers of the region, al-Bitruji, was a disciple of Ibn Rushd.) In the movie, *Destiny*, Chahine has our hero use a telescope with water lenses and mention Ibn al-Haythem (the great pioneer of optics), thus alluding to Ibn Rushd's astronomical knowledge and practice, though, of course, the whole concept of a telescope with lenses of any kind was not invented until the early seventeenth century.

In the same period he composed one of the most substantial of his earliest books, *Kitab al-Jawami' fil Falsafa* (*Compendium of Philosophy*), which also contained sections on astronomy (heaven and earth), physics and metaphysics. The other major work he produced during his early career in Marrakech was the very influential *Kitab al-Kulliyat fi-t-Tibb* (*Compendium of Medical Knowledge*). Shortly afterwards, he went back to Cordova and was appointed as a judge in Seville. There he must also have witnessed the erection of the city's famous Giralda Tower.

Becoming more and more famous for his encyclopedic mind and sharp intellectual insights, he started rubbing shoulders with the great thinkers of the town and the time. Two philosophers had appeared just before him, Ibn Baja (Avempace) and Ibn Tufail; they were to have substantial positive influences on him and his career. Legend has it that it was Ibn Rushd's grandfather who intervened with the Almoravid caliph to get Ibn Baja freed when he was imprisoned during the fundamentalist witch-hunt of that time.

Ibn Baja emphasised Aristotle, with a particular reading and understanding of the great Greek philosopher; Ibn Tufail emphasised the power of (pure) reason and introduced Abul Walid to the Caliph Yusuf ibn Abd al-Mumin, a meeting that was to lead to a highly fluctuating relation. Ibn Tufail became famous for his philosophical novel *Hayy ibn Yaqzan* (*Alive, Son of the Vigilant*), in which he asks whether a human being born on an island and having nothing (and nobody) but his mind and nature around him can deduce truths about his existence and about God; Ibn Tufail's answer was a resounding yes.

After the death of the caliph Yusuf, his son Yaqub succeeded him, and for a long time the new caliph made Ibn Rushd an even closer advisor, affectionately calling him 'my wizard'. But some circles, within and out of the palace and the court, and groups, who had both political and conservative religious motives, did not appreciate this close relationship and looked for ways to split it and control the caliph. The latter fell prey to the conspiracy, and so Ibn Rushd's 'hardship' (*mihna*) times began: He was expelled with his son and exiled not only to a small town outside of Cordova but his books were also burned in public, as famously portrayed in the movie *Destiny*. The film exploited this angle to draw a parallel between today's repression of liberal thought by fundamentalists and yesteryear's (attempted) suppression of the greatest rational intellectual Muslim mind of all time, Ibn Rushd.

It is no surprise then that Ibn Rushd had much more influence (and became famous) in Europe for centuries to come than in the Arab/Muslim world. His name was soon transcribed through Hebrew, so Ibn became Avn, then Aven through Latin, until it became Averroes through the series: Ibn Rushd/Roshd → Avenrosch → Avenros → Averroys → Averroes[2]. Alain de Libera, probably the best expert on Averroes today, has written simply: 'Among all the thinkers of the land of Islam, none other [than Averroes] will have had a greater influence on humanity's universal culture'[3]. Arab intellectuals echo this assessment; Mahmud Qassim, former dean at Cairo's College of Knowledge, starts an article on Ibn Rushd (published in 1963) in the following manner: 'Thinkers and educated people in the West are probably more familiar with this philosopher than the Muslim elite'. Indeed, his philosophy passed to Jewish and Christian medieval thinkers so that a whole school of Latin Averroism emerged, and he was never to be heard about again in the Islamic world until the twentieth century! Sadly and quite revealingly, a number of Ibn Rushd's works today exist only in their Hebraic and Latin translations[4]. Majid Fakhry, the great specialist of Islamic philosophy, has found 53 titles of works by Averroes in the literature, yet only 11 of them have been printed so far[5].

Today Ibn Rushd enjoys a golden admiration from everyone in the East and in the West. A few years ago, universities and official governmental institutions were competing to celebrate the 800th anniversary of his death, sometimes with hidden political agendas. However, his reputation and profile were not always sterling in Western writers' descriptions. Alain de Libera relates amazing stories that were written about Averroes in medieval times. One such myth had that he got tortured and killed his 'rival' Avicenna, who had presumably come to Cordova. (Never mind that Ibn Sina had died almost a century before Ibn Rushd was born!) Other, less risible fables portrayed him as an enraged fanatic, an arrogant megalomaniac who tears up books in fits of rage and, as in the startling last scene of Chahine's movie, throws his own books into the fire – apparently signalling that he had lost hope of any enlightenment in his people.

In more modern times a pernicious thesis emerged about Ibn Rushd's philosophy, the so-called 'duplex veritas' (double truth) thesis, whereby philosophy and theology allow one to adopt two contrary positions, for instance, about the creation of the world (philosophically eternal and theologically created). Needless to say, Averroes's philosophy is far from such warped logic. On the question of the eternity of the world, Ibn Rushd simply stated that although Revelation very clearly declares that God created the world, it does not say how or when, and so it is by philosophy (only) that one can explain whether the universe was created a finite or infinite time ago; then true to Aristotle, Averroes argued for the eternity of the world.

Unfortunately, this 'duplex veritas' thesis was spread widely in the West in medieval and modern times; when it reached the Muslim world, it tagged the great philosopher with a hypocritical or even heretical reputation. It took decades for his real philosophy to be rediscovered and properly appreciated, and even today among more conservative Muslims the mention of his views is always met with suspicion and the raising of guard.

It is thus important to note that what ended up polarising East and West over Ibn Rushd is a tortuous and decidedly incomplete understanding of his works. Indeed, although Averroes's commentaries of Aristotle, whom he admired without bounds and described as a pinnacle in the intellectual development of humans, were highly influential on thinkers like Maimonides, Thomas Aquinas and Albert the Great, they were often disputed. This led to a movement 'against Averroes' (the title of one of Aquinas's books). The duplex veritas description of our philosopher gradually turned into a simplistic depiction of him as 'the rationalist'. This explains how Ernest Renan, the famous French philosopher and historian, described Averroes as a 'free thinker' and positioned him in opposition to the conservative Muslim scholars.

In the land of Islam, most of Ibn Rushd's works had remained not only unknown but also the basic idea prevailed that he had 'dared' tear al-Ghazzali apart[6]. Many centuries later, the clash between the two still dominates debates on various issues related to Islam, with al-Ghazzali enjoying the support and following of every conservative and orthodox Muslim out there, and Averroes's approach claiming the minds of the (few) Muslims who try to uphold reason as an essential tool of analysis in most, if not all, subjects. In the West, Averroes's rationalist approach ended up prevailing and being even taken to extremes he undoubtedly would have fought very hard, but in the East a conventionally fideist, if not fundamentalist, Islam came to dominate.

truth cannot contradict wisdom

One of the main reasons that led to such a deep misunderstanding of our Muslim thinker (for he is thoroughly Muslim in his thinking) is the simple fact that two of the three books that contained his essential views and philosophy were not translated into Latin or later to European languages until the twentieth century. Only his commentaries on Aristotle's and other great Hellenistic works were known to the West until recently.

Indeed, the indispensable *Fasl al-Maqal fi ma bayna-sh-Shari'a wa-l-hikma mina-l-Ittisal* (*The Definitive Discourse on the Harmony Between Religion and Philosophy*) is a brilliant little book that contains enough clarity of exposition to dispel any 'double truth' thesis from anyone's mind. In his introduction to *Averroes – Discours Décisif*, Alain de Libera writes: 'Of all the texts by Ibn Rushd, none is more representative of the man, of his body of work and of those times than Fasl al-Maqal'. In it Ibn Rushd presents his simple theory: No true statement of religion can contradict any true statement of philosophy, if one makes sure to reach truth from each side. By religion, Abul

Walid means the irrefutable statements one finds in the Qur'an or those holy verses and prophetic statements that can be interpreted in a correct manner. By philosophy, Ibn Rushd means the conclusions that Reason can reach by careful methods. In his simple and timeless words, 'Truth (Revelation) cannot contradict wisdom (philosophy); on the contrary, they must agree with each other and support (stand with) each other'.

In this superb 'Discourse', Ibn Rushd refers to religion and philosophy as 'bosom sisters', such that 'injuries [to religion] made by people who are related to philosophy are the severest injuries, coming from close companions [. . .] the enmity and quarrels which such [injuries] stir up between the two, when they are in fact mutually loving friends by nature and essence . . . ' In Chahine's film, the Caliph (symbolising Islamdom) has two sons: Nasser, the disciple of Averroes, thus representing Reason, and Abdullah, who is seduced by the fundamentalist discourse, and thus represents a cartoon version of Religion. Ibn Rushd is the wise Muslim scholar, who can balance both sides in his mind and life. At some point, Chahine has him give a short speech taken verbatim from the *Fasl al-Maqal*: 'Divine Law combines Revelation with Reason. It is to be understood on the basis of causes, means and goals. The Revelation is completed by the element of Reason, and Reason is completed by the element of Revelation'. Indeed, Ibn Rushd starts his discourse by emphasising the importance of Philosophy as a way to reach divine truths and shows by logical steps that it must then not only be an exercise that is fully allowed for Muslims to practice but in fact it must be an obligation upon at least an elite group, for God has called upon humans to seek truths using the mind and the senses.

But what is one to do when confronted with a situation where, to all appearances, the Revealed text contradicts the conclusions reached by Reason/philosophy (and for us later, science)? Ibn Rushd addresses this issue squarely and clearly: whenever there is such an apparent contradiction, the (religious) Text must be allegorically understood and subjected to interpretation by those whom the Qur'an calls 'rooted in knowledge' (Q 3:7), namely the philosophers, who in Averroes's view are those who wield the highest methods of knowledge. (We shall explain shortly the three different approaches to knowledge, according to Ibn Rushd.) He then goes on to discuss the methods of interpretation that will stay true to the principles and the spirit of Islam's foundations.

Still, one must be careful not to think that Averroes subjects the Revealed Text and the necessity of its interpretation to the truths that one reaches by way of Reason/philosophy. Another important principle that our master

thinker expounds in the 'Discourse on the Harmony' is the different levels
of analysis and understanding that people can undertake with regard to
religious issues. He gives an illuminating example (pun intended): When
God describes Himself in the Qur'an as 'the Light of the heavens and the
earth', common people must understand it in those terms, and there is
nothing wrong with that first-degree reading; but if wiser minds wish to find
a deeper insight or grander theological meanings by interpreting the verse
so as to fit a metaphysical world view, they are not only welcomed but also
feel obligated to do so. Indeed, the Qur'anic verse that Ibn Rushd repeated
most was 'Reflect, o ye who have been endowed with clairvoyance' (59:2).
Alain de Libera tells us clearly: 'The *Fasl al-Maqal* is not a typical, scholarly
philosophical book . . . it is a text addressed to the public, not to any and all
public, but rather to an educated public. . . . That is why it is so relevant
today'.

the spirit of Averroes

It is becoming clear now how such a simple, balanced and fully genuine philosophy can be adopted and become a guiding spirit in contemporary science
and religion discourses. We will not go as far as Charles E. Butterworth
goes[7] and declare Averroes to be a precursor of the enlightenment, rather
we will content ourselves to having the spirit of Ibn Rushd (his principles
of philosophy, religion and their mutual relation) guide our exploration of
issues at the intersection of modern science, religion (particularly Islam) and
modern thought.

What Averroes achieved was revolutionary, not only for his times but for
ours as well: He showed that it was possible to unite reason with the core
tenets of the Islamic faith, and indeed of any similarly theistic creed or world
view. Any objective reader of Ibn Rushd, whatever his/her initial viewpoint,
will realise that a simplistic opposition or polarisation of philosophy/science
and religion cannot stand.

Let me explain this further by taking a brief look at the first paragraphs
of *The Definitive Discourse on the Harmony Between Religion and Philosophy*. After
the customary praise to God and prayer on the Prophet, Ibn Rushd starts by
defining the purpose of his treatise as 'to examine, from the standpoint of the
Law, whether the study of philosophy and logic is permissible, prohibited or
commanded, recommended or obligatory'. He then states clearly the following: 'We say: If the activity of "philosophy" is nothing more than the study

of existing beings, and the reflection on them as indications of the Artisan, inasmuch as they are products of art, for beings only indicate the Artisan through our knowledge of the art in them, and the more perfect this knowledge is, the more perfect the knowledge of the Artisan becomes. And if the Law has encouraged and urged reflection on beings, then it is clear that [any] activity [of the same kind] is either obligatory or recommended by the Law'. He then goes on to cite Qur'anic verses to this effect. Averroes is declaring any human activity, such as philosophy and science, to be at least commendable, if not obligatory, for the simple reason that it leads to greater knowledge and appreciation of God (the Artisan), through the study of His creation.

Ibn Rushd had obviously not adopted the inductive method of science, the one that Biruni had already espoused, contrary to Ibn Sina and other Muslim scientists. Both Ibn Rushd and Ibn Sina can be forgiven for being so 'Greek' in their approaches that they totally failed to see that the inductive approach was closer to physics (and thus to the world) than was philosophy. Still, concerning the debate on the sources of knowledge between religion and philosophy, Averroes makes an important and original division between three different modes of discourse: the philosophical (rational, logical and *objective* deduction of truths); the dialectical (debate between the experts, from which truths can emerge) and the rhetorical (persuading audiences using flights of eloquence that aim for the heart). Ibn Rushd discounts the rhetorical approach and attributes it to the simplistic religious scholars, who aim for the lay audience. The dialectical approach is the method used by the theologians (the *mutakallimun*, or experts in *kalam*, Islamic theology) and is not reliable in serious discussions. The only approach that is of value to those seeking to reach any truth is that of demonstrative reasoning (deduction).

What is important to realise is that people are not at all prepared for the usage of this high form of thinking and arguing, and that is why the Revelation has superbly accommodated all minds. Averroes points to the Qur'anic verse 'Invite to the Way of thy Lord with wisdom and with beautiful preaching, and argue with them in ways that are best and most gracious' (Q 16:125) as directly referring to those three methods. But this does not mean that thinkers must resort to the methods of the masses. On the contrary, he claims that the very reason why there are verses of apparent meaning and others with allegorical senses, and the reason why sometimes there seems to be contradictory meanings between some verses is indeed to make it clear to people that there is a need for an intellectual effort by those 'grounded in knowledge' to find deeper meanings that harmonise the whole text.

Averroes was very modern in his emphasis on a tolerant attitude towards all viewpoints and sects. He noted that intolerance and fights arise when people insist on only one possible understanding. 'One must not try to impose some kind of a consensus, which can only be illusory', he says, 'especially if such an attempted consensus is being built on dogmatic readings of texts'. Only the demonstrative reasoning can be objective, that is, independent of the author, reaching results that can be recognised as true and thus accepted by all.

Ibn Rushd's modernity is not difficult to detect in his writings. But above all, what makes him important is the flawless coherence and harmony he has achieved between his religious principles and his intellectual training. That is why I have adopted him and will use him as a model for a harmonious fusion of science, philosophy and religion in Islam today. That spirit of Averroes is what this book will try to capture and use to illuminate various topics of relevance.

yesterday

I grew up in a cultural environment, which, while firmly grounded in Muslim tradition, encouraged open explorations and varieties of learning sources. My father, like many boys in his days, memorised the entire Qur'an by the time he was a teenager; he went on to obtain doctorates in philosophy from Cairo University and Paris Sorbonne and later to become dean of the College of Religious Studies at the University of Algiers. My mother was more of a literary type and got a master's degree in Arab literature. Interestingly, all five of their children grew up to be scientists, medical doctors or science teachers, and all were deeply imbued with the rationalism of philosophy, the methodology of modern science, the strife for beauty in arts and literature, and the holistic world view of Islam/religion.

My family was one of cultural interface and synthesis. While my siblings and I excelled in mathematics and the natural sciences, my father was doing his thesis in Cairo on 'the concept of time in the philosophy of Averroes' and, almost simultaneously, his Sorbonne thesis on 'the concept of time in modern Arab thought', in the process surveying Western thought as well. Our home library, although scant on the hard sciences, was very rich in philosophy, religion and literature. Our education was bilingual (Arabic and French) from the start; English was added in our teenage years. No wonder then that during and after my formal education in physics (followed by research in astrophysics), I was always reading and sometimes reviewing philosophical books, especially those relating to science and religion.

The first topic to pull me into serious explorations in that field was cosmic design, fine-tuning and the anthropic principle. I must admit that early on I too was impressed by the 'science in the Qur'an' discourse, and it took me several years to recognise the deep flaws in that 'theory', something I will explain in detail in Chapter 5. Then in 1989 I discovered and read Barrow and Tipler's seminal book, *The Cosmological Anthropic Principle*, and was mesmerised not only by the wealth of data pointing to some metaphysical principle behind our existence in this finely tuned universe but also by the erudition and rigorous methodology displayed by the authors. And although I had already appreciated Paul Davies's *The Cosmic Blueprint*, to me *The Cosmological Anthropic Principle* was a quantum leap in both concept and scholarship. Similarly, Dyson's *Infinite in All Directions* showed me how a scientist can construct a personal philosophy of nature and existence.

In the following years I kept that strong interest; for instance, I reviewed for Arab readers Davies's *The Mind of God* and *About Time* and Dyson's *The Sun, the Genome, and the Internet*, among others, but I did not pursue such explorations in any systematic or academic way for almost a decade. Instead, I turned my attention to astronomical/scientific topics of societal and cultural relevance.

During the early nineties, while the conditions of life and work (in Algeria and then in Kuwait) were extremely difficult, I focused my attention on the practical – and socially relevant – problem of the observation of the new, thin lunar crescent, which determines the start of the Islamic months and the religious holidays of Ramadan, Eid and Hajj. This remains one of my proudest contributions to my society, for indeed I introduced[8] into the classical Muslim considerations new and modern astrophysical approaches that had been shown (by western researchers) to be highly useful concepts, such as the luminosity contrast between the crescent and the background sky, and ideas by Muslim scientists, e.g. the lunar dateline; these new approaches have now become standard in the scientific and religious treatment of the problem. To me that kind of work represented an ideal synthesis of modern scientific expertise, strife to solve societal problems and fusion of western methods and knowledge with Muslim topics, precisely fitting my general philosophy.

In the same spirit, I have collaborated with the United Arab Emirates' Ministry of Education and produced high school textbooks of physics, where for the first time chapters were included on astronomical topics of importance and relevance to the Muslim public: crescent observations, Islamic calendar, evolution of the universe, etc. The implicit and overarching principle in that work was indeed to instil in young Muslim minds the idea that science is relevant to their world view and evolves continuously . . . in an evolving world.

today

By the late nineties I began to take a more serious interest in the interface of science, religion and philosophy. First, I started to make systematic efforts at exploring the field of Western science, religion and philosophy, thus reading many essential works, such as those of Barbour, Peacocke, Polkinghorne, Ruse and others. I also turned to Muslim philosophy, from al-Ghazzali (1058–1111) and Ibn Rushd to Iqbal (1877–1938) and Nasr (1933–). Then, around the year 2000, my friend and collaborator Karim Meziane suggested that we undertake a three-way comparison between 'cosmology' as it has been defined and developed by (a) Muslim philosophers of medieval times (the 'golden age' of the Islamic civilization), (b) contemporary Muslim thinkers and (c) modern science. We thus wrote a short article (in French), which we circulated by e-mail among friends and acquaintances; the article drew some attention, so we started getting invited to workshops on science and Islam.

A common thread runs through all my intellectual and educational works. It consists of the following interweaving lines: (1) Science is important and relevant to Islam (and to other cultures, of course); (2) science can help make progress not only materially (this is obvious) but also intellectually, culturally and religiously; (3) science evolves, and theology should also progress; and (4) looked at properly, there is nothing (except pure materialism) that can oppose science and Islam.

The present book, in addition to both reviewing recent fruitful efforts and exposing dead ends in the science–religion–philosophy field, is meant to present the above principles more fully and systematically, first by building the basis for such a harmonious relation and then by giving examples of how this Averroesian model can be applied to several important topics, e.g. cosmology, design, evolution, etc. This approach, I believe, can help address

'Islam's Quantum Question', i.e. how to reconcile religious tradition with rational and scientific modernity, and how to be dual (quantum) without being schizophrenic.

tomorrow

Islam has in the past proven itself capable of completely integrating scientific knowledge into its own world view and producing important intellectual contributions in various fields and areas. The philosophers, from al-Kindi to Ibn Rushd, fully digested Hellenistic philosophy; the scientists, from Ibn al-Haytham and al-Biruni to al-Tusi and Ibn al-Shatir, did the same with Babylonian and Greek science (astronomy, in particular), and likewise for the physicians, from Ibn Sina (980–1037) to Ibn al-Nafis (1213–88). They did, however, go beyond their predecessors, uncovering their errors, discovering new facts, inventing new methods and tools and producing a genuine, organically cultivated civilization.

Islam achieved that great synthetic civilization when it confidently opened up and allowed thinkers to fully digest the sciences and the philosophical heritage of humanity and to explore new ideas and avenues without fear of any kind. Indeed, thinkers – temporal and religious ones alike – upheld the Prophetic statement: 'Whoever exerts intellectual effort (*ijtihad*) and succeeds (in reaching some truth) shall have two rewards, and whoever exerts effort but is not successful shall have one reward (for effort)'. Today, more than ever, the Muslim society needs to recover that spirit. It must recognise the multi-dimensional importance of science, readopt that open interactive attitude and intelligently mix modern ideas with its own principles. That is the mission Muslim scientists and intellectuals like me must pursue.

This book is one step in that direction.

introduction: Islam and science today

'Even before reaching enlightenment, a person knows that everything attests to the existence of God. But these virtuous feelings of innocence disappear as he begins to mingle with corrupt society, because worldly affairs hinder him from swimming in the vast ocean of insight.'

al-Ghazzali[1]

'We've arranged a civilization in which most crucial elements profoundly depend on science and technology. We have also arranged things so that almost no one understands science and technology. This is a prescription for disaster. We might get away with it for a while, but sooner or later this combustible mixture of ignorance and power is going to blow up in our faces.'

Carl Sagan[2]

technological vs. human development

Dubai has become the new Hong Kong: cosmopolitan, active and growing at breakneck speed – at least until the global economic crisis hit. I refer to it as the Manhattan of the Arabian Gulf and of the Middle East, for many reasons, chief of which is the large number of skyscrapers that have mushroomed over the past decade. Dubai has built the highest tower in the world (Burj Dubai) and the world's first seven-star hotel (Burj al-Arab), complete with underwater restaurant and other stunning features. Had the economic crisis not hit so hard, Dubai would soon have had six towers of more than 100 floors[3], the largest group of such skyscrapers in the world. Quite an extraordinary transformation for a city which only a few decades ago was a small poor village, where people lived by fishing, pearl diving and sea commerce. Since 2000, the population has doubled, both in Dubai and in the United Arab Emirates (UAE), testimony to the extraordinary growth in economic activity.

But it is not just the landscape that has changed so dramatically, from a scorching and barren desert to a modern city of glass buildings, huge malls with fancy entertainment centres (including an indoor ski resort), huge freeway systems with overpasses and tunnels and an under- and over ground metro. It is the complete way of life that has changed, from a sluggish pace regulated and slowed by the sun's position in the sky to a high-speed rhythm dictated by a lock to the global world. Life in Dubai is connected by cell and satellite phones, system supported by broadband Internet, and now landscaped by whole media, Internet and knowledge 'cities' (as they are called) built to host the world's largest and the most famous companies and institutions[4]. Every year Dubai hosts the Gulf Information Technology (IT) Exhibition (Gitex), one of the world's top-three IT fairs, which over four days attracts some 150,000 visitors to the Dubai International Convention and Exhibition Centre, where 2,700 companies present their innovative products over 30,000 square metres – the equivalent of six football fields. Wi-fi, Bluetooth and Blackberry now permeate and punctuate the lifestyle of this city's youthful and diverse population[5].

I live in Sharjah, which is adjacent to Dubai. With half a million people, it is about half the size of its more famous sister-city. It is also very different. Whereas Dubai thrives on commerce and high-tech activity, Sharjah is proud of its cultural outlook, with a dozen colleges, high institutes and universities, and at least another dozen museums, including the small science museum with its miniature planetarium. It is also rather a conservative city.

Ras Al Khaimah is an even smaller city and emirate[i] (state), some 80 kilometres east of Sharjah. It too has started a programme of fast economic

---

[i] The UAE is a federal system of seven states: Abu Dhabi, the political capital, Dubai the commercial and economical capital, Sharjah the cultural capital, and the small Ras Al Khaimah, Ajman, Fujeirah and Umm Al Quwayn city-states.

development, but with its size and level of activity it barely compares with a small town in Europe or North America. Still, Ras Al Khaimah has a branch of a major American university (George Mason University) since 2005; more stunningly, it has signed agreements with the Space Adventures Co. to provide a launch site for tourist and commercial space flights[6].

Abu Dhabi, the country's capital lying some 170 kilometres west of Dubai–Sharjah, presents another glittering landscape of glass, metal and concrete buildings spread out along a beautiful corniche. While it is far behind Dubai in terms of economic activity and global commerce, it has recently started an ambitious programme of cultural development, opening new universities[7], 'seven-star' luxury hotels to rival Dubai's and an amazing complex of several museums in the Saadiyat Island project, the first-ever branch of the Louvre outside France, an Abu Dhabi Guggenheim and several other art and performance centres.

The UAE in some ways resembles the USA – with major differences, of course. Both are federal nations, unified by some visionary founding fathers. Both are largely lands of immigrants, although the USA grants citizenship to and melts in its pot millions of people each year, while the UAE does not. Both have conservative and liberal states, though the UAE has no 'culture war'. Dubai, like New York, is certainly a 'blue' liberal city, and so is Ajman to some extent (akin to New Jersey). Sharjah is the very 'red' emirate state in the UAE, and Abu Dhabi is, perhaps, like Texas, the moderately conservative state. Dubai and Abu Dhabi are very rich, and it shows; Sharjah is modest in its revenues and projects; the other emirates have meagre resources.

Now, in contrast to that dazzling development, one finds the level of human and scientific development incredibly depressed – and depressing. It is true that the UAE is a very young country, having achieved its independence and unification less than 40 years ago, and that it had to build all its institutions, including education, literally from scratch. So one cannot expect it to have already produced social, cultural and intellectual progress anywhere comparable to that of developed nations; material progress is indeed infinitely easier to achieve than human development. However, as I have previously noted, the UAE is overwhelmingly inhabited by immigrants, a large portion of whom are well-educated professionals, from teachers to engineers, doctors and university professors. One would thus expect a certain qualitative or even quantitative measure of cultural and intellectual development, but that has been slow to come.

The story of stark contrast between material and intellectual development, and between technological and scientific progress that I am telling here is not about the UAE per se, but rather about the Arab/Muslim world more generally. Indeed, the conclusions would be quite the same if I were to transpose the narration from here to another Arab/Muslim country, although the contrast would not be so striking, for the pace of material development in the UAE is unparalleled in the history of the world, but the level of intellectual underdevelopment would pretty much be the same.

science and the Qur'an in Muslim society today

Let me start by relating some recent examples to describe and highlight the state of scientific ignorance that prevails in the Arab/Muslim society today.

In December 2006, the 'Eighth Conference on Scientific I'jaz (Miraculous Aspects) in the Qur'an and the Sunna'[ii] was organized in Kuwait (at the Sheraton Hotel) by the World Authority on Scientific I'jaz in the Qur'an and the Sunna. Over several days, 'scholars' presented 86 papers (in parallel sessions) on the following topics[8]:

- The scientific I'jaz (miracle) in the destruction by the Mighty shout.
- The scientific I'jaz in the distinction between the urine of the female maid and that of the suckling boy.
- The code of human life before birth and after death: scientific signs in the Holy Qur'an.
- The scientific I'jaz in the prophetic Sunna regarding the stagnant water.
- Satellites bear witness the truth of Muhammad's prophethood (PBUH).[iii]
- The disease and the remedy in the two wings of the fly.
- The miraculous description of the (re-)creation of human bodies (and not the soul) from the tail bone on the day of resurrection.
- The miracle in the 'descent' of iron (from the sky).
- The scientific I'jaz in the Qur'an's description of the movement of shadows (the immobile shadow).
- Study of the effect of bloodletting on the molecular biology of hepatitis-C patients.

[ii] Sunna refers to the Prophetic Tradition, statements and deeds, as recorded in the Islamic literature.
[iii] It is customary in the proper Islamic tradition to offer peace and blessings upon the Prophet Muhammad (and other prophets as well) whenever his (their) name(s) is (are) mentioned – thus the traditional usage of the abbreviation PBUH (peace and blessings upon him).

- The superiority of the treatment of lower backs by prayers over the treatment by lasers.
- Glimpse into the scientific I'jaz in the Prophetic hadith regarding remedying by vinegar.

I shall examine this 'scientific miraculousness of the Qur'an' in detail in Chapter 5; suffice it here to note the incredible topics that are being 'investigated' nowadays[9] and the implications regarding the understanding (or lack thereof) of science among a large section of the Muslim elite today.

In similar fashion, in April 2007, a conference was organized in Abu Dhabi on 'Qur'anic Healing' at the seven-star Emirates Palace hotel. The media, which showed up en force, reported[10] that 1,300 people, including several dignitaries, attended on the first day. Although the conference had a scientific committee and was apparently professionally organized, I could not find a website for it, so I am only able to rely on the (extensive) media reporting[11].

The opening keynote speech was given by Prof. Zaghloul al-Naggar, a former geology university professor who for many years now has specialised in I'jaz discourse. In his overview, he decried the 'duality' of the higher education system in the contemporary Muslim world, a result – in his view – of having been culturally influenced and dominated by the Western materialistic civilization. What he meant by that duality is the fact that, on the one hand, medical training does not make room for Qur'anic healing approaches, and, on the other hand, the Islamic theology and jurisprudence curriculum do not include medical subjects. This then, he says, prevents doctors from appreciating the role and value of Qur'anic healing, while the Qur'anic healers are not knowledgeable enough about scientific methods and facts. He asked for more cooperation between the two fields and for a greater amount of modern technology in the practice of Qur'anic healing.

But this was just an appetiser. The pièces de resistance were yet to come. Let me present a few highlights.

One main theme in the conference was the effect of the recitation of verses of the Qur'an on water, which then heals many ills in patients who drink it. Some speakers, most of whom were university professors, explained the healing on the basis of 'the memory of water' effect (needless to say, a long-discredited idea); others explained it by some electromagnetic waves, which, carried by the 'vibrations' of the Qur'an being read, 'rearrange' the molecular structure of the water, giving it special 'energy' (an overly abused word in that conference as in many others); others invoked the concept of 'information content', which somehow gets passed on from the Qur'an to the water and then to the patient, especially if the Qur'an is read by quasi-saints (*salihun*); others invoked telepathy; and finally some even based their claims on the 'theory' of homeopathy (infinitesimally small concentrations of medicine in water, which, it is claimed, gives it special power). One speaker, a medical instruments technician, even brought with him a device that purported to extract the 'Qur'anic energy' stored in the water (the energy is transferred to the water by just placing one's finger in it and reciting verses aloud or in one's mind); the device then transforms the energy into digital information, records it and even sends it by the Internet to anyone needing it anywhere in the world. I should thus add that the device was hailed as 'a qualitative leap in the development of psychological immunity, and a quantum leap in the concept of 'technological *I'jaz*' (miraculous Qur'anic technology), the first invention that combines the Holy Qur'an with modern technology'[12].

What no one mentioned was the placebo effect or the mind's ability to trigger the release of medicinal chemicals, thereby leading to natural healing that looks quite miraculous sometimes – an important effect that I shall discuss in Chapter 10 ('Science and Islam Tomorrow').

Another main theme of the conference was the specific methods that should be used for the treatment of cases of exposure to magic, demon possession (*mas bi l-jinn*) and exorcism, and evil eye (*al-'ain*), which was explained as electromagnetic waves radiated by the eye, as the latter is connected to the brain, which can produce evil thoughts.

Reading such reports, I had difficulty reminding myself that all this was being presented in the twenty-first-century conferences and not in dark medieval gatherings.

The conference concluded by adopting a set of resolutions, particularly recommending to the authorities that government clinics be set up for the practice of the Qur'anic healing, where well-trained, fully educated practitioners are visited by patients. The conference called for the ministries of health and religious affairs to form a joint commission, whose task would be to issue certifications for the Qur'anic healers, who would – of course – have to undergo examinations[13].

In his book *Characteristics of the Noble Qur'an*[14] (now in its tenth edition), Fahd Ar-Rumi, professor of Qur'anic Studies at the College of Teachers in Riyadh, Saudi Arabia, gives us his position on Qur'anic healing. He says that first, the Holy Book prescribed preventive treatments for physical and psychological ailments by prohibiting pork, homosexuality and adultery, then he adds, 'but Qur'anic healing becomes truly miraculous when the Book itself becomes a remedy to sickness, not by prescribing medicine'[15]. A few pages later, he adds that reading the Qur'an can cure Muslims (and Muslim believers only) from many illnesses; he justifies this opinion with the following verses: 'And We reveal of the Qur'an that which is a healing and a mercy to the believers' (Q 17:82); 'Say: It (the Qur'an) is to those who believe a guidance and a healing' (Q 41:44) and 'O men! there has come to you indeed an admonition from your Lord and a healing for what is in the breasts and a guidance and a mercy for the believers' (Q 36:5); these verses are contrasted with 'There comes forth from within [the bees] a beverage (honey) of many colours, in which there is healing for people' (Q 16:69). It should be clear to any reader that the healing mentioned in the above verses is spiritual, not physical, for it is each time associated with 'mercy', 'guidance' and 'the breasts'.

To be fair, I should state that other mainstream Islamic scholars decry such hodgepodge approaches to scientific – or at least medical – topics by Muslims, but the fact remains that the above religion–science concoctions are widespread and even prevalent among the public and most of the educated elite. The 'quantum question' here has largely turned into a chaotic state.

science, methodology and education

In May 2007, my sons' school invited me to be a judge at their first-ever science fair; I dutifully accepted. Two teachers organized the event, coaching the hundred or so participating students (from grades 7th to 9th) for over a month and putting together what they hoped would be a great learning initiative. I cannot praise the two teachers enough for their truly heroic efforts. But the limited success of their initiative speaks volumes about our society, as I would like to explain here briefly.

First, although there were several excellent pieces of work and projects which showed that at least some students understood the core idea at the heart of science, the systematic investigation of a phenomenon or of a question in one's mind, most of the students, however, showed no understanding whatsoever of what science is about. For example, one student showed me his 'experiment', which consisted of mixing some milk, food colouring and liquid detergent and observing the patterns that appeared in the coloured milk. When I asked him, 'So what are you trying to figure out by this, or what do you learn from this?' he simply muttered: 'surface tension', but upon further probing I realised he had no clue what that meant. My more important observation, however, was that none of the school officials showed up at any moment, even though the fair took place in the school's gym and lasted for most of the day, and only a handful of parents attended the fair, despite everyone having been alerted about its date and place – by SMS at least once – and through the regular notices that we had received regarding the projects and the initiative in general.

Three days later, the same school organized a celebration of a competition for the memorisation of as many chapters of the Qur'an as possible. This time, practically all students participated; about 200 parents attended the prize-distribution event; the media were invited, along with some guests of honour, and a few hundred prizes were given out.

Now, I have nothing against the Qur'an memorisation; I can think of quite a few positive aspects of the exercise. My own son participated, and although he was nowhere near the top achievers, he still received a cash prize of about $55, which to him was a large sum. I should also point out that my son did take part in the science fair, and although I of course did not help him one iota in putting together his (and his teammates') project, and certainly did not try to influence the judging process in his favour, his team won one of the top prizes, and the boys received small trophies and gifts, which altogether amounted to about $25. I estimated the budget of the whole science fair to be about $1,000 and that of the Qur'an memorisation competition to be at least $20,000 (contributed almost entirely by sponsors).

What I decry in this whole affair is not just the fact that the Qur'an memorisation initiatives are given 20 times more importance, resources and attention than science fairs, it is the nature of the competitions themselves. None of the guests, from parents to teachers, school officials and guests of honour, were bothered by the fact that no students were asked whether they had tried to understand any of the chapters they had memorised, whether they had conducted any research on the material, such as comparing different interpretations (exegeses) or translations of the Text; after all, the school is bilingual and implements a dual Arab–British curriculum, it would thus have been a great exercise to research the Qur'anic vocabulary in the English language, or compare one classical exegesis of the chapter being memorised with a modern one- etc. No, the whole competition, which lasted for a couple of months (and is repeated year after year), consists solely of rote memorisation.

No wonder then that I get students at the university with weak, if any, analytical and critical thinking skills. No wonder that only less than 10 per cent of my students know, upon coming to my Introductory Astronomy course, the difference between a star and a galaxy or roughly how old the earth and the Universe are. No wonder that during the whole semester as I am teaching those students some basic astronomical knowledge and emphasising *how we get to know these things*, they keep challenging me with 'but the Qur'an says that the Sun moves towards a "resting place"' or "but the Qur'an says that the moon broke up and reassembled right in front of people's eyes during the time of the Prophet"'.

I will not be able to address all such assertions in this book – there are too many of them, and I am bound to leave a few, no matter how many I discuss – but I will certainly come back to the important issues of the nature of science itself, the nature of the Qur'anic text and its interpretation and the relation between the Qur'an and science. Indeed, these will constitute the main material of Part I of this book.

I do wish, however, to go back to that example of the 'moon splitting' that my students are fond of bringing up with me. On a recent occasion, one student raised the question in class and apparently was not satisfied with my answer, so a day or two later he e-mailed me a picture and some slides from a PowerPoint presentation he had found on the Web[16]. (Googling 'moon splitting' turns up 12,500 pages in English and more than 100,000 pages in Arabic[17]!) The picture is reproduced below, and the gist of the PowerPoint slides I summarise here.

This picture was distributed by NASA; it shows the rocky band which proves that the moon was split one day. [The text then goes on to relate the story of how the Meccans challenged Muhammad to show them a miracle as proof of his prophethood, and how the moon then got split into two parts; the Meccans then said, 'Muhammad has played magic on us'. But later, the slide notes, people from other villages confirmed their sighting of the moon splitting, and so did the Qur'an in verses 54:1–2[18].]

The slides then relate the following story:

In one lecture at a British university, Prof. Zaghloul al-Naggaar [the *I'jaz* specialist] said that the miracle of the Moon splitting had been proven recently and related the following story.

One English brother, whose name is Dawood Musa Pitcock, and who is now the president of the British Islamic Party and intends to run in the next elections under the banner of Islam, which is spreading in the West at large rates, said that while he was searching for a true religion, he was given an English translation of the Qur'an by a friend, and when he opened it he fell on the Moon chapter (54) and read the first few verses, so he asked himself (denyingly), 'Did the Moon ever split?', and he set the Book aside and did not open it again. Then one day, he was watching a TV program on the BBC channel, where the presenter was reproaching to three American scientists the fact that the US spends millions, or rather billions in space exploration projects while millions of humans die of hunger. So the scientists started to recount justifications of the space program, among which were the benefits in agriculture and industry. [. . .] Then one mission was mentioned with its high cost of $100 billion [sic], so the presenter asked them, 'All this money just to put the American flag on the Moon?' The scientists replied that one of the important things they had discovered while studying the similarities and differences between the moon and the earth was a band of transformed rocks spread from the moon's surface through to its center and out to its surface again [on the other side], and when we showed this to the geologists, they concluded that the only explanation to this evidence is for the moon to have been split and reassembled one day, as this band of rocks must be the evidence of the reassembly collision.

Dawood Musa Pitcock finishes his story by saying: I jumped up from my seat, and I exclaimed 'Muhammad's miracle in the desert 1,400 years ago . . . God has guided the Americans to confirm it to Muslims. No doubt that this religion is right . . . '

Prof. al-Naggar adds that this Qur'an chapter was the direct reason for Pitcock becoming a Muslim.

I should remind the reader that Prof. al-Naggar spent his professional life as a university professor of geology, so it is astounding to see him relate such an incredible story without checking the facts with NASA, the BBC or any other credible source, and not ask himself when in the past such a geological event could have occurred. My students must then be excused if university professors make such momentous (or should I say moon-splitting) proclamations.

The picture, interestingly enough, appeared as NASA's Astronomy Picture of the Day on 29 October 2002[19]. The website, as if to rebut the claims

of my students and countless others of my Muslim brethren, added the
following text:

> Explanation: What could cause a
> long indentation on the Moon? First
> discovered over 200 years ago with
> a small telescope, rilles (rhymes
> with pills) appear all over the
> Moon. Three types of rilles are now
> recognized: sinuous rilles, which
> have many meandering curves,
> arcuate rilles, which form sweeping
> arcs, and straight rilles, like the
> Ariadaeus rille pictured above. Long
> rilles such as the Ariadaeus rille
> extend for hundreds of kilometres.
> Sinuous rilles are now thought to be
> remnants of ancient lava flows, but the origins of arcuate and linear rilles are
> still a topic of research. The above linear rille was photographed by the
> Apollo 10 crew in 1969 during their historic approach to only 14 kilometres
> above the lunar surface. Two months later, Apollo 11, incorporating much
> knowledge gained from Apollo 10, landed on the Moon.

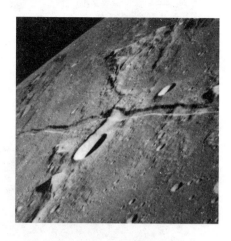

Ah! But then one would still have to convince my students and scores of
others that NASA astronauts did land on the moon. I will have something
to relate about this in Chapter 3.

Now, is there a rational explanation to this 'moon splitting' story, which
has fired up millions of Muslims lately, 'experts' and laymen alike? First, I
should point out that even some of those who strongly defend the 'scientific
miraculousness' of the Qur'an[20] have cautioned that the verses in question
may very well be referring to future events related to the Day of Judgment.
al-Jamili notes that there is no evidence that the moon ever suffered such a
split in its astronomical history, even though – he stresses – this does not
negate the possibility of a miracle for Muhammad. Many other writers[21],
however, have defended the moon splitting episode as really physical, supporting their position with the reported testimonies of people from around
Mecca during that night.

There is a similar story of a 'moon splitting' episode, this one in England in the year 1178. The story, known as 'the chronicles of Gervase of

Canterbury', tells of a group of men who on 18 June 1178, 'observed the upper horn of a crescent moon to have split with fire'[22]; it also relates that 'the group of five eyewitnesses saw the upper horn of the bright, new crescent moon "suddenly split in two"'[23]. There is also a report[24] relating a very similar episode recorded on poems and rock art by the Maori tribe in New Zealand at about the same time as the 1178 event in England; it speaks of 'a new moon with the lower horn of the moon split' (recall that New Zealand is in the southern hemisphere). The episode, which continues to puzzle historians and scientists, has been interpreted as possibly a huge meteoroid hitting the moon and producing a huge amount of rock ejecta; the 1178 episode has indeed been linked to the 22-kilometre Giordano Bruno crater on the moon.

Could our more famous (at least among Muslims) moon-splitting episode be the case of a huge meteoroid impact, the effect of which could have been described as a piece of the moon splitting from the main part? Sigismondi and Imponente (in a paper published in 2000) tell us that the new moon is one of the three favourable conditions[25] for such an event. That would explain why it would not have been seen and thus reported by many other people in farther regions, especially because the rare event would have produced a huge amount of energy. On the other hand, some scientists have remarked that the frequency of such huge impacts is extremely low; such collisions are expected to occur every few millions of years, not once or twice a millennium. But then again, we could be wrong about the whole case and its description.

more serious science–Islam discourse today

The examples I related in the previous sections – and many other similar ones – clearly highlight the problem of science and education in the Arab/Muslim society today. Not only is Ignorance of basic scientific facts hugely widespread, but more importantly the nature of science, its methodology, areas of validity and limitations are known to very select few. But I should certainly not misguide the reader into thinking that science in the minds of Muslims everywhere today is either a collection of high-tech devices that are used in one's life to make it easier and more fun or a number of ideas that one can cite and use, without any methodological constraints or underlying rules. There are serious voices that have for many years now been attempting to reconstruct the Muslim understanding of and interface with science. But these voices are too few, disconnected and dissonant to even make themselves heard, much less affect the general attitude towards science in the Muslim society today. Indeed, how many people today have heard of Seyyed Hossein Nasr, Ziauddin Sardar or Mehdi Golshani? It is true that the Nobel Prize winners Abdus Salam and Ahmed Zewail have become famous[26], but few know what they got their awards for, rarely is it emphasised that both did all of their research in the West, and hardly anyone has inquired about their philosophy of science and religion.

One prime objective of this book is to familiarise the public with these voices, particularly those who have expounded the philosophy of science and related it to Islam in a serious manner, and also to resurrect the discussion on the role and status of science in the Muslim society. Another chief objective of this work is to try to show that a modern harmonious synthesis of science and Islam can be attempted, in the form of a theistic version of science, where Islam can join hands with other monotheistic traditions, particularly Christianity, which has in recent times produced thinkers who have worked hard to try to find solutions to many of the problems I shall be bringing up and discussing. An intelligent discourse at the Islam–science interface is possible; the choice is not limited to either the secularists (Pervez Hoodbhoy, Taner Edis) or the mysticists (S. H. Nasr, Osman Bakar) and the neo-traditionalists (Muzaffar Iqbal). The quantum question can be resolved, or at least cogently addressed, without resorting to separationism and rejectionism (of either tradition or modernity).

this book

This book aims at presenting perhaps not only a brilliantly novel philosophy of Islam and science but also a coherent, viable and very modern synthesis of the main and valid principles of science with the core principles of Islam, through a reasonable (not overly liberal) reading. I will show that all we need to construct an evenhanded 'quantum' (dual) model of harmony is full knowledge of science and its philosophy and methods – not to succumb to its positivistic and materialistic doctrines – and a middle-ground approach to Islam, staying away from the traditionalist approach and resisting the ultra-liberal siren calls. A careful reconstruction will thus be needed.

This is why the first part of the book, made up of five chapters and collectively titled 'Fundamentals (First Things First)', starts with a chapter on Allah, and then one on the Qur'an, before moving on to discussing science (over two chapters) and then critically dissecting the *I'jaz* 'theory'. Indeed, these chapters are useful and even necessary not just for western readers who may not be very familiar with the Islamic core principles but also for the Muslim audience, who may need to be convinced that the present reconstruction and synthesis is done in a reasonable and informed manner (with respect to Islam). The two chapters on science are an absolute must, as the first one clarifies the fundamental methodologies – and limitations – of science, and shows how science is not to be confused with scientism (the doctrine that claims science to be the explaining system for everything there is); and the second one (Chapter 4) argues for the validity of 'theistic science' as a version of modern science, rigorous in every way, but enveloped in a theistic world view. The last chapter of this first part then draws the lines around this 'theistic science' in order to prevent it from being hijacked; indeed, the chapter shows how, what may at first have been a reasonable idea, namely, the 'science-infused Qur'anic exegesis', quickly turned absurd in the form of the '*I'jaz* (miraculous scientific content) of the Qur'an and Sunna'. After presenting a historical review of its development, I will show how this theory is both methodologically flawed and factually wrong on most, if not all, cases and how, if at all, it can be corrected.

All this epistemological ground-laying will be needed before embarking on an ambitious attempt to address the serious science–religion/Islam questions of the day, namely cosmology, design, the anthropic principle and evolution.

Part two of this book thus devotes a chapter to each of these weighty topics. In each case I will first review the current knowledge, largely from Western sources, but will make serious efforts to find and relate Islamic viewpoints on each sub-topic, thereby allowing me to construct for the reader a reasonable synthesis.

Part three ('Outlook') will be devoted to considerations of issues in the science–Islam field that Muslim thinkers will need to address in the near future. I will end with an Epilogue relating an encouraging long conversation I held with my students on various aspects of Islam and science, a conversation that leaves me cautiously hopeful for the future.

One last remark before taking off: I should stress that although somewhat linearly constructed, the book allows the reader to read most chapters independently from one another. Depending on their backgrounds and interests, readers may skip some parts, though it is hoped that even the most informed will be hooked enough to follow the whole story with its many quirky episodes.

This should be an exciting adventure for us, as I know that readers will find many ideas and examples quite surprising, for the topics are so diverse, and the field covers many centuries, cultures and thinkers that everyone is bound to find all kinds of new, interesting views here.

part i

fundamentals (first things first): God, the Qur'an and science

I

Allah/God, philosophy and modern science

هُوَ الْأَوَّلُ وَالْآخِرُ وَالظَّاهِرُ وَالْبَاطِنُ وَهُوَ بِكُلِّ شَيْءٍ عَلِيمٌ

'He is the First and the Last, the Evident and the Immanent; and He has full knowledge of all things.'

Q 57:3

لَيْسَ كَمِثْلِهِ شَيْءٌ وَهُوَ السَّمِيعُ البَصِيرُ

'There is nothing whatsoever like Him, and He is the One who hears and sees (all things).'

Q 42:11

the truest thing

Like all Muslims – and many non-Muslims who come across such Islamic texts – I continue to be mesmerised by the Qur'anic verses and other Islamic prose that describe God. The sheer magic beauty of the above verses, and many others like these, commands the attention of whoever comes to read them or listen to their recitation. Moreover, the portrayal of God as mysterious and obvious, transcendent and immanent, with both anthropic and unimaginable features, as pairs of opposite traits, makes one want to know God even while realising that this is an impossible goal.

This is not a book of theology in the sense of the study of God. This is a book about the relation(s) between science and Islam. But we will not understand the various issues stemming from those relations, and we will not be able to construct reasonable and fruitful views, if we do not first understand the core principles of Islam. And there is certainly no more central and solid core to Islam than God Himself, and no more important set of principles upon which Islam is built than its sacred book, the Qur'an. Moreover, these principles themselves have been largely affected by centuries of philosophical discourse and conceptual revolutions brought in by modern science.

All Islamic discourse begins with God. All Muslims are taught to start every action of their lives with 'In the Name of God, the Compassionate, the Merciful'. Parents are told to call the Name of God in the ears of their newborns so that it could be the first word a Muslim hears in his/her life. And every toddler is trained to memorise the following short chapter of the Qur'an: 'Say: He is Allah, the One! Allah, the Eternal, Absolute; He begetteth not nor was begotten; And there is none comparable unto Him' (Q 112:1–4). Indeed, so important are these few verses that the Prophet is reported[i] to have declared them to represent one-third of the Qur'an.

Another striking verse about God, the one known as *Ayat-ul kursi* ('the Verse of the Throne') is known and recited by heart by innumerable Muslims and hung on walls of living rooms in beautiful calligraphies. It reads: 'Allah! There is no god but He, the Everliving, the Self-subsisting, the Eternal. No slumber can seize Him nor sleep. To Him belongs whatever is in the

---

[i] al-Bukhari: Vol. 9, book 93, hadith 471; translation by the USC-MSA Compendium of Muslim Texts, available at: http://www.usc.edu/dept/MSA/fundamentals/hadithsunnah/bukhari/093.sbt.html. It should be recalled that al-Bukhari's hadiths are rated as genuine (*'sahih'*) *by all orthodox scholars.*

heavens and on earth. Who is there that can intercede with Him except as He permits? He knows what is before or after or behind them, while they encompass nothing of His knowledge save what He wills. His throne does extend over the heavens and the earth, and He is never weary of preserving them. He is the Most High, the Supreme (in glory)' (Q 2:255).

There is a famous book of Arab wisdom, *al-Mustatraf*, a 775-page compendium of wise statements from the Arab heritage, classified into 200 or so topics, which was written more than a thousand years ago and is still widely printed (free of copyrights) and read today. Here is how the very first chapter begins:

> You must know [dear reader] that God exalted is One; He has no partner or equal; He is unique; nothing resembles Him. He is Absolute, and nothing comes close to Him (in His power). He is Eternal, past and future; there is no beginning to His existence and no end to His eternity. He is Self-sufficient, and no eternity wears Him or changes Him. He is the First and the Last, the Evident and the Immanent; He transcends materiality; nothing is like Him. He is above everything, yet His height does not make Him far from His creatures; He is indeed closer to them than their jugular veins; He witnesses all events, and He is with you wherever you may be[1].

The book goes on with such statements about God for several more paragraphs and ends this introductory section with the following quote: 'The Prophet Muhammad, may peace and blessings be upon him, stated from on top of the pulpit that the truest statement ever uttered by the Arabs is that everything other than God is false'[2].

In this chapter's title I have used 'Allah' to refer to God, yet so far in this introduction I have simply referred to Him as 'God'. Is there any difference? For all intents and purposes, there is no difference; in fact I usually prefer to use 'God', if only to convey the essential fact that the God of Islam is, the way He is defined (or, more precisely, how He defines Himself), no other than what monotheistic believers and atheists mean when they use the term or refer to the concept. In fact, the word 'Allah' is nothing more than *al-Lah*, 'the God' in Arabic. Some scholars, however, insist that Allah is the name of God, the one He gave to Himself, and He therefore should be referred to by that name. Others add that because of some theological differences between the God of Islam and the God of Christianity, for instance, one should use Allah in order to convey one's beliefs more accurately. As far as we are concerned here, any such theological differences are non-consequential, since we are focussing on attributes (creator, sustainer, omnipotent, omniscient etc.) that are common to all monotheistic religions.

knowing Allah

In the Muslims' general consciousness, the existence of Allah is obvious, so obvious that hardly any discussion about it is ever conducted. For a long time, Muslim scholars adopted the position that humans know Allah by 'instinct' (*fitrah*), relying on Qur'anic verses to that effect[3].

Atheism has only recently entered the vocabulary in the Islamic discourse. The whole concept is practically non-existent in the Qur'an, and the closest the Sacred Book comes to addressing it is when it refers to a group of people who denied the afterlife ('And they say: There is nothing but our life in this world; we live and die and nothing destroys us but time; but they have no knowledge of that; they only conjecture' (Q 45:24)); these then came to be known in Islamic parlance as *dahriyyun* (from *dahr*, time/epoch, i.e. those who believe that only 'time' affects us). In contrast to this disregard for the concept of atheism, the Qur'an abounds in discourse with the polytheists, making it an important goal to prove the Oneness of God (*tawheed*) and to display His attributes. Only very recently have Muslim scholars started to develop arguments to 'prove' the existence of Allah, as we shall discuss.

In every standard treatise and textbook[4] of Islamic knowledge, the first chapter is always devoted to *tawheed*. This term literally means, 'unifying', and technically refers to the ways a Muslim is supposed to believe in and describe Allah. *Tawheed* is often divided into three parts: (1) *tawheed* of divinity (*rububiyyah*); (2) *tawheed* of attributes (*as-Sifat*) and (3) *tawheed* of 'worshipability' (*uluhiyyah*).

The first part emphasises the divine aspect of God, such as being the creator and sustainer, the giver of life and death, the lord and overseer, the omniscient and omnipotent, etc. Muslim scholars, especially orthodox ones, expound their greatest efforts on this part, trying to draw sharp lines between the acceptable and the unacceptable descriptions of God. This is the area where much theological debate took place in classical eras, particularly between the rationalist school of Mu'tazila and that of the more orthodox scholars, as I shall explain.

The second part describes God in the 'proper way', meaning the way it has come to us from God Himself, as revealed and recorded in the Qur'an or from the knowledge He inspired into Prophet Muhammad's mind. This too is an area of contention, not around the attributes themselves but rather about our perception and interpretation of them. A well-known *hadith* (Prophetic statement) declares God to have 99 'beautiful names' (*asma' Allah al-husna*) and promises paradise to anyone who memorises them all[5]; in another hadith the Prophet tells us that there is a hundredth one, which grants any person entry to Paradise who utters it[6]. The list of 99 beautiful Names was actually given by the Prophet himself in another hadith[7]; it can be found in numerous references, including on many online web pages[8], often with translations, explanations, discussions and further references. Many artists have produced beautiful works out of these 99 names (some of them can be found on the web as well[9]); many Muslim homes and mosques are adorned with hung carpets and calligraphic paintings of the 99 names. The Prophet has also mentioned that Allah has a 'Supreme Name' (*Ismu Allah al-'ADham*); he did not state it explicitly, but from the stories relating this it appears that the Prophet meant either some special way of addressing Allah or some particular combination of His beautiful Names[10].

The third part of *tawheed* stresses the fact that only God has features that make Him worthy of worship. As one author put it[11]: 'THIS creator Has Perfect Attributes [capitals in the original text]. He is the First, nothing is before Him, the Ever Living. To Him is the Final Return . . . '; therefore, man must 'turn to his Creator Alone'.

This conception of Allah sounds clear and straightforward until we start delving more deeply into the matter. Most contentious is the question of attributes, particularly those that have a definite anthropomorphic aspect. For example, how is one supposed to understand the many Qur'anic verses that speak of God 'seeing' and 'hearing'? Indeed, 'The Seer' and 'The Hearer' are two of God's beautiful names. Or hadiths[12] that speak of God's 'right hand', with which he 'spends' (gives), or those where God is said to speak to humans on the Day of Judgment 'with no interpreter between them nor a screen to hide Him'?[13] From such questions arose a serious theological dispute regarding whether humans will actually see God on the Day of Judgment[14].

This question of whether humans will actually see God 'with their own eyes' produced the most important rift in Muslim theological history. The Mu'tazilites, who emerged in Iraq in the eighth century and quickly became – for awhile – the dominant Islamic school, attempted to inject some (early) rationalism in their understanding of theological matters and insisted that God is so beyond anything we can picture that 'seeing' God as well as 'God seeing' must be understood figuratively. In fact, they made this one of their five seminal principles of theology, alongside 'God is One and Absolute' and 'God is (infinitely) Just'. But the fundamentalists rejected any such figurative and interpretative approach and insisted on upholding the Qur'anic and Prophetic statements in their literal meanings, even if sometimes we may have to declare our inability to understand them. It is true that we are limited in our understanding of all that God has revealed, but this approach leads to more complications than it helps solve the problem we have started with. This would, in fact, make us prisoners to either literal understanding of verses, more and more of which will become problematic as we come to discover the world, or to suspensions of human understanding, so often that knowledge just becomes meaningless . . .

I should also note that contemporary Muslim thinkers of the tradition known as 'perennialist' (more on this later), like Seyyed Hossein Nasr and William Chittick, readily acknowledge and defend the anthropomorphic conception of God. In his most recent book, Chittick puts it clearly: 'The Islamic God is anthropomorphic, because the Islamic human is theomorphic. If God is understood in man's image, it is because man was created in God's image'. He further justifies this viewpoint: 'Unless God is understood in human terms, a yawning gap will remain between the ultimate and the here and now. *Re-ligio* or 'tying back' to God is impossible without images of God and imagining God'[15].

a brief review of Islamic conceptions of God

The above presentation of Allah should not mislead the reader into think-ing that the Islamic conception of God is quite simple (or perhaps even simplistic). In fact, we find in Islam's heritage huge amounts of theologi-cal discussion about God, His nature, His 'characteristics', and what can be known about Him and how – the whole literature of *kalam* ('Islamic theology'), particularly from the classical period of the Islamic civilization.

The traditionalist (orthodox) Muslim theological position is personified by the great eleventh-century scholar Abu Hamid al-Ghazzali, perhaps the most influential Muslim thinker of all time, considering the many areas that he addressed, from his contributions to jurisprudence, theology and mysti-cism to his discussion of 'beneficial and harmful' sciences, and particularly because of the way he sharply criticised not just much of philosophy but also many areas of knowledge, and how he canonised certain approaches to Islamic issues.

After a long intellectual and spiritual journey, he came to conclude that questions relating to God and to many religious matters simply cannot be resolved through reason. Indeed, he could not even convince himself that his belief in God was not the product of psychological delusion. The usual religious arguments failed to convince him beyond doubt, for he realised that they were all based on pre-assumptions about the validity of scripture, the reality of revelation and other such beliefs. As to the theological (*kalam*) and philosophical discussions, arguments and counter-arguments, he found them to be no more than pure speculation. As Karen Armstrong explains: '[according to al-Ghazzali], it was absolutely impossible to prove God's existence beyond reasonable doubt. The reality that we call "God" lay outside the realm of sense perception and local thought, so science and metaphysics could neither prove nor disprove the *wujud* [existence] of Allah'[16]. He ended up concluding that real belief in God can only come from a truly religious experience; indeed, though he rejected some of the extravagant mystical ways and claims, he found that purification and contemplation were the correct ways to approach God. Once that feeling of God was experienced, knowledge of Allah could be obtained from the revelation.

The way the mystics (Sufis) approach God is highly spiritual, almost anti-rational. In that old tradition, perpetuated today in many 'orders' (*tariqas*, or 'ways') led by revered sheikhs, who establish their 'spiritual credential' among their followers, awareness of God is achieved by contemplation and recitation of chants of particular Names of Allah and by prayers, an exercise which must be preceded by physical and spiritual purification. These exercises, which vary somewhat from one order to another, are said to lead to an experience of God, ultimately reaching a 'dilution' or even 'annihilation' (*fana'*) in Allah Himself. There is, in the Sufi tradition, no other way to prove the existence and presence of God to oneself; of course, one can and must read the Qur'an for guidance, enlightenment and knowledge, but true awareness and feeling of Allah must come from direct experience.

A slight variation on this approach is the 'experiential proof' that the Turkish thinker Said Nursi (1877–1960), who is often admiringly referred to as Badiuzzaman (the "novelty, or marvel of the times") had advocated. His approach emphasised two ideas: the need to delve and let oneself drown in the Qur'an, and some soft mystically flavoured emotional and spiritual 'recognition' of Allah by the way of observation[17]. In fact, Nursi put great emphasis on looking at nature and letting it fill one with awe and love for the creator, while establishing connection with God through His revealed text.

The philosophers of Islam are some of the most maligned intellectual personalities of our history. They are often wilfully misunderstood and wrongly portrayed. To this day, they are mentioned with suspicion and often declared as heretics, so their ideas are simply discarded. What was their 'crime'? In a nutshell, most, if not all, adhered so strongly to Greek philosophy, its principles, methods and body of ideas so that they readily and brazenly applied it to Islamic issues. The fundamental mistakes that their critics make, even today, is to lump them together (from the extreme and least orthodox ones, e.g. al-Kindi and al-Farabi to the genuinely Islamic, like Ibn Rushd) and to put under the same umbrella their misguided speculations and views with their original, constructive and fruitful contributions, some of which we are in dire need of even today.

On the conception of God, the Muslim philosophers simply believed that the God of Greek wisdom ('Sofia') was the same as Allah[18]. Indeed, for them it went without saying that Allah, if He is the true God, must be the same for all true-believing traditions. Not only that, He must be compatible with the intellectual advances made by peoples in different ages. So they regarded the principles, methods and conclusions of (theistic) classical philosophy as no different from the Islamic approach; some of them even considered philosophy to be the 'superior way' of knowing God and everything that followed from that knowledge, at least for the elite. And that is why the religious establishment reacted strongly. It produced al-Ghazzali, someone who could fight the philosophers on their own turf and defeat them, at least until Ibn Rushd emerged.

All the philosophers, as far as we read from their writings, believed in God and even considered His existence as quite evident, but still required the development of proofs. What is important for us at this point is the conception of God that the philosophers upheld: Did God create the world at some point in the past, or is the universe infinitely old (both of these options raising some troubling questions)? Does God act in the world or was His 'act' limited to the creation part?

This is where the 'hellenized' Muslim philosophers ran into trouble. On the one hand, they wanted to adhere to the Islamic dogmas, and on the other they found the Greek philosophers' arguments compelling. For example, al-Farabi argued that it did not make sense to imagine that God, the Eternal, 'suddenly' decided to create the universe, coming up with a reason and a plan for all creation and particularly for human existence. Slowly, the philosophers slid along a slippery slope and started to argue not just that the world was infinitely old but also that resurrection must not be physical and that God does not control (and, in fact, He may not even know) the 'particulars' (the details) of events in this world, down to the atomic level, as that would limit the natural freedom of actions and events in our world. These last few issues are what led al-Ghazzali to declare the philosophers as heretics – and in his view, they were all guilty of the same crimes.

But a century after al-Ghazzali, our hero Ibn Rushd grew up in al-Andalus, who believed that only philosophy could undertake a systematic rational discourse on God and all other important issues. To him, theology was not deductive but dialectical, and classical religious discourse was rhetorical, which is even worse. He boldly declared that philosophers were the only class of thinkers properly equipped to set doctrinal principles and even to interpret the Qur'an; according to him, they are the ones referred to in the Qur'an as 'deeply rooted in knowledge'[19]. He thus advanced the following set of principles, which sum up his theological doctrine[20]:

- The existence of God as Creator and Sustainer of the world.
- The Unity of God.
- The attributes of knowledge, power, will, hearing, seeing and speech, which are given to God throughout the Qur'an.
- The uniqueness and incomparability of God, clearly asserted in the Qur'anic verse 42:9: 'There is nothing like unto Him'.
- The creation of the world by God, though we don't know how or when.
- The validity of prophecy.
- The justice of God.
- The resurrection of the body on the Last Day.

It is clear that these principles can be accepted by most if not all Muslims, including the traditional and orthodox laymen; in fact it seems to me that some of these principles (particularly the third and the last one) lack some of the sophistication and willingness to show novel interpretation that Ibn Rushd has displayed in many of his works. But we are far from the 'heretic philosopher' image that the traditionalists like to paint of such thinkers; indeed, on these doctrinal bases alone, he should be adopted by all as a model of genuine unification.

The (hi)story of the influence of the Spanish medieval philosophers (Muslims like Averroes, and Jews like Maimonides) on Christian Latin philosophers and theologians has been recounted in detail by many authors[21]. On the topic of God, one finds remarkable similarity[22] between Ibn Rushd's above eight statements of creed and Maimonides' following 13 articles of faith: (1) the existence of God; (2) the unity of God; (3) the incorporeality of God; (4) the eternity of God; (5) the prohibition of idolatry; (6) the validity of prophecy; (7) Moses was the greatest of the prophets; (8) the divine origin of truth; (9) the eternal validity of the Torah; (10) God knows the deeds of men; (11) He judges them accordingly; (12) He will send a Messiah and (13) the resurrection of the dead.

Similarly, in Medieval Latin Christianity we find St. Thomas Aquinas (1225–74), whose *Summa Theologiae* was a hugely important and influential synthesis of classical philosophy with Christian dogma. Aquinas had been particularly impressed by Ibn Rushd's explanations and commentaries of Aristotle, and he slowly came to realise that Christian doctrine could be 'proven', or at least argued, on rational grounds. His starting point was that one had to better distinguish clearly between God's ineffable reality and humans' descriptions and understandings of Him. He emphasised that God is the Necessary Being, and that this was pretty much all that one could state about Him without running into the difficulties inherent in humans' attempt to imagine and describe Him, especially in words.

Aquinas's most important contribution to Christianity and to the West, therefore, is his insistence that God must be considered as a Being that transcends everything, from our thoughts and words to the cosmos and our existence. Some have blamed Aquinas for making God a philosophical concept, much too remote from us and our lives. There may be some truth to this viewpoint, but I believe that it is preferable to err on the side of transcendence regarding God than on the side of immanence and anthropisme. Only the more abstract concept of God, as opposed to the concrete one, would have a chance to survive the onslaught of the Enlightenment and the scientific revolution, so in that sense at least, the Averroes–Aquinas theological approach was a watershed, even though their views had very little impact on the masses and on the orthodox scholars.

The Muslim reformers of the late nineteenth century and the early twentieth century, Jamal Eddine al-Afghani and his followers Muhammad Abduh and, to some extent, Mohamed Rashid Rida, came at a time when the Muslim nation was far behind the West in every respect. The Renaissance, the scientific revolution, the Enlightenment and the Industrial Revolution had taken place, and these had left Europe not only well ahead of the Muslim world in general development, capability and power but also in colonial control of vast Muslim regions, from India to Morocco. The reformers were fully aware that the Muslim world was in need of modernisation in every sphere of social life, starting from education. They were also painfully conscious that even the general understanding of religion was woefully inadequate and dangerous, as full as it was of superstitious beliefs, passive notions of *qadar* (divine destiny), and so on. And, of course, the mechanical and naturalistic model of the world brought in by Newton and Darwin had made the traditional Islamic understanding of phenomena as Allah's 're-creation' of the world at every instant seem totally unphysical if not mystical, and this realisation had forced a re-examination of the very understanding of God and his relation to the world and to humans.

While firmly insisting on the continued truth of the central core creeds of *tawheed*, prophethood, divine revelation, etc., al-Afghani adopted a modernist and rationalist stand, at times apologetic and at times aggressive. He presented his views mainly in two documents that he published: (1) a long letter in which he rebutted Ernest Renan, who had claimed in a Sorbonne lecture (later published) that Islam, and the Arabs in particular, had not produced any science – because Islam was inherently incapable of that; and (2) a book titled *The Refutation of the Materialists*, in which he tried to answer not only the modern naturalists (mainly the Western atheists) on a variety of issues but also aimed to describe Islam as a rational faith. al-Afghani was not a scholar or a thinker, he was a political activist who had realised that the Muslim world could only be awakened, and its civilization restarted, if its ways of thinking and behaving were reformed and modernised.

His disciple Muhammad Abduh was a religious scholar who, in his young Egyptian years, had been seduced by al-Afghani's message. Having worked with his mentor for years, many of them in Paris, where he was strongly impressed and influenced by the Western thought and science, Abduh devoted the rest of his life to a mission of modernisation of the religious educational system, which he identified as the central problem; indeed, as long as the Muslim world continued to produce worldly ignorant Imams and religious scholars, there was no hope of affecting any social change.

Abduh tried to rid Islamic teachings from superstition and miracles, and insisted that Allah's ways of convincing humans changed drastically with the advent of the Qur'an, where rational discourse replaced physical miracles. Indeed, in one of his great legacies, the (unfinished) commentary on the Qur'an, *Tafseer al-Manar* (*the Commentary of the Lighthouse*), many verses are given decidedly modernistic interpretations; for example, proposing that bacteria were the organisms described by the Qur'an as *jinns*, traditionally understood as spirits (mainly evil ones). He also defended the classical philosophers' attempt to fuse dogma with reason, but insisted on the need to uphold all true Islamic beliefs and injunctions[23], including, for instance, the order to stay away from idle speculations about the nature of God.

Unfortunately, I must close this brief historical review of the concept of God in Islam with the fundamentalists, who came to dominate much of the Islamic discourse in recent times, starting with the establishment of the Wahhabi school of Islam in the Arabian Peninsula. The school, led by and named after Muhammad ibn Abd-al-Wahhab, had resurrected and put on a pedestal the hardline tradition of Ibn Hanbal, Ibn Taymiyah and al-Jawziyah. This school rejects all 'extrinsic' elements of thought brought into Islam by philosophers and theologians and insists on deriving every element of Islamic life from fundamentals (Qur'an and Hadith), with as little interpretation as possible.

Ibn Abd al-Wahhab's main objective was the 'purification' of Islam from *shirk* ('associating partners' with Allah) and from *bida'* ('add-ons'). *Shirk* does not simply refer to bona fide polytheistic beliefs, but – in his view – also to actions such as the widespread popular tradition of visiting saints (alive or dead) and imploring them to intercede with God on one's behalf. *Bida'*, in the view of the Wahhabi school, includes any and all actions that were not performed by the *Salaf*, the early Muslims, who were the models to be emulated, for they were 'the closest' to the Prophet's Islam, both in time and in their understanding; for example, they condemned the new habit of celebrating the birthday of Prophet Muhammad. (The adepts of this school, often referred to as Wahhabis, prefer the denomination Salafis.)

The Salafis have also been very wary and critical of Sufism, often regarding it as suspicious in its practices if not downright heretical in at least some of its beliefs, such as the idea of 'emanation' and 'god-like' nature of man. And Salafis often nonchalantly dismiss scientific and other truths whenever they appear to conflict with their literal understanding of Islamic texts or with injunctions found in the Qur'an and in the Hadith. No effort at interpretation is ever made to reconcile such truths; the Texts come first – complete with the readings and understandings of the *Salaf*.

other traditions' conceptions of God

One will not be surprised to hear that Muslim and non-Muslim scholars of different philosophical affiliations tend to find more similarities than differences between the God of Islam and that of Christianity. Unfortunately, we have heard in the recent troubled years a number of fundamentalistic Muslims describe Westerners as infidels, whether they are deeply committed to Christianity or to Atheism, and we have likewise heard quite a few Christian preachers insist that Muslims do not believe in the same God as Christians, that Muhammad is a fraudulent leader (and we'll refrain from citing the more extremely insulting descriptions), and that anyone who does not accept Jesus as his saviour will be headed straight to hell, whether or not she/he believes in One God and does good in his/her life. But more reasonable voices are not difficult to find[24].

The more reasonable religious scholars have emphasised the fact that the common features of Muslim and Christian belief in – and description of – God are much greater and more important than the differences.[ii] The website Submission.org, an interesting site devoted to providing religious education about Islam on a variety of subjects, has an article titled 'God in the Bible and the Qur'an' in the form of a tabular comparison of agreements and disagreements. Interestingly, there are 11 agreements and 11 disagreements. But a closer look reveals the weight of the agreements compared to the lightness and triviality of the disagreements; indeed, among the features of God accepted by both Islam and Christianity, one finds supremeness, omnipresence, omniscience, omnipotence, justice, worshipability, creation and action in the world, etc. And among disagreements, one finds man being made in God's image in the Bible, anthropic 'personality' of God (jealousy, anger, revengefulness, etc.) in the Bible etc.

[ii] For example, in a program of the John Ankerberg Show devoted to 'Islam and Christianity' on 8 December 2002, Dr. Gleason Archer stated that 'Islam and Christianity are closer to each other than to any other religions'. He adds that even the words used for God are essentially the same: Yahweh and Allah, echad and ahad (for One), etc. In the same program, Dr. Jamal Badawi emphasized upon the three major areas of agreement between Christianity and Islam: (a) that God is in His essence beyond our imagination and description, but He has told us fundamental things about Himself through the Names/Attributes he has given to Himself, most of which are common to the two faiths; (b) the very concept of God, a loving and guiding creator; and (c) belief in God is not akin to accepting some dry principle but rather a commitment to a life of meaning and direction (transcript available at www.johnankerberg.org/Articles/_PDFArchives/islam/IS2W1299.pdf).

Pope Benedict, Islam and the 'Rational God'

On 12 September 2006, Pope Benedict XVI delivered a lecture titled 'Faith, Reason, and the University: Memories and Reflections' at his old University of Regensburg, in Germany. It wasn't entirely clear whether he was then speaking as the old theology professor or as the official pope of the Catholic world. Indeed, he started his 'memories and reflections' reminiscing about his old days in the theology department, the seminars and debates, and the fact that theology was then a bona fide academic discipline, largely accepted as a valid area of 'scientific' investigation. And he ended his lecture by calling upon academia to restart the tradition of dialogues about God and reason, which he had claimed to be one of the main areas of civilizational exchange. But on the other hand, he was wearing his official pontifical garb, and his speech was distributed to the press and placed on the Vatican's website[25]. And so the world ignored the academic aspect of his lecture and reacted to it much more politically then theologically. Indeed, within hours one of the worst crises between the Vatican and the Muslim world erupted, with the general public, the media and quite a few educated and official persons responding harshly to the pope and calling for apologies and boycotts.

What produced the strong, sometimes violent, Muslim reactions was a passage in the lecture where the pope quoted a Byzantine emperor who, in a dialogue with 'an educated Persian' (a Muslim scholar) in Constantinople in 1391, had made offensive statements about Prophet Muhammad and Islam to the effect that the God of Islam, was encouraging or at least condoning 'spreading the faith through violence' (jihad, as misunderstood by both the emperor and the pope). The pontiff then went on to half-subtly contrast the God of Islam, who seems to accept violence, which, the pope insisted, was unreasonable and thus contrary to the nature of God, and the God of Christianity who was fully rational and in total harmony with the Greek ideals with which Christianity had identified from its very beginning. Benedict found evidence in the first words of the St. John's Gospel, 'In the beginning was the logos'; the pope insisted that 'logos' meant 'both reason and word'. In contrast with this fundamentally rational God of Christianity, the pope referred to Ibn Hazm, the eleventh-century Muslim theologian, who declared God to be beyond our categories, totally free to act as He pleases, even to 'have us practice idolatry' if He wishes, according to Benedict.

It must be stressed however that, despite the momentous aspect of these declarations (all within the first page and a half of the speech), Islam's God was only a secondary target in the pope's speech. Indeed, his main goal was the modern positivist culture, which had pushed God outside of the realm of valid academic discourse, declared Him to be 'an unscientific or pre-scientific question', and reduced 'the radius of science and reason'. He decried the 'dehellenization' of Christianity and called for 'broadening our concept of reason and its application'.

Muslim scholars mounted a strong defence26 and some powerful counter-attacks.27. The 'Open Letter to His Holiness', signed by 100 Muslim scholars and dignitaries, pointed out factual errors in the pope's speech, stressed the importance of reason in the Islamic/Qur'anic discourse, and insisted on keeping the spirit of the Second Vatican Council's declarations alive[iii]. Aref Ali Nayed, after providing useful background to the pontiff's sudden slanted and superficial interest in Islam, undertook a point-by-point critique and rebuttal of the pope's lecture. As far as it concerns us here (the question of God and reason), Nayed remarked that Ibn Hazm and other Muslim scholars are correct in declaring God totally free to make decisions as He wills but that this does not mean that He grants Himself unreasonable freedom; after all, He has stated in the Qur'an that 'He has ordained mercy on Himself' (Q 6:12). (I may add that in a famous hadith qudsi, a non-Qur'anic divine pronouncement, He tells humans: 'I have prohibited injustice on Myself'.) By analogy, and although there is no specific statement to that effect in the Islamic texts, Nayed concludes with the following: 'Reason need not be above God and externally normative to Him. It can be a grace of God that is normative because of God's own free commitment to acting consistently with it'. He further warns against subjecting God to some human-defined 'reason' and stretching 'loaded words as logos and reason' and remarks that much of the medieval discourse depended on that kind of 'ambiguity-fueled leaping'. Finally, he adds, 'Although the pontiff strives to convince a secular university that theology has a place in that reason-based setting, he should not go so far as to make God subject to an externally binding reason. Most major Christian theologians, even the reason-loving Aquinas, never put reason above God'.

[iii] 'The church has also a high regard for the Muslims. They worship God, who is one, living and subsistent, merciful and almighty, the Creator of heaven and earth, who has also spoken to humanity. They endeavor to submit themselves without reserve to the hidden decrees of God, just as Abraham submitted himself to God's plan, to whose faith Muslims eagerly link their own. Although not acknowledging him as God, they venerate Jesus as a prophet; his virgin Mother they also honor, and even at times devoutly invoke. Further, they await the Day of Judgment and the reward of God following the resurrection of the dead. For this reason they highly esteem an upright life and worship God, especially by way of prayer, alms-deeds, and fasting' (Nostra Aetate, 28 October 1965).

proofs of Allah's existence

In the classical philosophical traditions of humanity, arguments for the existence of God have usually fallen into the following three to five categories (authors sometimes merge or ignore a few of these): (1) the cosmological argument; (2) the ontological argument; (3) the design/teleological argument; (4) the moral (within) argument and (5) the spiritual experience argument.

The cosmological argument, first introduced by Aristotle and later adopted by some of the major philosophical and theological figures in history (al-Kindi, Ibn Rushd, Thomas Aquinas and others), has several formulations, including the 'first cause' and the 'prime mover'; they all basically stipulate that the existence of the world can be traced back, from cause to cause, but this tracing back cannot proceed indefinitely and must stop at a first cause, God the creator.

The second argument (ontology), which was forcefully put forward by Anselm and Descartes, is a 'logical' proof, which essentially argues that because we can conceive of a God-Creator, that is, in our minds the greatest being that can be conceived of, either it exists only in thought or in reality; but as the latter is clearly a superior existence/possibility compared to the former, this greatest being must exist in thought and in reality.

The third classical argument for the existence of God is the simplest: It is impossible to observe all the glorious complexity and harmony (internal and mutual) of all the creation (both piece by piece and in totality) and not think that a supremely powerful and intelligent creator is behind it.

The fourth argument relates both to the existence of a 'moral law within us' and to the necessity of the ultimate triumph of the good (something that often is not fulfilled here on the earth), hence the necessity of the existence of a God who gives meaning and closure to our lives.

The last argument is the recognition of the mere existence of a widespread religious or spiritual experience that a multitude of humans report upon; in the view of some, this surely must point to the actual existence of a source of such a spiritual channel.

I shall come back to some of these arguments in later chapters, particularly the cosmological and the design arguments. Here, however, I want to focus on the ideas that Muslim authors have emphasised. Classical thinkers used to stick to the simple Qur'anic view that Allah's existence is obvious; it is instilled into us as *fitrah* (instinct); His presence can be glimpsed through His creation and action in the world; all that is needed is to 'clean up' one's soul to see that clearly and to correct any misconceptions about Him, His Oneness and His Attributes.

In his book titled *Islamic Dogmas*[28], the respected author and scholar al-Seyyid Sabiq starts by stressing that knowledge of Allah comes from two approaches: from rational and natural explorations, and from a proper understanding of His Names/Attributes. Let us focus on the first method, because that is closer to the idea of proof. By rational thought, Sabiq refers to the many Qur'anic injunctions to 'reflect' (*yatafakkarun*, Q 3:191) and 'comprehend' (*ya'qilun*, Q 2:242). He insists that belief must come from conviction and cites Qur'anic arguments against conformism, which was the attitude of unbelievers. He warns, however, that reflection must be about God's creation and signs (natural and revealed) and not about God Himself.

A major section of his treatise, however, Sabiq devotes to the 'natural' arguments (today referred to as 'design'); indeed, he starts with the repeated Qur'anic command to observe: 'Say: Consider/Observe (*'UnDhuru*) what is in the heavens and the earth' (Q 10:101); 'Say: Travel in the land and observe how He originated creation' (Q 29:20); and then cites one of the most direct passages from the Holy Book (Q 27:60–64):

> Or Who has created the heavens and the earth, and sent down for you water from the cloud; then We cause to grow thereby beautiful gardens; it is not in your power to cause the growth of the trees in them. Is there a god besides Allah? Nay! They are people who deviate.
>
> Or, Who made the earth a fixed abode, made rivers in its midst, set thereon firm mountains, and hath set a barrier between the two seas? Is there any god besides Allah? Nay, most of them know not.
>
> Or, Who answers the distressed one when he calls upon Him and relieves the suffering (or evil), and He will make you successors in the earth. Is there a god besides Allah? Little do they reflect!
>
> Or, Who guides you in the darkness of the land and the sea, and Who sends the winds as heralds of His mercy? Is there a god besides Allah? High Exalted be Allah from all that they ascribe as partner (unto Him)!
>
> Or, Who originates creation, then repeats it, and Who provides for you from the heaven and the earth? Is there a god besides Allah? Say: Bring your proof if you are truthful!

Sabiq dismisses the naturalistic (or materialistic) views on God and creation too superficially; for example, he claims that atheists consider the

cosmos to have come out of nothing with no 'originator' and that all creatures in nature are the result of chance transformations; in particular, he
blasts Darwin's evolutionary theory simply because, in his view, it rules out
purpose. Unfortunately, his discussion of these important issues remains
very cursory.

Sabiq lists other arguments in favour of the existence of Allah and briefly
discusses them:

- Human instinct indicates the existence of God; he refers to the following
  Qur'anic verse: 'And if misfortune toucheth a man, he crieth unto Us (in
  all postures) – lying down on his side, or sitting, or standing, but when
  We have relieved him of his misfortune he goeth his way as though he
  had not cried unto Us because of a misfortune that afflicted him' (10:12).
  He cites other similarly subjective arguments, which he considers to be
  separate from this one, though clearly they are not: (a) human worldly
  experience shows that God does indeed respond to our prayers, thus He
  exists; (b) history shows that God has intervened to support His good
  servants (prophets and others) etc. One might be tempted to relate these
  ideas to the aforementioned 'argument from religious experience', but
  this is much less; religious people may recognise in this feeling an innate
  seed of divine awareness, but non-religious people will readily dismiss it
  as illusion (or delusion[29]).
- Atheism presents no good argument/proof for its case; in fact, according
  to our scholar, statistics show that even scholars and scientists overwhelmingly (more than 90 percent of them, he says) reject atheism.
  Unfortunately, Sabiq gives no reference to his statistical figures, not even
  a date, when in fact we know that surveys from the whole twentieth
  century show far less religiosity among scientists, academics and highly
  educated people[30].

Another well-known religious scholar, Omar al-Ashqar, has expounded on
the same theme. The arguments he presents for the existence of God are the
following:[31]

- The 'argument from instinct', for which he cites a verse (Q 30:30), a 'holy
  statement' (hadith *qudsi*) and a Prophetic statement (hadith)[32].
- The argument of miracles shown by prophets, like Abraham, Moses and
  Jesus. Needless to say, this is a non-sequitur, as one has to believe in
  these prophets and their miracles to begin with, in which case the belief

in God is already there. Likewise for his argument about sacred books and prophets having brought very similar messages, here too, many people will disagree and point to the wars of religions as an indication of (extreme) disagreement rather than agreement.

- The argument of cosmic order, for which he cites the Qur'anic verses like, 'Behold! in the creation of the heavens and the earth, and the alternation of night and day, there are indeed signs for men who understand' (Q 3:190). This is the same as the argument from design, to which I shall devote a full chapter later, but al-Ashqar counts it separately; for this he cites the Qur'anic verse: 'He Who hath created seven heavens in harmony. You see no incongruity in the creation of the Beneficent Allah. So look again, can you see any disorder?' (Q 67:3)

- The argument of 'divine direction' for everything He created, for according to our scholar organisms and organs have to be 'directed' towards the functions they have been created for; he quotes the Qur'an, e.g. 'Our Lord is He Who gave to each (created) thing its form and nature, then guided it (to its goal)' (Q 20:50).

- Finally, he presents an argument akin to the anthropic principle (that the universe is extraordinarily well suited for complex life and intelligence, including humans – 'anthropos'), and he finds support for it in the Qur'anic idea of *taskheer* (putting at one's service) as presented in Q 31:20.

Other standard textbooks of 'Islamic knowledge' support and reinforce some of the above arguments and add a few new ones. Muslim and al-Zughbi[33], for example, strongly uphold the design argument, citing additional Qur'anic verses, but introduce the argument of providence, not only towards humans but for all creatures; they cite the following, among other verses: 'And thy Lord inspired the bee to build its cells in hills, on trees, and in (men's) habitations; then to eat of all the produce (of the earth) and find with skill the spacious paths of its Lord ... Verily in this is a Sign for those who reflect' (Q 16:68–69). They also support the 'instinct argument', adding new aspects to this approach; for instance, the 'spiritual longing' of humans (this is now much more like the argument from religious experience) as well as the feeling that because we humans are flawed and limited, there must necessarily be a perfect and complete Being, viz. God.

Finally, 'Azmi Taha al-Sayyid and his co-authors[34] insist first and foremost on what has classically been dubbed as 'the cosmological argument', that is the argument of the creator and prime mover. There are many verses to this effect; the two most striking ones are as follows: 'Is He then Who creates

like him who does not create? Do you not then mind?' (Q 16:17); 'And verily
ye know the first creation. Why, then, do ye not reflect?' (Q 56:62).

modern views

The Enlightenment (the French version, in particular) and the scientific
revolution had a profound impact on humans' conception of God, at least
among the elite.

First came the Protestant Reformation, an indisputable revolution, which
argued for a new conception of God, moving the emphasis from reason to
heart, from conviction to pure faith. God was no longer a Being which could
be proven and whose methods could be identified with those of reason.
The Augustinian viewpoint was brought back to the fore, making God an
unknowable entity, one which could only be felt emotionally and adhered
to by (unreasoned) choice.

Blaise Pascal (1623–62) was the first philosopher/scientist to conclude,
interestingly enough not by any rationalisation but rather by a spiritual
experience/crisis, that one cannot convince himself, let alone anyone else,
of the existence of God, and that the choice of belief was a deeply personal
one – the famous 'wager' and leap into the dark which ends up bringing
moral enlightenment. With this stand, Pascal had diametrically opposed
René Descartes' 'method', by which he rationally and systematically addressed any issue, including the existence of God, which he claimed could
actually be proven more easily than any other being. (Descartes' 'proof',
we shall recall, was the ontological argument mentioned in the previous
section.) It took some time for Pascal's modernist attitude to spread itself
widely, for in the meantime came the great Isaac Newton (1642–1727), who
unlike Descartes used a bottom-up approach, starting from the world and
working his mind up to God. Universal gravity, for example, Newton's greatest discovery/invention, was for him the evidence not only of God's creation
of the world but of His existence and continued sustenance of the cosmos, for
without God's perfect disposition of celestial objects, these would attract
each other and fall to the centre. This then was proof of God's existence,
power and action in the world. Newton then went on to relate God's attributes (infinite, eternal etc.) from physical observations (existence of time
and space, motion of objects, etc.). But with his physical explanations of
how the world works, both in the heavens and on the earth, Newton allowed
others (Laplace and other physicists) to introduce a 'mechanical universe'

that had, in Laplace's famous retort to Napoleon's question, 'no need for the God hypothesis'. The only thing for which God was still needed was the initial act of creation. Thus came the new concept of 'deism', the belief in a God who created and then retired from the world, never to have any affect upon it thereafter.

The (French) Enlightenment movement was based in large part on the sidelining – if not total erasure – of God and particularly on the discarding of revelation and its moral laws. Instead, one had to construct the world and its rules from scratch and do it as simply and justly as possible. The philosophers of the Enlightenment did not necessarily reject the concept of God; they only wanted to redefine religion on the basis of reason – and only reason – and establish a new 'humanist' social order. In fact they stressed that the idea of God was indeed needed and had to be created, if necessary; Voltaire himself called atheism 'a monstrous evil', particularly if the governing and educating elite fell prey to it and influenced the masses; he regarded it as another form of superstition and fanaticism. The philosophes had a problem much more with religion than with God, whom they claimed could even be proven by science and reason.

The fusion of religion and the Enlightenment was best personified by Baruch Spinoza (1632–77), a Dutch philosopher of Spanish-Portuguese descent, who had become dissatisfied with the orthodox Jewish theology after joining a circle of freethinkers. Spinoza, contrary to how he tends to be presented in popular literature nowadays, did not reject God, although he did get 'excommunicated' from the synagogue for his beliefs. Spinoza simply redefined the concept of God, away from the traditional description to a very modern one. Einstein, for example, the twentieth-century scientist par excellence, later adopted Spinoza's definition of God, which was an identification with the laws of the universe, or perhaps the underlying principle that gives birth to (or produces) the laws of nature. God's action in the world was then replaced by the mathematical rules that governed how things behaved. God then had to be the greatest, most perfect Being/Idea imaginable and the order to which He gave rise truly inspired awe and wonder in the heart of Spinoza and in the mind of Einstein. Everything we come to know and feel ultimately derives from Him; everything then 'emanates' (as the mystics say) from Him; in this conception, that kind of individual and deep personal realisation is tantamount to a revelation.

Modern science brought at least three hugely important theories, which had significant impact on our conception of and belief in God: Darwin's theory of evolution (explained by natural selection); the cosmological Big Bang theory and Quantum Mechanics (QM) as a description of the microphysical world.

I shall be reviewing the first two quite extensively in Chapters 9 and 6, respectively, and will then discuss their impact on theology, i.e. the very belief in God and the nature of the creation process, for the universe as a whole, and for organisms (and humans) in particular. QM, however, will not be discussed in this book, except in the last chapter, where I briefly review the modern propositions for divine action in the world. It is thus fitting to mention it here in relation to God, in terms of his attributes, as Creator and Sustainer of the world(s), and in terms of his relation to (and action in) the world.

Quantum Mechanics cannot be explained simply to a non-scientist. It is doubtful, in fact, whether it can be explained by anyone to anyone. As Richard Feynman famously put it, 'I think I can safely say that nobody understands QM'. The reason for that is because the world of particles, atoms and molecules it describes (so well) turns out to behave in a very strange way compared to what we are used to in the macroscopic world. In fact, even using the word 'particles' is misleading, since 'quantum objects' (my preferred phrasing) are neither particles nor waves, but things somewhat resembling this and that but in the end acting like neither. To succinctly summarise QM, one can say that quantum objects are ruled, first and foremost, by various uncertainties (known as Heisenberg's), which make it fundamentally impossible for us to determine all their properties (position, speed, energy, spin state, etc.) precisely. The best mathematical description that has been constructed for the description of the behavior of quantum objects is the 'wave function' (first introduced by Schrödinger, then also used by Dirac, when he came up with a relativistic version of the theory/equation). The wave function carries all possible information about the quantum object, including all possible positions, speeds, energy values and other parameters it can be found in, and it is only when a measurement is made that the quantum object 'takes on' a specific value for the physical quantity being measured.

There are two important interpretations to this strange behavior: One is the widely adopted so-called Copenhagen interpretation (for the school of physicists, led by the Danish great Niels Bohr, who came up with and publicised it) and one is Anthony Leggett's parallel branching of the universe into different options. The Copenhagen interpretation stipulates that each property of the quantum object is some kind of probability cloud, having no specific value and evolving through time; only the act of measurement 'collapses' the cloud into specific states with 'real' physical quantity values. This is where the popular books usually bring up a famous discussion that took place between Einstein and Bohr, the first one strongly opposing this 'indeterministic' description of the world, and the other, of course, defending his probabilistic interpretation. Over a heated dinner discussion, Einstein told Bohr: 'God does not play dice', to which Bohr replied: 'Albert, don't tell God what He can or cannot do. . .' As Witham[35] comments on the encounter: 'Einstein operated out of ancient Middle Eastern monotheism, which clashed with Bohr's ancient Asian precepts that defined Reality as a godless flux. . . Thus, no argument by Bohr could persuade Einstein, who could not abandon Spinoza's God'.

Various great scientists later took one side or the other and adopted various philosophical positions on the thorny question. Heisenberg, in particular, stressed that his uncertainty principle had allowed for a new acceptance of the 'natural language' of religion. As he put it: 'After the experience of modern physics, our attitude toward concepts like mind or the human soul or life or God will be different'. Most importantly, Heisenberg realised – and told Bohr – that many people would move to use the new indeterminacy as 'an argument in favor of free will and divine intervention'. Still, others were very cautious; Eddington had remarked to the 'religious reader' of his 'Nature of the Physical Universe' that 'I have not offered him a God revealed by the quantum theory, and therefore liable to be swept away in the next scientific revolution . . .'

As to Muslims, I have yet to come across a cogent integration of QM principles into Islamic beliefs. Searching the literature, one can find a number of cocktails, from QM-fueled Sufi mysticism to metaphysical reinterpretations of QM; some[36] have also stressed the non-materialistic nature of QM. Let me take just one such example, an article titled 'Quantum Islam: Does Physics Confirm the Qur'anic Worldview?'[37] – no less. The author, O'Barret, is impressed that 'the quantum description of Reality puts an end to the age of materialism, by requiring the existence of nonphysical Mind behind the physical brain', something he identifies with *ruh*, the Islamic term for spirit; he does not explain how he establishes the existence of that mind. For him, however, it is this *ruh* that makes the choices between the various possible quantum realities. Secondly, O'Barret wholeheartedly adopts Leggett's 'branching into parallel universes' interpretation of QM, instead of the more dominant Copenhagen interpretation. Why? First, because the existence of many worlds in a realm hidden from us corresponds (in his view) to the Qur'anic concept of *'alem al-ghayb* (the unseen world), which the author explains as 'the concealed dimension of reality'. And secondly, because 'Muslims . . . have always known Allah as *rabb al-'alamin*, the Lord of the worlds'. As one can see, the religious resonances that writers have generally tried to find with QM are not very profound or convincing. There remains the question of how QM has been seen to provide a channel for God's action in the world, and this we shall review in Chapter 10.

the creator and sustainer

Seyyed Hossein Nasr explains[38] that being a creator was a logical necessity for a One God defined as Infinite, Absolute and Utterly Good. This viewpoint has the merit of rendering moot the famous question, 'And who then created God?'; indeed, if God, – the – Creator, is not only a logical necessity, then there cannot be another creator, also infinite and absolute, who would have created the one we just defined as the creator. In addition to this, one quickly infers that God must be outside of time, for if He were temporal, He would have some of the qualities of worldly things, which is contrary to our starting definition. (Recall that we must always uphold the Qur'anic principle of 'Nothing is like Him' (Q 42:11).) And if He is Infinite and First, then He must be self-sufficient and thus also be everlasting. And with such a definition and logical properties, it becomes obvious that such a Being must be One.

Nasr adds another reason for God to be a Creator, one that had been advanced later by St. Augustine and many others: It is in the nature of the good to give of itself, thus the utterly good could not but create the world (out of Himself).

Several of the Names/Attributes (among the 99 previously discussed) refer directly to God's power as creator. Among those are the Creator (*al-Khaliq*), the Maker (*al-Bari'*), the Fashioner of forms (*al-Musawwir*) and the Originator (*al-Badi'*). Indeed, the Qur'an makes the act(s) of creation the first argument that should impress humans in recognising God and His supreme powers: 'Were they created of nothing, or were they themselves the creators? Or did they create the heavens and the earth? Nay, they have no firm knowledge' (Q 52:35–36).

The Qur'an also stresses Allah's omnipotence in creation: 'To Allah belongs the dominion of the heavens and the earth. He creates what He wills (or pleases)' (Q 42:49). And, of course, the Qur'an attributes to God the power to create out of nothing ('ex nihilo'): 'And Our word unto a thing, when We intend it, is only that We say unto it: Be! and it is' (Q 16:40).

In Islam the power to create is so exclusively associated with and reserved to God that in many religious schools, drawings, paintings and sculptures are prohibited, especially if the works are to portray living creatures. The many orthodox scholars who adopt this position rely on a literal interpretation of Prophetic hadiths like the following: 'The painters of these pictures will be punished on the Day of Resurrection, and it will be said to them, "Make alive what you have created"'[39]. I should note, however, that other scholars later interpreted these hadiths as applying only to those who may ascribe to themselves a creator attitude ('Allah said, "Who are more unjust than those who try to create something like My creation? I challenge them to create even the smallest ant, a wheat grain or a barley grain"'.[40]). They thus concluded that those who draw or paint (sculpture has mostly remained a taboo in Muslim society) for constructive purposes should not fear such prohibition or punishment. Still, the underlying idea remains strong: Creation is an act reserved for God!

Classical Muslim theology goes even further than defining Allah as the Creator; He is also considered the Sustainer of the world(s): 'Allah is Creator of all things, and He is Guardian over all things' (Q 39:62); 'It is Allah Who sustains the heavens and the earth, lest they cease (to function): and if they should fail, there is none – not one – can sustain them thereafter: Verily He is Most Forbearing, Oft-Forgiving' (Q 35:41); 'There is no creature on earth but its sustenance is upon Allah' (Q 11:6).

This in fact takes us to a long and unresolved debate among Muslim theologians regarding the way God acts in the world or controls it. Some have argued that He sustains the world by way of the laws he built in nature; in Islamic parlance, God uses 'divine habits' (*sunan ilahiyah*). Others have insisted that God acts directly, that he 'recreates' the world at every instant. Opponents of this position refer to it as 'occasionalism'. Rationalists tend to subscribe to the first view; traditionalists tend to adopt the second position.

Going back to the question of creation itself, I must emphasise one idea that is strongly upheld in the Islamic world view, that of purpose. There is no doubt in Islam that the world was created for a purpose; it was not just left to follow random transformations and produce whatever creatures may come about. This important point of contention between materialistic scientists, in particular, and religiously inclined thinkers will surface again in our discussions later, especially when we come to biological evolution; we shall encounter it also when I review the physical evolution of the universe, cosmology, fine-tuning and the extraordinary appropriateness of the laws of the cosmos for life, intelligence and humans to appear. It is thus important to emphasise from the start that in Islam purpose is a fundamental aspect of creation.

On the other hand, the purpose of creation itself is not obvious for humans; our mental abilities are limited, and the revealed scriptures have presented us only with spiritually oriented answers ('I created the jinn and humans only that they may worship Me' Q 51:56).

A different and important perspective on creation is the one proposed by some of the medieval thinkers and espoused by Nasr and a few like-minded contemporaries; it can be summed up in one world, 'emanation', and it represents the 'cosmology' of the old perennial mystical philosophy. Indeed, Nasr relies on a rather disputed hadith *qudsi* (divine statement made outside of the Qur'anic corpus) where Allah (presumably) stated: 'I was a hidden treasure. I loved to be known. Therefore, I created the creation so that I would be known'[41]. The cosmos is then viewed as a 'theophany' (*tajalli*), a 'primordial revelation' from Allah, building on the fact that the word *ayat* (signs) is used in the Qur'an to refer both to the revealed scriptural verses and to the phenomena of nature. In earlier writings[42] Nasr had explained that the regularity of natural and cosmic phenomena is a confirmation of all creatures' subservience to God, in fact by no choice of theirs. He also finds in the harmony between religious rites (of Muslims in particular) and natural phenomena (solar-timed prayers, lunar-timed fasting, etc.) further evidence of that divine common origin and connection (between humans, nature and God).

Nasr also interprets the verse 'We will show them Our signs in the horizons and in themselves, until it becomes manifest unto them that it is the Truth . . . ' (Q 41:53) to mean that human beings themselves are 'like the macrocosmos'[43] and therefore constitute a revelation too. Nasr finds similarities between some divine attributes and some human characteristics, particularly those of intelligence, will and speech[44], of course with the obvious difference that all of God's attributes are infinite and absolute, while human features are limited and relative. Our Sufi philosopher goes further and declares man to be a 'theomorphic being', one that has been made 'in the image of God'; humans thus are – or can be – God-like and, by refusing the 'trust of faith' can also 'play the role of a little deity and deny God as such'[45].

summary: who is God?

We have seen that the concept of God in Islam can be considered with various degrees of sophistication. The general conception presented by the traditional religious scholars and taught to children in schools is simple and rather satisfactory: God is One; He is the Creator and Sustainer of the universe; to Him we shall all return to account for our actions; He acts in this world and answers our prayers; He is beyond our imagination, yet He is 'closer to us than our jugular veins'. He can be known from the way He portrayed himself in the Qur'an and from the way His Prophet described Him.

But on another level, we do find in the Qur'an itself descriptions of God that are beautiful but mysterious. One famous and remarkable verse that many Muslims know by heart and recite while admitting not to have much understanding of what it means is the Verse of the Light:

> Allah is the Light of the heavens and the earth. The parable of His Light is
> as (ka) a niche within which is a lamp; the lamp is enclosed in glass; the glass
> as it were a brightly shining star; lit from a blessed olive-tree; neither of the
> east nor of the west, whose oil would almost glow forth (of itself) though no
> fire touched it. Light upon Light! Allah does guide whom He will to His
> light. Allah does set forth allegories for men; and Allah is Knower of all
> things. (Q 24:35)

Many pages have been written in commentary of this one verse – and we can imagine why. As Karen Armstrong remarks, there is strong emphasis on the allegorical aspect of this description, starting with the usage of the word 'parable' and of the participle *ka* (Arabic for 'like'), 'a reminder of the essentially symbolic nature of the Qur'anic discourse about God'[46].

Another researcher, Qamar-ul Huda, reminds us that 'the Qur'an is not a treatise about God and his nature, instead it is a reminder for humankind of God's infinite mercy; God is Creator, sustainer of the universe and of human beings, and in particular the giver of guidance to his creation'[47].

But if the Qur'an is not a treatise about God, and if its verses are often highly allegorical, how do we get to know God? What about the Hadith? We have seen that they emphasise knowledge of God by way of his 99 Names, and that was a path conducive to much understanding about the Omniscient and Omnipotent Lord and Creator (to mention just a few of His attributes); however, we also found that several hadiths adopted anthropomorphic descriptions of God, which later led to theological disputes.

We also found that the philosophers and the mystics adopted different viewpoints about God's nature and His relation to the creation (the cosmos and humans), as well as different 'proofs' of His existence, sometimes emphasising phenomena in the physical world and sometimes stressing the personal spiritual experience, each of which supposedly pointing us to God.

Last but not the least, I reviewed – somewhat briefly – the effect of modern philosophy and science on the concept of God:

- The Protestant Reformation made belief rest less on reason and more on 'pure faith'.
- The Enlightenment made God an unnecessary assumption and, if present, disconnected Him from the world, hence the concept of deism.
- Darwinism showed that features of all creatures could be neatly explained in a natural way, thus doing away with the idea of a Designer.
- Modern Cosmology seemed to bring reintroduce the idea of a moment of creation, thus apparently the need for a Creator.
- Modern physics (QM) destroyed the determinism of classical physics and opened the door for other levels of reality and for divine action.

Now, while recognising the impossibility of understanding and properly describing God, as well as the multiplicity of paths that one can take to Him, and acknowledging the rational constraints that philosophy and science present around the concept of God, can we still attempt some kind of synthesis, or at least some common denominator that most, if not all, believers (including Muslims) can adopt?

This is perhaps where the emphasis on the Creator attribute becomes crucial; indeed, it is not only a necessary aspect of anyone's definition of God but one that automatically encompasses other fundamental attributes (Infinite, Absolute, Omnipotent, Omniscient etc.); it also makes the world contingent upon Him, and it further implies purpose, which I have proclaimed to be a central tenet of the Islamic doctrine. On the other hand, it leaves open the question of how creation was performed, how the world, life and humans evolved, and it keeps open the nature of the purpose behind creation.

Clearly that the question of God is intimately related to science (and philosophy), we shall thus return to these issues in various ways in future chapters.

2

the Qur'an and its philosophy
of knowledge/science

'Islam is a culture grounded on a book, the Qur'an.'

Massimo Campanini[1]

the Qur'an in the Islamic culture

'In the beginning was the Qur'an[2].' With this smart take-off on the opening sentence of St. John's Gospel, Fr. Georges C. Anawati,[i] an Egyptian Dominican scholar and specialist of Islamic thought, aptly and pertinently sums up the status of the Qur'an in Islam.

Many other authors have expressed in similar ways the extraordinary position the Qur'an occupies in the Muslim culture. Sachiko Murata and William C. Chittick, for instance, refer to it as 'is undoubtedly one of the most extraordinary [texts] ever put down on paper'[3]. More recently, Reza Aslan adopted Kenneth Cragg's phrase, calling the Qur'an 'the supreme Arab event'[4]. Furthermore, Aslan refers to the common Muslim view that the Qur'an was Muhammad's God-given miracle, noting that '[i]n Muhammad's

---

[i] Fr. Georges Chehata Anawati (1905–94) was the Director of the Dominican Institute of Oriental Studies in Cairo, Egypt, a leading center for the Christian study of Islam and for Christian–Muslim dialogue.

time, the medium through which miracle was primarily experienced was neither magic nor medicine,[ii] but language'.

In their daily lives, Muslims treat the Qur'an as a sacred book, both in the text it contains and as a package. Among the many examples one could cite in this regard are the facts that the Qur'an is always supposed to be placed on top of a pile of books, that it must never be put on the ground, that no Muslim would carry even fragments of it into a bathroom and that one is supposed to touch the Qur'an only after having performed ablutions (the rites of cleaning that one undertakes, usually with water, before prayers).

Westerners are often surprised by the constant referral to Qur'anic verses by both Muslim scholars and laymen on any topic of importance. In fact, referring to the place that the Holy Book occupies and the role it plays in Muslims' lives today, Suha Taji-Farouki notes that 'millions of people refer to the Qur'an daily to justify their aspirations or to explain their actions' and considers the scale of that kind of direct reference to have in contemporary times reached levels that are 'unprecedented in the Islamic experience'[5].

Seyyed Hossein Nasr has remarked that even though the Qur'an can be compared to the Old and the New Testaments, the more proper analogy to be made is not with the Bible but rather with Jesus Christ himself. Indeed, both the Qur'an and Jesus can be defined as God's logos, which He sent in similar forms to Muhammad and to Mary. Furthermore, in Christianity both the spirit and the body of Christ are sacred, in the same way that I have noted with regard to the Qur'an, that is as text (in Arabic), as meaning(s) and as an object. In fact, many Muslims regard the Arabic language itself as sacred in a way, for it carries the original word of God. And that is why Nasr rejects the 'rationalist and agnostic methods of higher criticism'[6] that secular scholars propose to apply to the Qur'an as a text, just as Christians would object to having the remains of Jesus (were they to be found miraculously intact) dissected and subjected to modern medical techniques, with the aim of determining whether Jesus was born miraculously or was the son of Joseph.

One defining feature of the Qur'an is the otherworldly quality it exhibits in its original Arabic version. In his standard textbook *Characteristics of the Noble Qur'an*[7] (now in its tenth edition), Fahd Ar-Rumi, professor of Qur'anic Studies at the College of Teachers in Riyadh, Saudi Arabia, takes at least 40 pages to review just the style of the Qur'an. One may summarise the many characteristic features that the author presents[8] in the following: rich in

---

[ii] By 'magic' and 'medicine', Aslan is alluding to the miracles of Moses and Jesus, respectively.

vocabulary (which is said to be about five times more diverse than a typical book in Arabic), with some words carrying one, two or multiple meanings; striking in its musical tone and rhythm; effective in addressing laymen and elites alike across ages and eras; balanced in addressing both the mind and the heart; simultaneously using a literary and a scientific style[iii]; concise in expression and yet full in meaning(s)[iv] and extensive in the usage of imagery and metaphors.[v] But, the gap in quality which separates the Arabic Qur'an from any and all translations is so huge that non-Arabic speakers are often both confused by the text in its translated form and bewildered by Muslims' claims that it is the most uniquely beautiful text to ever appear in the history of humanity.

Aslan reminds us that to this day, 'Muslims of every culture and ethnicity must [in prayer recite] the Qur'an in Arabic, whether they understand it or not', recalling that 'the message of the Qur'an is vital to living a proper life as a Muslim, but it is the words themselves – the actual speech of the one and only God – that possess a spiritual power known as *baraka*'[9]. Murata and Chittick abound in the same sense: 'Only the Arabic Qur'an is the Qur'an, and translations are simply interpretations'. They even go so far as to state, 'The Arabic form of the Koran is in many ways more important than the text's meaning'[10].

The Qur'an then plays a central role in defining the beliefs, the lifestyle and the world view of Muslims. Kenneth Cragg notes that Muslims regard it as 'the groundplan of all knowledge'[11]. He draws the following essential conclusion, one which I will use as an important principle in our discussion: The Qur'an 'is always the arbiter to which verdicts must appeal and whose support they will assume. *We may say that if Muslims are to be assured on any and every issue, they will need to be Qur'anically persuaded*, however variously they invoke it'[12] (emphasis added).

It should be clear, however, that the Qur'an couldn't be viewed as a classic book that expounds a given philosophy. Indeed, the Book describes itself as guidance for humans, one that is both spiritual and temporal. Nevertheless, as Anawati rightly notes, the Qur'an does present a certain philosophy

---

iii Examples are the verses 'And [man] makes comparisons for Us, and yet forgets his own creation', He says, 'Who can give life to bones when they have rotted away?' Say: He will revive them Who produced them at first, for He is well versed in every kind of creation!' (Q 36:78–79).

iv An example is the verse 'There is nothing whatsoever like Him'(Q 42:11).

v Example: 'and they put behind them a heavy day' (Q 76:27) and 'Allah has not made for any man two hearts within him' (Q 33:4).

of nature. Muzaffar Iqbal goes further and points to entire cosmological sciences, with 'the Qur'an contain[ing] a significant number of verses that describe the origin of cosmos and life'[13].

interpretation of the Qur'an

Considering the fundamental importance of the Qur'an in Islam, the art and science of interpreting the Text is one of the most central intellectual activities and areas of debate in the Islamic tradition. Muslims have developed a suite of 'sciences' around the Qur'an, collectively referred to as *'ulum al Qur'an* (the sciences of the Qur'an), which includes exegesis in its two forms, *tafsir* (literal interpretation) and *ta'wil* (allegorical interpretation), occasions of revelations (of each verse), chronology of revelation, readings (as there are slightly different ways of reading some verses), the art of recitation, etc.

Now, while the Qur'an is one Book, it is diverse and multiple in the meanings it presents. Murata and Chittick emphasise the importance of the variety of interpretations that can be produced from the text[14]: 'One of the sources of the richness of Islamic intellectual history is the variety of interpretations provided for the same verses'. These and other Muslim thinkers often quote the Prophet to the effect that every verse of the Qur'an has seven meanings, starting with the literal sense and ending with the seventh and deepest meaning, which 'God alone knows'. Murata and Chittick then add, 'The Prophet's point is obvious to anyone who has studied the text carefully'[15]. Mohamed Talbi, an important contemporary historian and scholar of Islam, concurs: 'There is not *one* reading key for the Qur'an, but rather *several* keys, all at the same time subjective and objective'[16] (emphases in the text).

A strong confirmation of this possibility of multiple readings and interpretations of some of the Qur'an's passages can be found in the Qur'an itself. Indeed, Muhammad Asad, the highly respected Muslim scholar[vi] from Eastern Europe, who converted from Judaism, refers to the following verse as 'the key-phrase of all its key-phrases'[17]: 'He it is Who hath revealed unto thee (Muhammad) the Scripture wherein are clear revelations – they are the substance of the Book – and others (which are) allegorical. But those in whose

---

vi Muhammad Asad was born as Leopold Weiss in today's Ukraine in 1900, converted to Islam in 1926, and died in 1992; among his most celebrated books are *Road to Mecca*, *The Message of the Qur'an*, and *Islam at the Crossroads*.

hearts is doubt pursue that which is allegorical, seeking (to cause) dissension by seeking to explain it (for their own goals). None knoweth its explanation save Allah. And those who are grounded in knowledge say: We believe it; all of it is from our Lord' (Q 3:7)[18]. Asad then comments: 'Thus, the Qur'an tells us clearly that many of its passages and expressions must be understood in an allegorical sense for the simple reason that, being intended for human understanding, they could not have been conveyed to us in any other way'[19]. He explains the reason for the allegorical form given to many of the Book's verses: 'All truly religious cognition arises from and is based on the fact that only a small segment of reality is open to man's perception and imagination, and that by far the larger part of it escapes his comprehension altogether'. He adds: 'How can we be expected to grasp ideas which have no counterpart, not even a fractional one, in any of the apperceptions which we have arrived at empirically? The answer is self-evident: By means of loan-images derived from our actual – physical or mental – experiences'. He supports this view-point with Zamakhshari's[vii] commentary on the Qur'anic verse 13:35[viii]: 'Through a parabolic illustration, by means of something which we know from our experience, of something that is beyond the reach of our perception'.

Mohamed Talbi has gone somewhat beyond the reformers' modernist approach to the Qur'an; he has described his methodology as 'intentionalist', which consists of an attempt at finding in the Revelation God's underlying intentions on various topics, general or specific. He insists[20], however, that 'the intentionalist reading of the sacred text is not a hasty new invention, for it has its helpful [intellectual supports] in the past; these today need new momentum and spirit. [The analogical reasoning – *qiyas* – employed by the legal scholars was one such support.] The intentionalist reading of the sacred text, though, goes beyond this analogical reasoning – which many legal scholars in the past rejected – and I prefer the intentionalist reading to the analogical reasoning, without rejecting analogy in all circumstances'.

Finally, no discussion of approaches to the Qur'an today can be considered complete enough without the mention of Mohamad Shahrour, the

---

vii Zamakhshari (d. 1144), the great classical Qur'anic exegete who belonged to the rationalist Mu'tazila school, wrote the famous commentary *al-Khashshaf*.

viii Qur'an 13:35: 'The parable of the Garden which the righteous are promised: underneath it flow rivers, perpetual is the enjoyment thereof, and the shade therein; such is the reward of the Righteous; and the requital of Unbelievers is the Fire.'

controversial Syrian thinker who burst on the scene in 1990 with a highly original book[ix] that was no less than an earthquake and that may one day be considered the start of a revolution in Islamic thought. Indeed, Shahrour has already been called 'a Martin Luther of Islam'[21] and an 'Immanuel Kant in the Arab World'[22]; his groundbreaking book has been considered as potentially the Muslim equivalent of Martin Luther's *95 Theses*[23].

In his important book, Shahrour produces a very original reading and interpretation of the Qur'an, one that is based on the principle that each word in the Qur'an has a precise and unique meaning (there are thus no synonyms in the Holy Book). Indeed, he dissects the text in an attempt to unlock a 'semantic code', one which then allows him to find connections between verses that produce a new, heretofore unexpected meaning. However, as Andreas Christmann remarks in one of the best reviews of Shahrour's work: 'The problem that lies in such a literalistic and essentialist approach to meaning is [. . .] that it prevents him from acknowledging any symbolic or metaphorical meanings, inasmuch as it does not allow any appreciation of the different usage of one term within the Qur'an . . . '[24] One of Shahrour's main principles is his differentiation between 'the permanence of the textual form' of the Qur'an and 'the movement of its content'. For him, the Holy Book's uniqueness or miraculousness (*I'jaz*) lies in the dialectical relation between the permanence of the text's form and the movement of the text's content, which then allows not only for new readings to be made and meanings to be found by humans of all ages but also for gradually obtaining a larger and larger share of the divine knowledge.

In another revolutionary move, Shahrour declares such interpretations or hermeneutics (*ta'wil*) to be open to everyone, specialists or not, Muslims or non-Muslims, Arabic speakers or non-Arabic speakers. Christmann comments: 'Shahrour holds this view, which overturns everything that has previously been prescribed as the prerequisites for *ta'wil*, which has led him to the conclusion that "those who are deeply rooted in knowledge" (*al-rasikhuna fi-l-'ilm*), are not, as conventionally assumed, the most learned and devout among the *ulama* [religious scholars] and *fuqaha* [Muslim jurists], but "scholars and philosophers who occupy the most eminent place in society"'. Christmann further explains: 'Among the examples of such *ta'wil* which Shahrour cites are Newton's theory of gravity, Darwin's theory of

---

[ix] Mohamad Shahrour's first and most important of his four books was *al-Kitab wal Qur'an: qira'a mu'asirah* ('The Book and the Koran: a modern reading') (Damascus: al-Ahaly li-Tiba'a wa-Nashr wa-Tawzi', 1990).

evolution and Einstein's theory of relativity'[25]. It is indeed a very novel and revolutionary approach to the Qur'an.

the Qur'an's philosophy of knowledge

'And were every tree that is in the earth (made into) pens and the sea (to supply it with ink), with seven more seas to increase it, the words of Allah would not come to an end. . . . '

Q 31:27

Typical to the standard Islamic literature and teaching on the subject, Fahd Ar-Rumi tells us that the Qur'an comprises all kinds of knowledge, which he summarises in the following main branches[26]:

- *Theology*: Proofs for the existence and unicity of God, and description of His attributes.
- *Linguistics*: The rich and complex vocabulary and syntax of the Qur'an has allowed scholars to build Arabic on a clear, solid and common foundation.
- *Literature*: The Qur'an has contributed much to poetry: metrics and music, as well as metaphors and idioms.
- *Ancient history*: At least partially constructed from the many stories related about prophets and nations of old times.
- *Jurisprudence*: From principles to detailed rulings, based on the many injunctions and explanations given in the Qur'an for different cases and circumstances.
- *Natural science*: Due to the many verses (some 750) dealing with natural phenomena, and due to the needs of the Muslim life (prayer times related to the sun's position in the sky at various times, month of fasting related to first-crescent visibility, etc.).

Now, as many Muslim thinkers have emphasised, the Qur'an presents a complete world view: a philosophy of knowledge, a philosophy of nature and perhaps (in the view of some) a philosophy of science as well. A few contemporary authors have given this topic much thought and written about it with erudition and passion: Mehdi Golshani, Ghaleb Hasan, Muntasir Mujahid and Muzaffar Iqbal. I shall make extensive usage of their works, particularly Golshani's *The Holy Qur'an and the Sciences of Nature*[27] and Hasan's Arabic book

The Theory of Knowledge/Science in the Qur'an[28], where I have rendered the Arabic/Qur'anic fundamental concept/term of *'ilm* as 'knowledge/science', as indeed the term is ubiquitous[x] and all-encompassing in the Qur'an, and deciphering its meaning(s) is the start of any construction of a philosophy of knowledge/science in the Qur'an.

The first principle that the Qur'an presents in its philosophy of knowledge is that man has been endowed with the capacity to learn and comprehend. Indeed, this is what makes him God's *khalifah* vice-regent (or viceroy or deputy) on the earth. The first act of God towards humans shortly after their creation was to teach Adam 'all the names' (concepts) and to ask him to restate them, which he did successfully, thereby proving to the angels that humans had a distinct capacity that made them superior to all other creatures and thus worthy of carrying God's mission on the earth. Indeed, the concept of reasoning appears in the Qur'an 49 times, always in the active form, not as an abstract idea or passive human ability. (I am rebutting the view that conservative religious scholars often put forward, namely that the original meaning of *'aql*, reason, was 'restraint'.) Man can thus learn anything – in principle. Conversely, this means that nature can be understood. Furthermore, knowledge is vast and encompasses many fields.

Golshani insists on the fact that, contrary to what many religious scholars have proclaimed, *'ilm* as described in the Qur'an is much wider than – and is not limited to – the religious fields that may be more obligatory upon Muslims to know about. In fact Golshani simply rejects the traditional classification of knowledge into religious and non-religious; he notes that in the Qur'an (e.g. Q 39:9, 96:5, 16:70) *'ilm* is presented in its most general sense; he further refers to the Prophetic statement to 'seek knowledge even in China' and remarks that 'the Prophet couldn't have been asking Muslims to seek *religious* knowledge in China'[29]. Finally, he adopts the twentieth-century Iranian scholar Murtaza Mutahhari's principle: 'Islam's comprehensiveness and finality as a religion demands that every field of knowledge that is beneficial for an Islamic society be regarded as a part and parcel of the 'religious sciences''[30].

Other writers have pointed to the concept of *hikmah* (wisdom), as in the verse 'and Allah has revealed to you the Book and the wisdom, and He has taught you what you did not know, and Allah's grace on you is very great' (Q 4:113) as implying that wisdom is 'the science that embraces any possible

---

[x] Hasan states that the term "*'ilm*" and its derivatives are used 900 times in the Qur'an; Golshani says there are 780 such occurrences.

knowledge'³¹. Muzaffar Iqbal identifies *'ilm* with science but considers it to be of very wide applicability and relevance: 'Unlike Greek and Latin, Arabic does have a specific word for science: *al-'ilm*. This word as well as its derivatives frequently occurs in the Qur'an. It is used to denote all kinds of knowledge, not just the knowledge pertaining to the study of nature'³².

In the many Qur'anic verses, which relate to knowledge and its acquisition, one finds a variety of terms pointing to a hierarchy of methods, words such as listening (in the sense of understanding), observing, contemplating, reasoning, considering, reflecting, etc., most of which occur a dozen times or more.ˣⁱ

In the closing sections of his book with Mohamed Talbi, Maurice Bucaille develops this idea more fully: The Qur'an, he notes, uses a different vocabulary each time it calls upon humans to observe or notice a particular sign in nature, depending on whether the phenomenon is obvious or subtle; indeed, depending on the situation, the reader is asked to watch, listen, think, reflect or exhibit wisdom, these being gradually higher and higher functions of the mind. Bucaille writes: 'Oftentimes the expression "haven't you seen . . . " comes back in the Qur'an, each time addressing people on items that offer themselves to our observation'. He then quotes verses that refer to 'people who listen', to 'people who reflect' and to 'people who remember', and adds, 'One degree further and it's the call to reason which completes the preceding calls'. Bucaille concludes: 'Reflections on the observational data are asked of both "people of understanding" (Q 3:190, 39:21) and "people of wisdom" (Q 20:53–54). [. . .] Verse 10:5 tells us that God "exposes His Signs in detail for those who know"; this idea is repeated in 6:97 in similar manner'³³.

Golshani divides the sources of human knowledge according to the Qur'an into three parts: (1) the senses, which allow humans to conduct observations and measurements; (2) the intellect, which allows men of ability to ponder, reflect, reason etc. and (3) divine revelation, which either directly brings knowledge unknown to people (e.g. news of ancient peoples) or helps minds to reach truths, by means of parables, intuition, sudden enlightenment etc. According to Golshani, such God-assisted acquisition of knowledge can occur by way of meditation, intellection or normal observation that does not miss crucial information. Our author, however, reviews many Qur'anic verses

---

ˣⁱ Examples include: 'It is He who has made for you the night that you may rest therein, and the day to make things visible (to you). Verily in this are signs for those who listen' (Q 10:67); 'And He has subjected to you, as from Him, all that is in the heavens and on earth: Behold, in that are Signs indeed for those who reflect' (Q 45:13).

but cannot exhibit any clear and explicit one to show that this channel is not restricted to prophets and saints but can extend to simple humans; all he can find is 'That is God's grace; He grants it to whom He pleases' (Q 62:4). He then supports his view by referring to the idea of 'inspiration' (*kashf* or *ilham, hadas*), something that many thinkers and scientists have claimed to exist as a nonrational way of reaching some truths, including scientific ones; Golshani quotes Avicenna and Charles Townes (the Nobel prize winning inventor of the laser), both of whom emphasise the importance of intuition in making discoveries.

One can also infer from Qur'anic verses different levels of knowledge, each characterised by a different term, such as believing, doubting, thinking, understanding, envisioning, realising, ascertaining etc.[xii]

I must also note here that the Qur'an draws attention to the danger of conjecturing without evidence ('And follow not that of which you have not the (certain) knowledge of . . .' Q 17:36) and in several different verses asks Muslims to require proofs ('Say: Bring your proof if you are truthful' Q 2:111), both in matters of theological belief and in natural science. In many instances, the Qur'an/Allah argues with the unbelievers and cites examples/arguments to try to convince them; for instance, 'Or, Who originates the creation, then reproduces it, and Who gives you sustenance from the heaven and the earth? Is there a god beside Allah? Say: Bring your proof if you are truthful' (Q 27:64).

But what constitutes proof in the Qur'an's philosophy? Ghaleb Hasan extracts the following methodology from the Qur'an[34]. First, 'proof' in the Qur'anic context means convincing evidence or argument; it must be clear and strong. Secondly, proof is not achieved by relying on tradition or forefathers' views (what we today call the argument from authority): 'And when it is said unto them: Come unto that which Allah hath revealed and unto the messenger, they say: Enough for us the ways we found our fathers following. What! Even though their fathers had no knowledge whatsoever and no guidance?' (Q 5:104)[35]. Thirdly, both assertions and rejections require proof; e.g. 'O mankind! Verily there hath come to you a convincing proof from your Lord: For We have sent unto you a light (that is) manifest' (Q 4:174).

[xii] Examples are: 'To Solomon We inspired the (right) understanding of the matter: to each (of them) We gave Judgment and Knowledge' (Q 21:79); 'Nay, were ye to know with certainty of mind, (ye would beware!)' (Q 102:5); 'If you obeyed most of those on Earth, you would be misled far from Allah's way. They follow naught but conjecture, and they do but guess' (Q 6:116).

Finally, it is important to note that the Qur'an distinguishes several groups among humans with regard to their abilities and willingness to learn, to understand, to penetrate to deeper truths or conversely to stubbornly reject truths even in the face of proofs. According to Hasan[36], the Qur'anic discourse addresses five categories of people: (1) the most general audience; (2) the community of Muslim believers; (3) the people of knowledge (as in Q 30:22, 29:43)[37]; (4) the righteous people, i.e. those who go beyond faith to an eagerness to live, behave, and think according to the laws that God has decreed and (5) people of certainty, those who do not doubt.

Golshani does not present a clear categorisation and hierarchy of audience but nonetheless divides people according to the attributes they are labelled with when addressed by the Qur'an, e.g.: the believers; the pious; the mindful (or God-conscious); the listeners to the truth; the meditators; those of understanding; the learned; those who have purified their intellects; the wise and the people of certainty. It is also important to note that Golshani puts much emphasis on the moral dimension of the 'people of knowledge'; he insists on the purification of one's intellect as an important criterion for achieving high knowledge and wisdom, and for this he cites the Qur'an (Q 2:282) and Prophet Muhammad ('No servant [of God] devotes his full forty mornings to (the service of) God but the springs of wisdom flow from his heart to his tongue'), and quotes Ali[xiii]: 'One who does not render his heart clean does not benefit from his intellect'[38].

Golshani also lists the 'impediments of cognition': pollution of the intellect (as in the above paragraph); lack of faith (Q 63:3, 30:53, 10:101) and subjectivity (Q 45:23), which can be due to ambition, arrogance or prejudices of various sorts; acceptance of conjectures and past beliefs without proof or following authorities (as we've explained above).

And last but not the least, the goal of such knowledge, as one can start to infer from the above categorisations, is to move up the ladder from mere belief to knowledge, then righteousness, and certainty in one's awareness of God. This overarching goal of knowledge and truth in the Islamic world view is stated and emphasised by all Muslim thinkers and built upon Qur'anic references; for instance, 'And truly are God-conscious among His servants those who are possessors of knowledge' (Q 35:28).

[xiii] Ali (ibn Abi Taleb) was the Prophet's cousin and son-in-law; more importantly, he became the fourth caliph of Islam, the last of the 'rightly guided caliphs' and is the first of the Imams of Shiism. For Shiites he is, after Prophet Muhammad, the most important human ever; for Sunnis he is one of the most important ones, but without any special religious significance.

the Qur'an's philosophy of science

I have not yet defined 'science'. This will be discussed in detail in the next chapter, where an overview of the philosophy of science will be presented as well as a review of the various critiques that have been levelled at the scientific enterprise. For now, let us limit ourselves to the study of the natural world, that is leave aside the study of the human species and society, and thus define science as 'the methodical study of nature in the aim of understanding its phenomena'.

In his latest book, Muzaffar Iqbal makes the following assertion: 'The Qur'an itself lays out a well-defined and comprehensive concept of the natural world, and this played a foundational role in the making of the scientific tradition in the Islamic civilization'[39]. In the end, Iqbal could not support the first part of his statement with convincing evidence, he barely gave a few verses of some relevance; the second part, however, was exemplified by al-Biruni's writings, which Golshani similarly cited, and which I quote in the next paragraph. Iqbal, however, adopts a Nasrian view of Islam, nature and science, which considers 'the Qur'anic view of nature [as] characterised by an ontological and morphological continuity with the very concept of God – a linkage that imparts a certain degree of sacredness to the world of nature by making it a Sign (*ayah*, pl. *ayat*) pointing to a transcendental reality'[40]. I should stress that this view, although upheld and strongly expounded and publicised by a few prominent thinkers, represents a school of thought that is marginal in the general intellectual landscape of Islamic thought; I shall come back and describe this school and its views more completely, in Chapter 4.

Muhammad Iqbal (the illustrious nineteenth-twentieth century Muslim poet–philosopher, whom we shall meet closely in Chapter 7) considered[41] the Qur'an's methodology and epistemology to be empirical and rational. The Holy Book discusses nature at length and considers the amount of information that can be gleaned from its study and description to be (metaphorically) infinite. According to many Muslim authors[42], one can count some 750 verses (out of about[xiv] 6,300) in the Qur'an dealing with natural phenomena. In many of the verses, the exploration and study of nature is highly

---

xiv The number of verses in the Qur'an is not universally agreed upon, as some do count the 'Basmallah' ('In the Name of Allah, the Merciful the Compassionate'), which starts in 113 of the 114 'suras' (chapters), some do not; some divide some verses and count them as two, etc.

encouraged and recommended[43]. Golshani remarks: 'In fact the main reason that our great scholars, in the glorious period of Islamic civilization, paid attention to foreign (especially Greek) sciences was due to the Qur'an's emphasis on the study of nature ... al-Biruni has explicitly stated that the motive behind his research in the scientific fields is Allah's Words in the Qur'an: "Those who reflect on the creation of the heavens and the earth (and say): Our Lord! Thou hast not created this in vain! Glory be to Thee'" (Q 3:191)[44]. Similarly, the illustrious astronomer al-Battani (Albategnius 850–929) has written: 'By focusing attention, observation, and extensive thought on astronomical phenomena, one is able to prove the unicity of God and to recognize the extent of the Creator's might as well as His wide wisdom and delicate design'[45].

From the above remarks, two main interrelated principles emerge as a Qur'anic philosophy of science: (1) The exploration of nature, from mere observation to full scrutiny, should clearly point out the order and purpose of the cosmos; and (2) the study of nature should point to a certain unity and thus lead to a (greater) faith in the Creator. Golshani emphasises the theistic objectives of the study of nature according to the Qur'an: 'The Qur'an does not approve of such cognitions which aim at nothing except satisfying one's own curiosity. On the way of understanding nature, one should not busy oneself with the means and forget the ultimate end'[46].

Ghaleb Hasan further extracts some important philosophical principles of science from the Qur'an; he summarises them in the following three points: (1) unity, (2) generalisation and (3) prediction. He cites various verses in support of his view; for instance: 'For you shall not find any alteration in the ways (laws?) of Allah; and you shall not find any change in His ways' (Q 35:43). He adds that 'science' in the Qur'anic philosophy is meant as the act of interpreting the observed signs (*ayat*) of God, just as – one may add – exegesis is the 'science' of interpreting God's written verses (also *ayat*). Regarding prediction, he notes that the Qur'an points out the regularity in the phenomena of nature and further explains that the computability and predictability of such phenomena is for human benefit: 'It is He Who made the sun a shining brightness and the moon a light, and ordained for it phases that you might know the computation of years and the reckoning (of time)' (Q 10:5). Mujahid finds the concept of cosmic laws in the following Qur'anic verses: 'And the sun runs on to a term/resting-place determined for it; that is the decree of the Exalted in Might, the All-Knowing. And the Moon, We have ordained for it mansions/stages till it becomes again as an old dry palm branch' (36:38–39), where the terms 'determined', 'decree' and 'ordained' are understood to imply a 'natural law'[47]. Mujahid draws the same conclusion from the verse: 'Verily We established Zulqarnain's power on earth, and We gave him the ways and the means to all ends' (Q 18:84), highlighting the words 'ways and means'[48].

Muzaffar Iqbal identifies the three Qur'anic concepts of *tawheed* (unicity), *qadr* (measure) and *mizan* (balance) as 'not only central to the teachings of Islam but. . . . [a]lso of immense importance for understanding the relationship between Islam and science'[49], a statement and position to which I subscribe wholly.

Iqbal stresses further that 'God's ways and laws are unchanging', citing the Qur'anic verse 'That was the way of Allah in the case of those who passed away of old, and you will not find for the way of Allah any changes' (Q 33:62), and adds: 'Thus the entire world of nature operates through immutable laws that can be discovered through the investigation of nature'[50].

Campanini has also pointed out that the Mu'tazilite (rationalistic) theologians 'linked the arranged structure of the universe by God with the exactness of demonstrative proofs'[51]; he refers to 'Abd al-Jabbar (ca. 1024), one of the school's most powerful and influential thinkers, who held that God operates according to rational laws. On this basis, the Mu'tazilites went so far as to severely limit the occurrence of miracles as 'irrational occurrences' that could only be performed by God in order to vouchsafe the claims of his prophets (Moses, Jesus, etc.)[52].

Other thinkers of different philosophical propensities adopt a very different view, however; Campanini reminds us that the Egyptian thinker Sayyid Qutb[xv] (1906–66) 'denies, on principle, any interaction between faith and technical knowledge; the Qur'an is all about the route to salvation, while science deals with other issues'[53]. That is why Campanini calls Qutb's position 'galilean, for [Qutb] is adamant in distinguishing the religious dimension from the scientific'[54].

Hasan then summarises the general objectives of science according to the Qur'an as follows[55]:

1. Satisfying our human desire to understand what is around us. Here Hasan cites Prophet Abraham's experience, when the Patriarch kept

---

[xv] Sayyid Qutb is more famous today for being the pioneer of radical political Islam, one who was tried and executed by the Egyptian powers in 1966, but he should be known for his impressive literary exegetical work *Fi Dhilal al-Qur'an* ('In the shades of the Qur'an').

misidentifying God with the great celestial objects (sun, moon, etc.); Muntasir Mujahed also refers[56] to the human 'pre-disposition' (*fitrah*) for knowledge and the soul's tendency and capacity to fulfil itself with contemplation of the cosmos, seeking – among other nourishments – beauty[57]; in fact Mujahed makes it a duty on humans to search, pointing to Qur'anic verses and citing the famous Egyptian writer and historian Abbas al-'Aqqad, who wrote a whole book titled *Thinking/Reasoning: an Islamic Obligation*.

2. Improving the world, for the Qur'an insists that nature has been put at man's service, and the cosmos can be perfected, as it has been prepared to be developed further.

3. Identifying the 'First Principle' (the cause of this glorious universe), in other words reaching and connecting with God.

Golshani also insists on the utilitarian aspect of science in Islam. To that effect he cites Prophetic statements ('the best fields of knowledge are those which bring benefits') and quotes Ali ('there is no goodness in knowledge which does not benefit people')[58].

Finally, both Hasan and Golshani and many other Muslim thinkers insist on the ethical dimension of science in the Qur'an (and in Islam more generally). Hasan starts by noting that the Qur'an presents the act of creation as mercy ('The Most Merciful (Allah)! He Who has taught the Qur'an. He has created man and taught him expression. The sun and the moon follow courses (exactly) computed. The stars and the trees prostrate in adoration. And the firmament He raised high, and He set up the balance' (Q 55:1–7). He then picks the following verse as the nexus between science and ethics in the Qur'an: 'And follow not that of which you have no (definite) knowledge; verily the hearing and the sight and the heart, all of these (you) shall be questioned (responsible) about' (Q 17:36).

for a hermeneutical approach to the Qur'an

In the book's prologue and after I introduced our iconic figure and thinker Ibn Rushd, I briefly stated what can be considered the core principle of his philosophy: that religion and philosophy (which he refers to as *hikmah*, wisdom, and which can thus be enlarged to any truthful knowledge, including science) can never be in contradiction, for they are 'bosom sisters' and express the same truth in different ways. But there do occur frequent instances of at least apparent contradictions between the religious proclamations on a given issue and the results derived from profane knowledge; what does one then do? Averroes, as we have seen, then explicitly called for metaphorical interpretation of the religious text[59], strongly basing himself on the Qur'anic verse 3:7.

For these reasons, Campanini calls Ibn Rushd 'one of the fathers of philosophical hermeneutics [*ta'wil*, interpretation of religious texts] in Islamic thought'[60]. This Averroesian specialist further explains that '[our philosopher's] idea of *ta'wil* does not refer to a secret, concealed and esoteric metaphysical level which is beyond literal meaning, as for the Shi'is, but to a linguistic level in which the concept, or even the being, shows itself in words'[61]. Qur'anic verses, therefore, carry various meanings, some of which are literal, and others are symbolic in nature. Campanini also emphasises the role of the interpreter in the hermeneutical process: 'The role of the interpreter is central: it is the interpreter who decodes the meanings of the text'.

Two contemporary Arab thinkers have also insisted on the multiplicity of readings and meanings in the Qur'an: Mohamed Talbi and Hasan Hanafi.[xvi] I have already quoted Talbi, who explicitly wrote, 'There is not *one* reading key for the Qur'an, but rather *several* keys, all at the same time subjective and objective'[62] (emphases in the text). Similarly, Hanafi stated, 'There is no true or false, right or wrong understanding. There are only different efforts to approach the text from different motivations.... There is no one interpretation of a text ... An interpretation of a text is essentially pluralistic'[63].

Also referring to another Arab thinker, Nasr H. Abu Zayd,[xvii] who has been insisting on hermeneutics, Campanini concludes that 'if religious texts are normal linguistic texts and accordingly may undergo hermeneutic analysis, ipso facto the Qur'an cannot be interpreted *literally* as a scientific text'[64] (emphasis by the author). He adds, 'The biological or cosmological hints of

---

[xvi] Hasan Hanafi (1935–) is a professor of philosophy at Cairo University who is often identified with the 'Islamic left' movement; he is also an expert on Islamic thought and a prolific writer (his CV lists 24 books, at least three of which deal primarily with exegesis: http://www.ispionline.it/it/documents/CV_Hanafi.pdf).

[xvii] Nasr Hamid Abu Zayd (1943–) is described in Wikipedia as 'an Egyptian Qur'anic thinker and one of the leading liberal theologians in Islam. He is famous for his project of a humanistic Qur'anic hermeneutics': http://en.wikipedia.org/wiki/Nasr_Hamid_Abu_Zayd (September 2009).

the Qur'an are not properly scientific. However, they can be symbolically interpreted by a shrewd textual hermeneutics'[65]. And further, 'The linguistic system of the Qur'an constitutes a fixed text, but the intentions of the interpreter can disclose a plurality of meanings: philosophical hermeneutics may perceive the foundation of science in the Qur'an without arguing that the Qur'an is a scientific text'[66].

This is indeed a philosophical and religious position I fully subscribe to.

summary and conclusions

The first major idea I have tried to emphasise in this chapter is the extraordinary place and influence that the Qur'an occupies in the lives and minds of Muslims. That strong emphasis aimed at the following objectives: (a) to explain why the Muslim discourse on science and religion (and other social and political issues) tends to often be filled with references to the Qur'an; (b) to show that for a credible and reasonable discourse on science and Islam to have a chance to be well received by the public as well as by the elite, it must at least ensure a Qur'anic acceptability (or non-objection) of the ideas being put forward, if not fully compatibility and (c) to explain that this can be achieved by a hermeneutical approach, which is indeed part of the Islamic tradition.

The second major idea that we have uncovered is that one must note that the Sacred Book repeatedly draws one's attention to the general predictability of the world's physical phenomena. Furthermore, the Qur'an continually encourages people to reflect and search. Mohammad H. Kamali[67] considers 'scientific observation, experimental knowledge and rationality' as 'the principle tools that can be employed in the proper fulfillment of [Man's] mission [on earth]', i.e. the 'vicegerency (khilafah)'. Referring to Muhammad Iqbal (the Indian philosopher), Kamali adds that the Qur'an leads to the birth of the 'inductive intellect' by making it 'an obligation therefore of every Muslim to master the inductive method to uncover the laws of nature and society'[68].

Nevertheless, the concept of science in the modern sense cannot easily be found in the Qur'an or indeed in most of the classical Muslim heritage; rather the concept of knowledge is developed. The confusion between the two concepts has often been made by Muslim thinkers and educators; indeed, the word 'ilm is today routinely used for 'science', although it is quite certain that it originally stood for knowledge, perhaps even religious knowledge (as opposed to knowledge of the world). This has led to strong disagree-ments, essentially along the traditional versus reformer lines, regarding the possibility (or not) of building a case for a Qur'anic basis for science,[xviii] with the latter taking several possible definitions ranging from 'sacred science' to 'traditional knowledge', 'Islamic science' and (a theistic version of) Western science, as we shall see.

The position I advocate is simply a rejection of all extreme positions. The idea of 'scientific content' in the Qur'an is to be rejected, for a variety of reasons that I shall expose in Chapter 5. Instead, I have emphasised and promoted a multiplicity of readings (with multilayered nuances) of most, if not all, of the Qur'an, an approach which allows for an intelligent enlightenment of one's interpretation of Qur'anic verses, using various tools, including scientific knowledge, at one's disposal. I have argued that this approach meshes well with that of some of the most intelligent scholars of Islam, from Ibn Rushd (Averroes) to Mohamed Talbi. The latter has written that 'reading, interpreting and reflecting upon the Qur'an in the light of the sciences that we have here and now is [. . .] an enduring tradition within Islam'[69]. Furthermore, 'each approach to the Qur'an must take into account the fact that it is, by the continuous reading of the signs that it asks to undertake, a constant revelation that is incessantly being disclosed to us concurrently with our discovery of the Universe. The Qur'an asks us to observe and to read. Yet, how could one observe or read . . . without science?!'[70]

We have also seen how Mohamad Shahrour has broken with traditional rules of interpretation, constructing not only a very original and radical new approach to the Holy Book but also opening the gates of interpretation to anyone who has the intellectual capacity to do so, including non-Muslims. Furthermore, he views the inclusion of modern science and philosophical theories into our reading of the Qur'an as one way to expand some of the Book's potentials as well as to help a given society mesh its knowledge with God's truths.

To summarise, while the Qur'an cannot be turned into an encyclopaedia of any sort, least of all of science, one must keep in mind the fact that if the Qur'an is to be taken seriously and respectfully, one must uphold the

---

[xviii] When translating the verse (50:11), i.e. 'Allah will exalt those of you who believe, and those who are given knowledge, in high degrees', the Qur'anic interpreter Yusuf Ali adds the word 'mystic' between parentheses before 'knowledge'; other commentators could have easily replaced 'those who are given knowledge' by 'men of science'.

Rushdian principle of 'no possible conflict' (between the word of God and the work of God). In practice this principle can be turned into a 'no objection' or 'no opposition' approach, whereby one can convince the Muslim public of a given idea (say the theory of biological evolution), not by proving that it can be found in the Qur'an but rather by showing that at least one intelligent reading and interpretation of its verses is fully consistent with the scientific theory in question.

3

science and its critics

'The ideas of Ibn al-Haytham, al-Biruni and Ibn Sina, along with numerous other Muslim scientists, laid the foundations of the "scientific spirit" within the worldview of Islam.'

Ziauddin Sardar[1]

'It doesn't matter how beautiful your theory is; it doesn't matter how smart you are; if it doesn't agree with experiment, it's wrong.'

Richard Feynman

the moon-landing story

When I first started receiving emails claiming that the National Aeronautics and Space Administration (NASA) never put a man on the moon, I dismissed them as another example of the online junk that comes our way every day. That was in 2001, so 'spam' hadn't entered our everyday vocabulary and become a major nuisance, consuming time and resources. And yet the Internet had already become a catalyst for conspiracy-type discourse. What made me pay special attention to the no-moon-landing emails, however, were two things: (1) many of them were coming from students and academic colleagues; and (2) they seemed to be supported with 'proofs', in the form of attachments containing images and diagrams with arrows and captions pointing to the 'flaws' in NASA's story of 'one small step for Man, one giant leap for mankind'. It was then that I decided to address the issue by giving public talks, which I titled 'Did NASA fake the moon-landing . . . or are we miserably failing to educate the public?' It was clear to me that what we were missing was a correct understanding of how science works, how we get to know what we know and how sure we are of whatever we know.

On a similar level, some of the most interesting and challenging courses I teach are those which attempt to impart some understanding of science: Conceptual physics, which is offered to non-science majors; and Introduction to Astronomy, which is mainly for non-science students, but is also open to engineering and science majors as a free elective. Both courses are part of the science requirement of the curriculum, aiming to educate non-science students in the methods of science. The reason I find these courses fascinating to teach is because they constitute a window to the collective mind of my society and give me a chance to correct misconceptions in at least a small group of people. For example, to this day, some of my astronomy students invariably bring up the no-moon-landing 'proofs' and forcefully argue against the 'NASA claim', and many are still convinced that astrology is at least partially true, basing their convictions on personal experience. And needless to say, most of my students strongly argue against evolution, but that's a story for another chapter.

The dilemma I find myself facing, when giving talks upholding the scientific method and results to general public and especially while teaching these courses to university students, is how simplified (or even cartoonish) an image of science I should present. What I mean by this is that I could give a stern 'science is objective, universal, progressive, testable' description and try to negate any doubts in the minds of my listeners, or I could be more realistic in my portrayal of how science really works (including the subjective, paradigmatic aspects of it) and risk feeding the audience with even more suspicion about the certainty (or lack thereof) of any claims made by scientists. As Ziauddin Sardar (a cultural commentator of very wide interests, whom we'll shortly be meeting and interacting with more closely) said when describing Thomas Kuhn's dilemma: 'Kuhn was particularly concerned to preserve the public face of science. Whatever the "internal" problems and "truth" of science, the public's belief in science as Good and True had to be

defended, for the social consequences of not doing so could be devastating'[2].
(More on Kuhn later.)

What I have come to realise is that the general public as well as most of the
elite, unfortunately, can be characterised by two traits: (a) a widespread lack
of appreciation of how science works and how many results must today be
considered certain to a very large extent; and (b) a high credibility attitude
towards almost any type of claim, as long as it comes from some source, be
it a newspaper, a television show or some respectable person. In *The Demon-
Haunted World*, Carl Sagan mentions the Chinese proverb, 'better to be too
credulous than to be too skeptical,'[3] and pits it against the scientific attitude;
I believe the same case can be made with the Arab-Muslim society in general.
Indeed, one of the main quarrels some of the modern Muslim thinkers[4]
have expressed with regard to Western epistemology (the philosophy of
knowledge) is that it emphasises skepticism too much. This may be the case
to some extent, but there is no doubt in my mind (from my many years of
teaching and practicing science in the Arab-Muslim society) that the latter
errs far more in the other direction.

To begin with, we scientists, science educators, in particular, and society
at large (particularly the media and the publishers) have done a very poor job
at explaining what science is about and what it entails. The most successful
area of human endeavor is paradoxically the least understood and appreciated
human story! People are not only largely ignorant of how science makes
progress but also lack basic scientific information.

When I teach the Introduction to Astronomy course (almost every
semester), I give a pre-quiz on my first encounter with the students; the aim
of that informal test is to highlight, for me and for the students themselves,
the basic astronomical information they do not know, and facts and concepts
I will clear up and emphasise during the course. You would be surprised at
how basic some of the questions are; let me give a few examples:

- Who was Copernicus? Less than 10 percent of my students will have heard
  of him.
- What is the difference between a planet and a star? One gets a mixed bag
  of answers.
- What makes the earth go through seasons during the year? Most, if not
  all, students answer that this is because the distance between the earth
  and the sun varies during the year (a true fact, but not the reason for the
  seasons).

- How old is the universe? How old is the sun? How old is the earth? How old is humanity? To these questions, I get anything from 'thousands of years' to 'I have no idea'; very few – if any – students give ages in billions (for the first three).

Let me now list some of the basic questions on which the public is largely confused:

- Is science a bunch of facts?
- What are 'facts'? Do scientific facts change over time? (One year Pluto is a planet and the next it is not.)
- How do we know that a scientific result is true, will stand and will not be contradicted and annulled later?
- What is a scientific 'theory'? Was Newton's theory of gravitation not re-placed by Einstein's? How do we know that Darwin's theory of evolution will not be thrown out in the future?

Most importantly, the public has no understanding, whatsoever, of the falsification criterion for considering a hypothesis as scientific; likewise for the peer-review and wide checking process that is used to confirm or reject a result or a discovery. And last, but not the least, our society has no appre-ciation at all for the role that science has played in reinforcing rationality and strengthening critical thinking in our thinkers' reasoning, whether it be about economic, social or religious issues.

In this chapter my aim is to review the status of science, with its various processes and methods as largely adopted by the scientific community, but also from the perspectives of many critics, who as we shall see have chipped at the edifice of science bit by bit for valid or not so valid reasons.

Let me thus start by reviewing for the reader the foundations of the scientific process. Many of the questions that we will be posing, e.g. 'what is science?' and 'how does science work?' are not at all easy to answer. To some extent, the public can be forgiven in not being clear about these issues. But only to some extent, as I believe that everyone should under-stand why science is fundamentally different from philosophy, art, politics and religion. Unfortunately, we scientists, educators and communicators have failed in our missions, or else people should at least be able to tell that man did set foot on the moon and that astrology is fundamentally wrong.

what is science?

One of the main sources of the problems I described above is the absence of a clear definition of science, or rather any consensus on it. Moreover, there is usually an implicit assumption that what one means by 'science' is modern science, but that assumption turns out not to be always justified. And even in that case, one would still have to reach a consensus on what modern science is, when it actually became modern, and what to make of the earlier forms of science. Just like the concept of time, one may actually know what science is but may not be able to define it when asked about it. Indeed, when asked for a definition, Rutherford (the discoverer of the atom's nucleus) is known to have replied: 'Science is what scientists do'. Needless to say, one could still retort 'and what's a scientist?', especially because the term 'scientist' has entered the vocabulary relatively recently.

When Richard Feynman (one of the greatest twentieth-century physicists) gave a talk titled 'What is Science?' to the National Science Teachers Association in 1966, he confessed that when he started to think about the nature of science, he was reminded of the centipede in the following anonymous little poem:

> *A centipede was happy quite, until a toad in fun*
> *Said, 'Pray, which leg comes after which?'*
> *This raised his doubts to such a pitch*
> *He fell distracted in the ditch*
> *Not knowing how to run.*

In my introductory science courses, one of the first things I need to do for my students is to define science. I do this in several steps. First, I state that science is our human attempt to construct objective explanations for the world around us. In fact, this is pretty how Samir Okasha starts out his short *Philosophy of Science* treatise: 'Surely science is just the attempt to understand, explain, and predict the world we live in?'[5] This indeed is not a bad start, especially because it contains several important aspects of science: (1) the attempt to make the explanation objective (and this also signals that it may not be easy to reach that goal); (2) the fact that science is constructed by humans, which highlights the possibility of error or bias and (3) the attempt to describe the world, as widely as possible. But this first-draft definition may be substantially lacking, at least in some respects: (1) It does not declare whether science is limited to the natural world or is to be extended to all observable phenomena, including psychological, social, religious, historical etc., and this leads us to a grey area where the label science in its hard meaning may be difficult to stick; and (2) it says nothing about the nature of the enterprise of constructing objective explanations.

Ziauddin Sardar provides us with a more precise definition of science; he writes[6], '[Science is] an organized, systematic and disciplined mode of inquiry based on experimentation and empiricism that produces repeatable and applicable results universally, across all cultures'. Indeed, this has the virtue of emphasising the goals of objectivity (repeatability and universality) and testability (experimentation and empiricism); it also implicitly restricts the fields to the natural sciences, which is what we will content ourselves with here, as our issues are complicated enough not to have to add to them questions like 'is psychology not (objective) science?', or 'what about information science?', not to mention library science, political science etc.

So the essential and perhaps defining characteristic of science may very well consist in the method and process it has developed, almost canonised, and required everyone who practices it to adopt. That is what we call 'the scientific method' and what we uphold to our students as the glorious aspect of science, in fact much more than the marvelous array of results that scientists have achieved until now. And what exactly is the scientific method? In its simplistic, almost cartoonish formulation, it is the successive acts of (a) observing a phenomenon, and recording as much data/info about it as possible; (b) mentally constructing an explanation/hypothesis for it, one that must be based upon knowledge of nature; (c) assuming that the explanation is testable, that is, it leads to specific consequences (or predictions), checking whether the hypothesis is correct, in that its predictions are found to be true; (d) fine-tuning the hypothesis until its predictions are all found to be correct or throwing it away and replacing it by a new one if its results are strongly contradicted by experiments and observations.

Well, anyway, this is what we tell our students and people in general. This is not actually wrong, but it is a very incomplete picture of how science is conducted. Two important aspects are missing from this simplified description: (1) It leaves out the human element altogether (which data/info are used to construct a hypothesis, do any personal/subjective elements come into play when this is done, at what point does one declare a hypothesis/model to be erroneous and call for its replacement etc.); and (2) it does not emphasise strongly enough the scientific community's role in the process, in particular the fact that data gathering, both in the initial study of the phenomenon and in the checking of the experimental predictions of the hypothesis, is done by many scientists independently, and so is the internal (logical, mathematical etc.) consistency of the proposed theory/model; this is known as the peer-review process. This second aspect is actually what has given science its robustness and allowed it to filter the correct stuff from the junk, or at least the erroneous claims and results; it is this part that is often greatly underappreciated by the public at large, including the highly educated but non-scientific groups. The other aspect (the subjective human element in the scientific enterprise), however, is its Achilles heel, and few people, including the foot soldiers of the scientific community, are fully aware of it. Indeed, the philosophy of science subject, which deals with such issues (and a lot more), is hardly ever taught even to students who major in science fields, and so in the end we produce hordes of technically advanced scientists who, very much unaware of the limitations and internal problems of science, fight valiantly to defend the objectivity of science against the attacks that often come from the traditionalists, the cultural relativists and other post-modernists.

Indeed, traditionalists like William Chittick dismiss the claim that modern science carries any true objectivity; he considers it simply as 'a vast structure of beliefs and presuppositions'[7]. He insists, 'modern scientists, intellectuals and scholars have acquired all their knowledge by imitation, not realization. They take what they call 'facts' from others, without verifying their truth.'[8] He adds: 'In effect, everyone has to accept empirical verification on the basis of hearsay'[9]. For Chittick, 'what people call "science" is strikingly similar to what in [previous] times was called "sorcery". Certainly, the goal is exactly the same'[10]. Finally, he rejects any usefulness of science: 'People have no idea that all this information is irrelevant to the goal of human life. [. . .] The more 'facts' they know, the less they grasp the significance of the facts and the nature of their own selves and the world around them'[11]. Needless to say, this is an overly harsh description and assessment of science, one that is, in my view, based on an erroneous understanding of the scientific enterprise, its foundations, its methods (which are far from rigid and simple) and its claimed results, which are often open to interpretation and re-evaluation. The bulk of this chapter is devoted to precisely clearing up and correcting such misunderstandings.

Let me thus try to address these issues of relativism and limitation by giving the reader (if she/he needs it) a crash course in the philosophy of science. But before I do that, there is one (minor) issue that needs to be settled, if we can, as it will have some important consequences later: Is science, as we have now defined/described it, a universal and old human enterprise, or is this a modern and western development? In other words, can we say that as humans have since the dawn of consciousness been trying to understand and explain natural phenomena around them, they have been performing science? Well, a minute of reflection will lead most people to conclude that clearly the early mystical view of nature, filled with spirits and demons, and the explanations of the cosmos as the realm and battleground of gods, cannot be assimilated to any kind of science. But what about the science that was developed and practised by the ancient but not primitive civilizations, for example, the Chinese, the Maya, the Indian, the Babylonian, the Greek, the Roman and the Arab-Muslim societies, do we call their attempts to develop explanations of the world 'science', or is this term limited to the modern enterprise? That is far from an easy question to answer, for while in those civilizations the study of the world clearly carried in it many of the features we identified (and defined) as part of the scientific enterprise, for example the experimentation part of it and sometimes even the attempt at universality, other essential aspects were often absent, such as the need to insist on a naturalistic aspect of explanations (as opposed to mystical or metaphysical ones) and also the falsifiability of any hypothesis, i.e. that it be a subject to checking and verification, leading to confirmation or refutation.

But how important is this, actually? Is trying to decide whether explorations in the Babylonian or Arab-Muslim eras qualify to be called science in our understanding merely an academic question? Or does it have any later repercussions in our discussions? Suffice it to say that I would not have brought it up if it were only an academic issue.

when did 'science' really start?

Western writers tend to consider the Renaissance, more specifically the Copernican revolution and the Galilean overthrow of the Aristotelian world view, as the great turning point. Muslim (and some non-Muslim) critics counter with claims that essential elements or even foundations of modern science clearly existed within the Arab-Muslim civilization.

One typical example is Okasha, whose aforementioned book *Philosophy of Science* is otherwise excellent. He reviews the history of science (in its modern sense) in the first few pages of his book and states, 'The origins of modern science lie in a period of rapid scientific development that occurred in Europe between the years 1500 and 1750, which we now refer to as the scientific revolution'. He quickly adds: 'Of course scientific investigations were pursued in ancient and medieval times too – the scientific revolution did not come from nowhere. In these earlier times the dominant world-view was Aristotelianism . . . But Aristotle's ideas would seem very strange to a modern scientist, as would his methods of inquiry'. In other words, modern science started with Galileo, as earlier science did not conform to the scientific method as we have defined it. Okasha goes on to briefly review the major milestones in the development of modern science: the discoveries of Kepler (his laws of planetary motion), Galileo (the overthrow of the Aristotelian world view, in the cosmos as well as on the earth, and his emphasis on the usage of mathematics as the language of science), Descartes (natural/mechanical description of the world), Newton (unification of heavenly and earthly physics), Darwin (natural explanation of the diversity of biological organisms), Einstein etc.

Let me immediately switch to a strongly opposing historical description of science. In an essay titled 'Muslims and the Philosophy of Science'[12], Ziauddin Sardar starts to build his case that Muslims developed at least the foundations and the 'spirit' of modern science by highlighting the repeated call of the Qur'an on humans to observe and reflect upon natural phenomena as a way to know God, which many understand as knowing the laws of God. Sardar then recalls the fact that many theologians, especially those of the Mu'tazilite rationalist school, were highly interested in physics (or understanding nature) as it allowed them to decide upon certain theological propositions (God's knowledge of particulars, God's action in the world etc.). Then Sardar, after briefly reviewing the works and statements of some of the great Muslim scientists of the classical era, who are known to have carried a truly scientific spirit and methodology, makes some bold statements: 'The ideas of Ibn al-Haytham, al-Biruni and Ibn Sina, along with numerous other Muslim scientists, laid the foundations of the "scientific spirit" within the worldview of Islam'. He then adds: 'The "scientific method", as it is understood today, was first developed by Muslim scientists'. He justifies this by claiming that the classical theologians 'placed a great deal of emphasis on systematic observation and experimentation'. He adds: 'The insistence on accurate observation is amply demonstrated in the literature of astronomical handbooks and tables'. And in another article ('Science Wars: A Postcolonial Reading'[13]), Sardar writes, 'From Islam, Europe learned how to reason logically, acquired the experimental method, discovered the idea of medicine and rediscovered Greek philosophy. Most of algebra, basic geometry and trigonometry, spherical astronomy, mechanics, optics, chemistry and biology – the very foundation of the scientific renaissance of Europe, came from Islam'.

Ziauddin Sardar (1951–) may be surprised at this characterisation, but in my view he can be considered as a modern-day version of Ibn Rushd,

albeit a decidedly unconventional one. Indeed, Sardar would be surprised because he is an admirer of al-Ghazzali, a giant in his own right, but one who was forcefully and masterfully rebutted by Ibn Rushd.

Like Averroes, Sardar is well read and versed in the knowledge both of his times and of his classical culture. He too has sharp analytical skills and refined writing talents, and uses both to dissect and shred the views of various schools of thought. Unlike Ibn Rushd, though, Sardar has been endowed with a fine sense of humor.

Ziauddin is something of a contemporary cultural prodigy. Indeed, rarely does one find a person who is so well versed in such a wide spectrum of subjects, from science and technology to philosophy and religion, cultural and postmodernist theories, arts, future studies, policies and politics and more. Moreover, Sardar is a prolific writer; now in his mid-fifties, he has published over 40 books and innumerable articles, not to mention a number of documentaries. Unfortunately, it seems that only one book of his (*The Touch of Midas: Science, Values and Environment in Islam and the West*) has ever been translated to Arabic, and that too became possible when the then crown-prince of Jordan read it, liked it and funded its translation.

Sardar continues to write regular columns in the British newspapers and weeklies (*The Observer, The Independent, New Statesmen*); in past times, he wrote extensively (and worked for awhile) for the high-level science journal *Nature* and the influential weekly *New Scientist*. Most amazing about Sardar is the fact that he is equally at ease and at home with several cultures at once: the Muslim, the British, the Pakistani and the South Asian; indeed, he can use the whole intellectual heritage of these cultures as well as the large palette of tools they can put at one's disposal (books, journals and magazines, TV broadcasting, universities and research institutes etc.); and yet he is critical of all of them at the same time. Belonging almost equally to two worlds is never an easy feat; Sardar has described himself as 'an enigma in both spheres'.

The man has already led a very fulfilling life and has been very influential; indeed, he had achieved that much (fulfilment and influence) by the time he was barely thirty. For a full measure of the variety of experiences he has gone through, one should read his autobiography (*Desperately Seeking Paradise*), where he describes his intellectual and spiritual voyages from his teenage years in Britain to his trips in the Muslim world (in the seventies and eighties) trying to survey the state of science and technology there, to the years he spent in Saudi Arabia (from 1975 to 1980) when he tried to help modernise some aspects of Islamic life (Hajj, for example), to his intellectual battles within and on behalf of his Ijmali school of thought, to his failed attempts to help make Malaysia the 'new Andalus' (the spirit of Ibn Rushd surfacing here again).

Sardar has always been interested in the future, particularly that of Muslim societies. For that reason he examines the present. And to understand the present, he sometimes needs to delve into the past. He is convinced that no progress can be made by any society if it tries to apply artificial solutions, such as those that consist of emulating the West or implementing its ready-made techniques and solutions. Furthermore, his critique of modern science, of hard-line rationalism and of colonial (classical and neo) methods has not earned him the love of proponents of the West and its civilization. Because of this stand, he has often very mistakenly been considered a traditionalist. On the other hand, the neo-traditionalist camp (Seyyed H. Nasr and his followers) have criticised Ziauddin for his little usage of the spiritual dimension of Islam; in their view, he seems mainly interested in ethical, environmental and Third World (post-colonial) critiques of science, not in its fundamental metaphysical bases.

But if such is the extent of the Arab-Muslim civilization's scientific legacy to the West, how come it is never recognised and acknowledged? Sardar explains his theory in a long diatribe against the Western history of science, a paragraph that is worth quoting at length:

Clearly, a Europe perceived to be far superior to Islam could not admit a deep intellectual debt to the inferior, barbaric civilization of Islam any more than it could acknowledge the existence of an Islamic science that was at par with anything that Europe produced. Thus began the first and the original science wars: the war of European science against the science and learning of Islam. This war had three main functions. First, to sever the Islamic roots of European science and learning. Second, to make the history of Islamic science all but invisible. Third, to deny the very existence of science in Islam. In the initial stage, this was done consciously as an integral part of the Orientalist scholarship . . . When the sixteenth- and the seventeenth-century Orientalists looked at Islamic science they found it to be trivial and, in many cases, to be not science at all, but simply a rag-bag of superstitions and dogma. The fiction was created that Muslims did little more than translate the works of Greeks and were themselves unable to add anything original to them. Thus was born what I call the 'conveyer belt' theory: the Muslims preserved the heritage of Greece and, like a conveyer belt, simply passed it on to its rightful heir: the western civilization[14].

It is a bold and quite extreme assessment. Few would disagree that there are elements of truth in this view, but few, if any, would subscribe to it in full. It is not hard to see why Sardar, a well-read cultural critic and commentator with a strong background in science and Islam, would be driven to construct such a harsh judgment of the western history of science. Indeed, few of the western contemporary works dealing with the history and philosophy of science or with the questions of science and religion make any reference to Islam, when it is quite obvious that one should expect to find such topics in the history of Islam as much as in Christianity and the West. A thinker like Denis Alexander, for whom I have the greatest respect and admiration, has written a voluminous tome titled *Rebuilding the Matrix: Science and Faith in the 21st Century*, a major and impressive opus that took him many years and yet – despite having lived for a number of years in Muslim countries and cultures like Turkey and Lebanon – did not make almost any reference to the contributions of the Islamic culture, science, philosophy and civilization on any of the topics he addressed. Indeed, while devoting at least four full chapters to 'the roots of modern science', including one to 'from the Greeks to the scientific revolution', it devotes a total of four lines to the contribution of the Muslim world and era: 'The story of how the ancient texts of Greek science were gradually translated and filtered into Europe is a familiar one. Preserved by the Muslim world, transcribed texts, some of them very poor, began to be translated from Arabic into Medieval Latin from the late 10th century, and translation became a major scholarly activity from the 12th century onward . . . '[15]

I should not let any wrong impression develop here that only such opposite viewpoints are expressed in the literature on the history of science. Let me mention another important Muslim voice in the field of science, philosophy and religion: Mehdi Golshani (see sidebar). This important, though discrete Iranian physicist and philosopher who has written several books and many articles in this domain cites several authors who state, sometimes very explicitly, that modern science is a legacy of the Islamic civilization. He approvingly cites Briffault[16], who declared that what we call science today, what developed in Europe as a result of the modern world view, that is the new approaches to investigation, the observational and experimental method and the advanced mathematical laws, was all fuzzy among the Greeks; and yet such a new methodology was indeed transferred to Europe by the Arabs, so modern science must be considered as the main contribution of the Islamic civilization. Similarly, Golshani cites Randall[17] as follows: '[The Arabs] in medieval times possessed that type of scientific thinking and technological life that we tend to attribute to modern Germany. I believe that contrary to the Greeks, they did not look down upon the laboratory and the patient experimentation; furthermore, they subjected science to the benefit of life in a direct manner in such fields as medicine, mechanics and other technological areas, instead of considering science as a goal in itself; Europe indeed inherited from the Arabs what we like to call the Baconian spirit, which aims at developing and enlarging the domain of mastery of nature by man'.

Although generally very careful and moderate, Abdekader Bachta[18], a Tunisian contemporary philosopher of science, finds important aspects of the modern scientific method – or at least the scientific spirit – in the works of Ibn al-Haytham, al-Biruni and Ibn Rushd, and to some extent in those of al-Kindi and al-Farabi. The elements of the scientific spirit that he finds are the usage of mathematics (arithmetic and geometry, in particular), the reliance on measurement (and thus on quantification) and the insistence on experimentation – to various degrees for each of them. Bachta remarks that Ibn al-Haytham – and the other great Muslim scientists – did adopt the inductive method, generalising limited observations that they or others had made before them, and that they often started with a given explanation (today we would say 'hypothesis') and set up experiments to try to confirm it. In the case of our icon, Averroes, Bachta analyzes one of his medical works[19] and shows[20] how he refutes Galen using additional observations and arithmetic calculations (averages). The author then makes the following general assessment:

> All of this refutes the widespread view that the experimental approach was absent from the Arab-Muslim civilization, but it should not make us think that it was identical with the modern one either. [. . .] The Arab-Muslim thinkers were far from such constructed experiments; they applied their

studies directly on nature, either raw or by means of what we can only consider today as elementary instruments. History allowed them only that much. Consequently, one must be careful when describing the experimental approach and induction of the Arabs and Muslims. One must always take into account the historical context as well as the exact content of the manuscripts. One must rather emphasize the Arab-Muslims scientists' physical realism, which they in fact found in the Greeks' works [. . .] What the Arabs did was to extract that approach from the multitude of ideas and schemes that could be found in the Hellenistic legacy and to refine it and apply it in their natural philosophy and research. Hence they built a bridge toward the modern scientific spirit[21].

So are we to conclude that because one can make a case for a serious scientific spirit and method of investigation among a number of Muslim scientists of the golden era that modern science emerged within the Islamic civilization? Isn't that another kind of cultural centrism? What, if someone else makes a good case of serious scientific investigations during the Aztec, the Babylonian or the Chinese eras? Some have actually done precisely that. The first half of the twentieth century saw the production of two monumental and revolutionary studies in the history of science: (1) George Sarton's *Introduction to the History of Science*, in several tomes, published between 1927 and 1948; and (2) Joseph Needham's *Science and Civilization in China*, also a colossal multi-volume production, published from 1954 onwards. Sarton's work, largely devoted to science during the Islamic era, made a detailed case that the contributions of Muslims were truly enormous, both in quantity and quality, much more in fact than had hitherto been realised; he concluded that no history of science, call it modern or western, can be envisaged without taking full measure of that contribution. Needham made the same case, perhaps even more forcefully and compellingly, about science in (ancient) China. And of course, one can examine other societies and cultures as well and search for signs of scientific explorations. One, for example, learns that the Aztecs, long ago, developed a calendar that was far better than that of the European civilization which invaded it, considered it to be primitive and barbaric and destroyed it, even though its knowledge of astronomy was more advanced. The Indian civilization made inroads in arithmetic (famously invented the concept of zero) and quantitative science. And yet, one must be careful not to call every kind of investigation of nature, and every successful invention or discovery, science.

Another controversial issue is the claim by a growing number of thinkers that science appeared and prospered in monotheistic cultures (western writers say 'Christian', while Muslim commentators insist on the precedence of the Islamic civilization in this regard), and that the belief in divine providence and design, which implies the existence of order and laws governing the world, was the main drive behind the development of the scientific enterprise. Some Muslim thinkers (e.g. Nasr and Golshani) have stressed the centrality of the Divine Unity in all Muslim intellectual activities, from religious acts to artistic, philosophical or scientific endeavours, all of which aimed at underscoring the Oneness of God. Even fields such as astronomy, which had obvious practical applications for Muslims (e.g. the determination of prayer times, fasting and pilgrimage dates etc.) were considered first and foremost for their spiritual values; indeed, that is what is emphasised in the Qur'an and in the books of exegesis.

Lastly, I should note that Chittick finds it logical and culturally consistent that 'modern science' did not arise in the Islamic civilization. For him, the key feature of modern science, the one that makes it a dangerous 'anomaly', is its rejection of teleology. Furthermore, unlike the (traditional) 'Islamic science' he upholds, modern science makes a clear (he calls it 'brute') separation between subject and object and, according to him, refuses to 'admit that consciousness and awareness are more real than material facts'. In the end, modern science disregards 'the ultimate and the transcendent', which is why it could not have arisen in Islam[22].

What is clear, however, is that science (as we now see it) is, in comparison to the whole history of mankind, a very recent phenomenon[23]. Science certainly could not develop until reason had become an important characteristic of human behavior. On this everyone agrees. And science was perhaps not possible before superstitious and polytheistic beliefs began to disappear, and monotheistic religious thought appeared and flourished. On this, however, there is quite a bit of disagreement.

Let me repeat the central idea I have developed so far: Science is the systematic, objective, quantitative, falsifiable set of methods by which one studies the (natural) world. It is clear that some civilizations, particularly the Islamic one, had introduced and implemented some features of this general program, but it is not clear whether any society had previously developed the understanding we have today of what science is and how it should proceed. Furthermore, it should be stressed that even today's science is far from being a completely developed objective program, the results of which are to be safely trusted. As I have mentioned previously, the human and subjective elements of science may be hidden from most people's eyes, but they are certainly there. Let us thus examine the general philosophy(ies) of science in order to get a better understanding of what can and must be trusted – and thus taken fully into account in philosophical and religious consideration – and what must be viewed with some skepticism.

a (very) basic review of the philosophy of science

What do we mean by 'philosophy of science'? This refers to the critical examination of the methods of inquiry and exploration that science (or scientists, to be more accurate) adopt and implement. It is a vital part of the human quest for understanding nature, for without such a critical examination of our methods, we may never be sure that what we discover is not actually a reflection of our biased and unconscious beliefs. Indeed, one often sees what one wants to see, and conversely one becomes unconsciously blind to what is actually there but is not to one's liking.

Introductory science textbooks often credit the thirteenth-century English philosopher Roger Bacon with the development and canonisation of the scientific method, at least the one based on experimentation and empirical approaches (*scientia experimentalis*), although later studies[24] showed him to be highly interested and versed in alchemy and astrology. His modernity has been somewhat exaggerated.

Historically, the first major shift in the general scientific enterprise was the realisation that the method of deduction, which was actually inherited from philosophy, had to be replaced with induction. Indeed, it took a long time for scientists (before the term and specialised profession existed) to realise that in science, except for mathematics, one couldn't start from axioms and deduce results by various methods (logic, experimentation or whatever else one may devise). Instead one has to start from 'facts', with whatever certainty one could have about them, and try to construct (by generalisation, i.e. induction) a law that both fits and explains the data. This too is often credited to Bacon, although one can show that al-Biruni made explicit statements to this effect, adopted this approach and debated it with Ibn Sina (who remained a 'deductionist'), two or three centuries before Bacon.

Philosophically, however, the first major development in the construction and delineation of the scientific enterprise consisted in the positing of 'methodological naturalism' as a foundational base of science. By this is meant that only natural explanations will be accepted within the scientific enterprise, and therefore supernatural explanations (e.g. divine interventions) and arguments from authority (e.g. 'Aristotle says') are all evacuated from the scientific discourse. This may appear to be a clear demarcation, and implementations of it should be easy, but that is not always the case; for example, oftentimes the views of leading scientific experts (one could say 'high priests of the field') are given credence purely on the basis of their authors' status. And indeed, a school of thought referred to as 'critical rationalism' holds that a strict implementation of the above rigid rule is not possible and instead calls for upholding the falsifiability criterion to sift through all scientific propositions, however they may have been formulated or obtained.

Indeed, it quickly became evident that the manner in which hypotheses or explanations were to be constructed and formulated was by no means rigid and prescribable. One should, in principle, adopt the inductive approach and stick to methodological naturalism as honestly and consciously as possible, but beyond these general guidelines, anything goes. In the eighteenth century, William Herschel had started to emphasise the role of the imagination (or, more generally, creativity) in constructing models and theories, at least in the early stages of formulations; it is the testing of such theories that will give them credence or send them to the dustbin. Shortly afterwards, William Whewell, one of the leading figures of the nineteenth-century science who is given too little importance today for various reasons, showed that the history of science indeed supports the important role played by insight and creativity in the discovery and formulation of scientific models, results and laws.

In the twentieth century, three philosophers of science made seminal contributions to the understanding of how science works and how to be cognisant of its limitations and of any undue credit and authority it may be given in many cases. These philosophers are Karl Popper, Thomas Kuhn and Paul Feyerabend. Very recent times saw the emergence of some severe critiques of science, and I shall mention them later as well.

It would not be a wild exaggeration to state that Sir Karl Popper introduced what may now be considered the defining characteristic of a scientific hypothesis, what differentiates between science and non-science. Simply put, Popper insisted that for a hypothesis to be scientific, it must be falsifiable, that is it must make some predictions or lead to consequences that can be tested and shown to be true or false. Any hypothesis that cannot be tested or disproved is simply not scientific! Popper thus leaves open the way a scientist comes up with a hypothesis, as others before had noted the role of the imagination, insight or creativity in this regard, but he puts the onus on the end part of the proposed explanation.

It was soon pointed out, however, that this definition of the scientific nature of a hypothesis is not foolproof; for example, there are a number of entities in nature the existence of which are predicted (quark, dark matter etc.) but which cannot be directly shown to exist, yet no one will claim that any physical theory that proposes their existence is not scientific, as its predictions cannot be proven. And here we get into murky grounds. Indeed, what about the cosmological theories, which predict the existence of other universes that cannot be confirmed experimentally? Many scientists argue that if indirect evidence can be found for the theory, then its other claims can be accepted, including the existence of whole universes that no one will ever have any evidence for, others are more skeptical. Suffice it to say that the criterion of fasifiability, while a very important and perhaps a defining one for scientific hypotheses, can in some cases be faced with difficult judgments to make.

The next philosopher of science to make an important contribution in these discussions is Thomas Kuhn, who in his seminal book *The Structure of Scientific Revolutions*[i] explained how science moves ahead and, in particular, how science comes to discard theories. In a nutshell, Kuhn explained that science goes through periods of 'normal science' in which scientists follow a paradigm – a generally accepted state of affairs in a given field, complete with the dominant theories of the day and the main questions and problems that are in need of investigation. The geocentric cosmology was a paradigm for many centuries until the heliocentric model overthrew it. Similarly, Newton's theory of gravitation was a paradigm; it was later replaced by Einstein's more general theory. And so on.

Now, Kuhn focused on how one paradigm is replaced by another. He explained that the revolution, which achieves this, takes place when too many anomalies are found in the prevailing paradigm/theory. So far, it sounds reasonable and seems to indeed conform to how things have occurred in the history of various fields of science. But Kuhn's view ran into strong opposition when he argued that these transformations are very subjective, it often depends on the scientists' 'appreciation' of the anomalies and people's

---

[i] Thomas Kuhn's *The Structure of Scientific Revolutions*, first published in 1963, was so influential that it is often considered as one of the most important books of the past half-century or even of the whole twentieth century. It has sold more than a million copies and been translated into more than 20 languages.

readiness to cling to or discard the old theory. Indeed, as Alexander explains, 'It is simply not true that scientists give up their theories that easily after they have set up tests for the theory which have turned out not to support it'[25].

Kuhn seemed to expose as a myth the objective nature of science, especially when he insisted on the 'act of faith' by which a scientist 'converts' to a new theory. Furthermore, he described as 'peer pressure' the collective move to the new paradigm. He says: 'As in political revolutions, so in paradigm choice – there is no higher standard than the assent of the relevant community'. So the argument from authority is still present – albeit hidden – after all. Kuhn also criticised the 'normal' enterprise of science as being highly conservative; in his view, 'normal science does not aim at novelties of fact or theory, and when successful finds none'[26]. Sardar sarcastically describes the role of scientists in Kuhn's normal periods as that of 'puzzle solvers within an established worldview' instead of being 'bold adventurers discovering new truths'[27] as one tends to imagine them. Sardar further comments that 'Kuhn showed that far from being the pursuit of objectivity and truth, science was little more than problem-solving within accepted patterns of beliefs'[28]. Moreover, he refers to science's claims of objectivity and universalism as no more than a mirage. The whole concept of 'facts' and 'truths' is thus shaken by the Kuhnian view of how science proceeds.

Imre Lakatos is a philosopher of science who was born in Hungary. He was first incensed by Kuhn's description of science and scientists, particularly the purported conservative and irrational clinging to theories, the blindness in the face of contrary evidence, the religious-like 'conversions', and the 'mob psychology' of the community as it establishes a paradigm. His major point of criticism towards Kuhn was his insistence that paradigms are not identical to individual theories and thus scientists have few chances to cling to their preferred ones. It is a collective process, according to Lakatos, and that is what ensures the objectivity of the enterprise. Lakatos thus insisted on replacing the concept of paradigm with that of 'research program', which indeed starts from a set of accepted facts or assumptions and moves on to obtain new results; so long as the research program is successful in that goal, it continues with those assumptions, otherwise it shifts to new ones. In other words, the same Kuhnian process that describes the evolution of science through history is upheld, but without the strong negative features of subjectivity and 'faith' that Kuhn attached to scientists.

Paul Feyerabend was an Austrian philosopher who started out as an advocate of Popper's views on science but gradually developed a personal view that stood out in sharp contrast. In one word, Feyerabend's philosophy can be described as anarchist; it is however often stated in the two words that he has made famous: 'anything goes'. His magnum opus is *Against Method: Outline of an Anarchist Theory of Knowledge*, published in 1975. His viewpoint is simple and sounds very reasonable in fact: No single approach to the pursuit of knowledge should be canonised as the scientific method; every idea and theory, including what others would consider non-scientific, such as creationism and astrology, should be thrown into the arena and allowed to seek confirmation and acceptance. In short, there is no boundary that delimits the scientific enterprise, whether in methodology or in content.

postmodernism and science

In the wake of the above fundamental critiques of the (standard) enterprise of science, it is no surprise that the general cultural movement of postmodernism pushed the criticism further by applying its general philosophy to science. Indeed, postmodernism is often identified with relativism, a cultural philosophy that gives no special status or importance to any field, trend, school or program. This means that science must be seen as one approach to seeking knowledge and to viewing the world – among many other approaches that are all equal in principle. Postmodernism defends multiculturalism and gives non-rational views, including mysticism and the occult, an equal opportunity. One of postmodernism's main arguments is that knowledge is intimately related to and based on language, which is always related to a society's culture and world view and can never be shown to describe any objective or universal truth about the world.

The age of postmodernism (the sixties and the seventies) saw the emergence of social movements that were also going to have some bold things to say about science. Environmentalism attacked not only the harmful applications of science but also its current methods. The first big blow of this kind was Rachel Carson's *Silent Spring*, published in 1962, which described for the public the great harm science had been doing to the world (the environment), particularly in the usage of pesticides, DDT and such. A zoologist and marine biologist, Carlson made a huge environmental impact in the USA with that book and other publications; indeed, she had always had great writing skills and made utmost use of them. So important was her legacy that she was posthumously awarded the Presidential Medal of Freedom.

A few years later (1971) Jerry Ravetz, an important contemporary philo
sopher and historian of science, published another seminal contribution: *Scientific Knowledge and Its Social Problems*. In it he developed the theory of 'post-normal science', which applied 'when facts are uncertain, values in dispute, stakes high, and decisions urgent'[29]. Ravetz's main thesis was that scientific research does not lead to immediate discoveries of facts, rather truths are *produced* by societies over long periods of time, merging facts with other social ideas. In his book, he dissects, for example, the way a fact is 'discovered' and 'processed' by the community (scientists, writers and teachers), and then taught – in a simplistic version – as 'absolute truth'. Sociologists picked up Ravetz's views, created the new 'sociology of science' field, and soon replaced the term 'discovery' with 'construction'. The most famous work on that theme is *Laboratory Life: Social Construction of Scientific Facts*, written by Bruno Latour and Steve Woolgar in 1979.

Ravetz went further and exposed the modern 'pathologies' of 'industri-
alized science': (a) entrepreneurial science, the game of getting grants and publishing having replaced any honest objective search for truths; (b) shoddy (sloppy) science, a consequence of the previous ill; (c) reckless science, a consequence of the previous two ills as well as of being under the financial mercy of industrial powers; (d) dirty science – harmful applications not being given any second thoughts. Recently, Ravetz added new 'structural contradictions' to the list: global pollution and climate change; knowledge and power; knowledge vs. ignorance; elitism vs. democracy; corruption in research; image and audience; societal context and reality; violence vs. safety and more. Finally, I should note that Ravetz is a close friend and collaborator of Ziauddin Sardar; the latter has referred to him as 'my Maimonides'.

There were other important criticisms of contemporary science and its methods. Everyone now knows about the negative products of technology (and, by corollary, of science): pollution of air and water, pesticides, acid rain, deforestation, disruption of biological life to the point of species extinc-
tion, chemical poisoning, toxic and nuclear wastes, exponential proliferation of cancer, global warming and many others. Not everyone, however, knows that roughly half the scientists work at least part-time for the military[30]. Sardar explains: 'American science [. . .] is tightly controlled by an alliance of the military, powerful multinationals and research universities. American militarism directs American science while American science propels American militarism: they define each other'. He goes even further: 'De-
vices and other equipment developed for the military not only provide new instruments for science but also shape the conceptual categories and toolkits

of scientists. The computer, rocket launchers, the laser, microwave communication technologies and satellites were all developed for the "security interest" of the United States and then became key research instruments in shaping and directing science'.[ii] In Sardar's (and others') views, the connection between the military and science spreads even to undergraduate teaching, because the classroom material and the projects and theses that professors teach and assign are directly related to what they work on, which, in turn, is controlled (through funding) by the military.

Now another dimension of science's ugly side has been uncovered and strongly publicised lately: The inhuman methods that are allegedly employed in some fields of research. Sardar, citing *Time* magazine, thus tells us: 'In the United States, in the late 1940s, teenage boys were fed radioactive breakfast cereal, middle-aged mothers were injected with radioactive plutonium and prisoners had their testicles irradiated – all in the name of science, progress and national security . . . '[31] Sardar adopts Ashis Nandi's description of science as 'a theology of violence' and explains, 'It performs violence against the subject of knowledge, against the object of knowledge, against the beneficiary of knowledge and against knowledge itself'[32]. If our thinker sounds extreme in his indictment of science, it is partly because this is his personal style of writing (direct and blunt) and partly because he has his own agenda to promote, as I shall discuss in detail in the next chapter.

In the past few decades, a new critique of science has surfaced in the west: The accusation of systematic gender discrimination and male domination in the scientific enterprise. In fact the feminist critics do not simply state that the number and prominence of women in science is much too low, they further argue that the content of science itself is largely sexist. Indeed, feminist scholars are not simply calling for an equal or at least a fairer, participation in the scientific research program, they are calling for a complete revamping of the programmes themselves, of the assumptions and even of the very basic concepts.

Finally, perhaps the most recent line of criticism that has been opened against science is that of cultural-centrism, that science today is deeply rooted in the Western social norms, precepts and methods. The claim is simple and strong: The science one produces is a direct result of how one

---

[ii] Sardar: *Islamic Science: the Way Ahead*, op. cit., p. 167. Sardar quotes Stuart Leslie, who adds to the list of scientific instruments that started as or were the outcome of military projects: the Hubble space telescope, the supersonic wind tunnels, the high-power klystrons, MIT's whirlwind computer, the gravity probe B, transistor and integrated circuits, etc.

investigates problems, which is a product of one's view of the world, which, in turn, is affected by one's cultural identity. This idea was first introduced and slowly developed by S. H. Nasr, who has been emphasising the fact that the very definition of science, the very conception of the world and nature, is strongly related to the cultural and metaphysical principles that underlie the society where such science is developed. He made that point when he described in detail the Islamic cosmologies of the classical era, cosmologies that bear little resemblance to today's conceptions of the cosmos, for the simple reason that the very definition and notion of a 'cosmology' is quite different in the classical Islamic culture from that of the western science. Nasr made that point forcefully again in *The Encounter of Man and Nature*[33], where he argued that the western concept of 'nature', as an object that can be dissected or even tortured for the benefit of man, is fundamentally different from the Islamic view (of nature as a sacred entity) and those of other cultures. Similarly, the concept of time can be different from one culture to another; for example, time is cyclic in eastern traditions, whereas it is linear in the West and in other monotheistic cultures. Sardar also likes to mention that western (Aristotelian) logic is binary (A vs. non-A), whereas in Hinduism it is four-fold.

In some instances, the criticism is made even bolder by linking the fast development of modern science to the imperial adventures of the western powers and by showing that even today, the science gates are guarded by western 'men in white' against too much intrusion and participation by Third World scientists. Sardar claims that there is 'truly monumental evidence for deep ideological, political and cultural fingerprints in modern science', but the only evidences he can cite are the following: 'The British needed better navigation so they built observatories and kept systematic records of their voyages'[34]; 'the subordination of blacks in the ideology of the black "child/savage" and the confinement of white women in the cult of "true womanhood" emerged in this period and are both a byproduct of the Empire'[35]; and 'racist and androcentric evolutionary theories were developed to explain human behavior and canonised in the history of human evolution'[36]. On the basis of this (less than convincing) evidence, Sardar concludes that 'modern science has become an instrument of control and manipulation of non-western cultures, marginalised minorities and women'[37]. The second charge, that to this day the scientific enterprise is deliberately kept largely closed to non-Western participants is made on the basis of the ratio of the journals (3,000) that are indexed in the West (by the Institute for Scientific Information (ISI)) to that of the number of journals in the whole world (between 60,000 and 100,000), which translates into a similar (or even smaller) ratio of papers that are available to the Western researchers (and thus read and cited) compared to the total number of papers published in the world. What Sardar does not stress is the huge difference in quality between the ISI-indexed journals and the rest.

To make a long story short, there has developed a new trend among historians, philosophers and critics of science to both expose the 'Eurocentrism' of Western science and to develop a new vision of how a fair and equitable post-colonial science can be developed to do justice to all cultures and their contributions, past and present.

The latest episode in this tug of war between the hard conservative scientists, who believe in the objectivity and universality of the scientific enterprise, and the liberal philosophers and sociologists of science, who believe in the relative and social nature of the scientific research and its outcomes, came in 1995 when a physicist at New York University, Alan Sokal, tired of both the postmodernist attacks and the cultural relativist portrayals of science, wrote a hoax-paper and submitted it to the journal *Social Text* with the title 'Transgressing the Boundaries: Towards a Transformative Hermeneutics of Quantum Gravity'. The paper duly appeared in the spring/summer issue of the journal, and a few weeks later Sokal sent a short article to the journal *Lingua Franca*, revealing the hoax, by showing the numerous ridiculous statements it contained, and statements, which any student of physics or science would have caught. He explained that these postmodernist philosophers and sociologists of science not only understand little of science to begin with but are more impressed with a long list of references that hardly relate to the contents of the paper, as long as it mentions all the high priests of postmodern criticism. As Sardar concluded, 'Sokal's hoax proves what many scholars already suspected: cultural studies has become quite meaningless, and anyone can get away with anything in the name of postmodern criticism'[38]. The war of 'two cultures' had started anew, with the latest battle won by the scientists.

metaphysical bases of science

In the previous sections, I brought up many issues pertaining to science, from its very definition and nature to its hidden processes, limitations and flaws. I did, however, leave aside (temporarily) one of the most important questions that ends up dividing philosophers, thinkers and scientists, at least those who concern themselves with the implications of what science seems to be telling us; that question is the 'metaphysical' basis (or bases) of science.

Now, it may sound a little strange to bring up 'metaphysics' in a discussion of modern science, particularly because we are limiting ourselves to the physical sciences and are trying to keep as objective and universal a standpoint as possible. Indeed, the very word 'metaphysics' conjures up ideas of 'beyond this world', spirits, mysticism and anything outside of nature. But those are only erroneous images and (mis)conceptions; in fact, one simple dictionary[39] definition for metaphysics is, 'the underlying theoretical principles of a subject or field of inquiry'. So what is meant by 'the metaphysical basis' is simply the (often undeclared) foundational principles or accepted tenet upon which science (in this case) or philosophy (in another instance) or other fields of knowledge, will be constructed.

This is a fundamental issue, one that is not often recognised, particularly by scientists. It is a critical aspect of the whole discussion because one can indeed show that depending on the metaphysical bases that one chooses to built science upon, one ends up with a materialistic, a theistic or even an Islamic kind of science. The myth that science is a hard and cold field where everyone checks out his/her own philosophy and belief at the door and agrees to pursue an objective search for truth, and where nothing but naturalistic explanations and interpretations will be made, is a huge myth, and a very recent one too. In her very enlightening and balanced book, *The Fire in the Equations: Science, Religion, and the Search for God*[40], Kitty Ferguson gives the following summary points to describe the main principles upon which all practicing scientists agreed upon in the seventeenth century:

- The universe is rational, reflecting both the intellect and the faithfulness of its Creator.
- The universe is accessible to us.
- The universe has contingency to it, meaning that things could have been different from the way we find them, and chance and/or choice have played a role in making them what they are. Knowledge comes by observing and testing it.
- There is such a thing as objective Reality. Because God exists and sees and knows everything, there is a truth behind everything.
- There is unity to the universe. There is an explanation – one God, one equation or one system of logic – which is fundamental to everything.

Of course, these are not meant to represent the metaphysical bases of science in that era; they do represent, however, some unproven and generally accepted principles regarding science and nature. The reason I bring them up here is to show that today, much of the above, except perhaps the existence (or non-existence) of an 'objective reality', is upheld by science, only after any reference to God or a Creator is deleted. What has changed? Simply the general metaphysical bases, which have in the past few centuries, turned decidedly materialistic.

Mehdi Golshani (see sidebar) quotes the twentieth-century fundamentalist Pakistani Muslim thinker Abu al-A'la al-Maududi (1903–79), who here has made a rather reasonable discourse on science: 'Science moves on two planes; the first one consists of facts of nature or Reality, the other consists of what people do when they construct theoretical molds out of those facts and establish certain concepts. One must clearly distinguish between the two fields. Indeed, the first part represents a global endeavor, and has no relation to the nationality of the researcher. On the other hand, however, the Marxist mind, for example, will construct from those facts models that are consistent with the Marxist philosophy; indeed, you hear statements and expressions that mirror the communist philosophy, which has its own conception of man, of the world, and of history . . . Therefore, every child in the communist countries will learn science framed in the Marxist ideology. Similarly, in the (secular) West, scientists will present their theories in a way that suits their view of a universe that is devoid and totally free of any divine existence or role. Some scientists may well believe in God at their personal level, but they will take Him out of their considerations with regard to fields of science, philosophy and history. We thus conclude from this that ideology gives science and knowledge a special shape that conforms to its general precepts. In the past, Muslims Islamized all branches of science that they digested, that is they subjected these to the Islamic mindset'[41].

'scientific imperialism' (scientism)

Another metaphysical issue attached to science that has come to pit scientists, philosophers and theists against one another more and more is the epistemic limits of science, that is what science can rightfully be applied to and what, if anything, must be accepted as lying outside of its boundaries. A whole school, dubbed 'scientism', wishes to extend the application of science to everything in nature, including human life and society. Indeed, a number of philosophers[iii] consider science not only applicable (in principle) to every subject and field of nature (which, in their view, amounts to everything there is) but is also in the end the only valid path to knowledge. It is easy to see then why scientism has become such a pejorative term and why this conceptual program is sometimes referred to as 'scientific imperialism'[42] and as 'an ideology of arrogance'[43]. It has widely been recognised that the fundamental malignancy at the heart of scientism and naturalism is 'reductionism', which is normally a methodological principle that declares all fields of knowledge to be reducible to fields that lie under them; in particular, physics underlies chemistry, which underlies biology – at least in such a view – biology then underlies neurology, which determines psychology, and so on. Nomanul Haq has also identified another malignancy, the political one, which he defines as the tendency to place the scientists (or as he dubs most of them, especially in the Muslim world, the 'technicians') upon a pedestal, thereby intimidating the common citizen into thinking that scientists are geniuses who understand more and see deeper than anyone else, and therefore that what they say is absolutely correct and that they can – if given the resources and the freedom – invest in any other field and work wonders in it. In other words, scientism says, let's make science the central command for exploring everything that exists.

Other thinkers argue that 'clearly' some topics and fields will always lie outside the scope of science; yet, even the examples that are often offered, e.g. consciousness (a very 'different' and highly subjective phenomenon), are not always accepted as obviously beyond natural science and reductionism. Other examples, such as the religious/spiritual experience that many people report, are viewed with skepticism, if not outright denial. Indeed, one must be very suspicious of challenges to scientific explanation; history is full of such cases that science was surprisingly successful at shooting down, contrary to expectations, as well as of all kinds of subjective experiences and phenomena.

Still, philosophers argue that if one claims that any explanation can only be made on the basis of some first principles, which then must either be posited outside the scope of explanations, thus leading one to admit that

---

[iii] Okasha mentions the late Willard van Orman Quine, whom he considers 'arguably the most important American philosopher of the 20th century'.

some things (these principles) cannot be explained by science, or be explained by something else, thus continuing the series of searches. Even the advocates of the much-touted (future) theory of everything (if it ever comes to light) call upon some principle of beauty, logical consistency etc., to save their imagined theory from the need to further explanations. But then one is led to such questions as 'what is beauty that one should base full explanations of nature upon it?'; 'consistency of whose logic are we relying upon?'; and so on and so forth.

So we come to realise that depending on one's metaphysical principles, one can construct a very different flavour or even shape of science. What is most interesting is that although one's flavour/shape of science may be quite different from others', for pretty much all intents and purposes, the outward look of one's science may be quite identical to others', though this is highly contested by some, as we shall see later. So the difference in the end, and it is a big one, no doubt, will lie in the heart of science, in its spirit so to speak, and in its interpretations and considerations of its possible implications, much more than in the results one will obtain.

Let there be no doubt, however, that this foundation idea, that there are (often hidden) metaphysical principles underlying modern science and that there can be varied interpretations and rich discussions of possible implications of science, is very much the crux of the battle that rages on today, at least in the West, particularly in Europe. (In the USA and the UK, discussions of this kind have long been accepted as within the permissible philosophical reflections around science, religion etc.) In this context, in February 2006, I co-signed a letter[44] to the prime French newspaper *Le Monde* with a dozen high-calibre scientists and thinkers, including a few Nobel and other prize winners; in the letter we practically begged for tolerance towards those of us who dare consider ethical, metaphysical, spiritual or even religious implications that one may see or find in scientific discoveries or developments. The response from the (French) materialistic camp was blistering, accusing us of attempting a 'spiritual foray' into science and even waging a 'crusade' against materialism.

theistic science

Resisting the materialistic philosophy of science as it has been enforced in the recent times, and recalling that modern science owes much to the monotheistic religions and their principles, a growing number of

scientists and philosophers today have been arguing that without wreck-
ing the methods of science and without trying to introduce any dubious
approaches (e.g. mystical or subjective experiences) into the realm and practice of science, beliefs in a creator, designer and sustainer of the world should
not only be accepted (as they do not violate any methodological principles)
but may also be a more fruitful, more complete world view to adopt.

As I have mentioned above (in the general seventeenth-century paradigm
described by Ferguson), the belief in God and the scriptural commands upon
the faithful to explore the world and see the glory of God, all played a fundamental role in the development of science, at least as a field of investigation of
nature. Alexander thus emphasises this point: 'It is quite a surprise to peruse
the voluminous writings of the 17th-century founders of modern science and
find that, whether Catholic or Protestant, Italian, French, Dutch, German
or British, all are unanimous in their conviction that a study of God's "book
of nature" is both a duty and a delight. This defining paradigm did much to
set the tone of 17th-century science'[45]. For example, Alexander quotes part
of letter from Kepler to Galileo, where he refers to 'the many undisclosed
treasures of Jehovah the Creator, which He reveals to us one after another'[46].
With equal passion, Galileo wrote, 'The holy Bible and the phenomena of
nature proceed alike from the divine Word, the former as the dictate of the
Holy Ghost and the latter as the observant executrix of God's commands. A
hundred passages of holy Scripture [. . .] teach us that the glory and greatness of Almighty God are marvelously discerned in all his works and divinely
read in the open book of heaven'[47]. (Needless to say, one could cite hundreds
of similarly eloquent statements and passages by classical Muslim scientists
and philosophers, particularly the giants al-Biruni and Averroes.) And so one
is not surprised to read no less a towering twentieth-century philosophical
figure as Alfred North Whitehead explaining the development of modern science in the seventeenth century as having 'come from the medieval insistence
on the rationality of God, conceived as with the personal energy of Jehovah
and with the rationality of a Greek philosopher . . . My explanation is that
the faith in the possibility of science, generated antecedently to the development of modern scientific theory, is an unconscious derivative from medieval
theology'[48]. Alexander adds one last argument that Joseph Needham, whom
I introduced earlier as the scholar who presented to the world (the West)
the history of Chinese science and technology, explained the Chinese failure
to develop 'modern' science on account of their lacking the idea of divine
creation.

The same Muslim tradition and outlook continues today, with Golshani, for example making similar pronouncements. For instance, he cites the following Qur'anic verse: 'Say: Consider what is in the heavens and the earth; yet signs and warners do not avail a people who would not believe' (10:101)[49].

Mehdi Golshani (1939–) is a distinguished Iranian physicist and philosopher, who has been described as a modern Avicenna. What distinguishes him from many other Muslim thinkers is not the fact that

he has received formal education from the most reputable institutions (in his case, a PhD in particle physics from the University of California at Berkeley in 1969) or that he has been appointed at high-level positions (chaired the Faculty of the Philosophy of Science at Sharif University of Technology since 1995, headed the Institute of Humanities and Cultural Studies in Tehran), or has been a member of international organizations (senior associate of the International Centre for Theoretical Physics,

Trieste, Italy; member of the Philosophy of Science Association, Michigan, USA, the European Society for the Study of Science and Theology and the Center for Theology and Natural Science at Berkeley, California). What distinguishes Golshani is his international, cross-religious outlook: He is a Shiite, but he quotes abundantly from both Shiite and Sunni sources; he is a Muslim, but he participates in international ventures (he has been a judge on the prestigious Templeton Prize for Progress Toward Research or Discoveries about Spiritual Realities and was one of the earliest winners of the Templeton course program on science and religion).

Golshani has contributed much to the study of the relation and interaction between science and religion. In 1998, Golshani approached 32 high-level scientists, philosophers and theologians (only six of them were Muslims, the others being Christians of various denominations) with eight essential questions addressing the relationship between science and religion; he published their answers, along with a long commentary of his own, in a book titled *Can Science Dispense with Religion?*[50].

In addition to a number of significant papers[51], he has published several books[52], in Farsi (Persian) and/or English, some of which have been translated to Arabic. His articles and books clearly show a global outlook.

So today there is a new, growing trend to put God back into one's world view and approach to science (the investigation of nature). This trend is called 'theistic science'.

But what would this theistic science consist of in today's philosophical and scientific methodological landscape? First, theism is the belief in a God, who is not only the creator and designer of the universe, but a sustainer of it, one without whom the very existence of the universe at any instant would not be possible; this belief includes the idea that God may act in the world (or interacts with it) somehow. This is to be contrasted with deism, which is a belief that the creation of the universe required a God (some infinite power transcending this world), but that from that point on, laws of nature took over and steered the development of every aspect of it, with no need (some will insist on 'no possibility') of intervention or involvement on the part of this God. Just as the seventeenth-century scientists fully subscribed to a theistic philosophy, many enlightenment philosophers accepted such a deistic intellectual position. Today, there is among scientists a whole spectrum of philosophical and religious positions, ranging from mysticism to belief in and personal relation with an acting God, a theistic philosophy, a deistic position, agnosticism and finally atheism.

And why would such philosophical and religious positions, which are in principle purely personal views that anyone is free to uphold, have any bearing on our scientific pursuit? Alexander answers this fundamental question by stating that, first 'theism is as much a metaphysical belief system as atheism'[53], and then adds, 'on balance theism as a model appears to "fit" better the properties of the world that we observe'[54]. He explains that the comprehensibility of the world, which so mystified Einstein, or the haunting and perplexing beauty of the equations that describe nature, which led many modern physicists (e.g. Einstein, Dirac, Schrödinger, Weinberg etc.) to believe in their truths, all this (and more) fits better if one frames it in a theistic world view. Another 'fit' is sometimes seen in the existence of conscious creatures, and sometimes in religious experience as well.

Mehdi Golshani has focused much of his discourse on the concept of theistic science, advocating a shift from 'secular science' and showing that it fits the Islamic philosophy of science as well as that expounded by some of the western thinkers. In an article titled 'Seek Knowledge Even if It Is in China'[55], he writes, 'One can learn from the Qur'an some of the things that humanities and physical sciences cannot provide: a solid metaphysical base for all sciences'. There are many other realities in the world that we cannot comprehend or do not have access to; this forces us to adopt a much broader world view than what is presently ruling over secular scientists.

Golshani eloquently expresses the Muslim viewpoint on theistic science; he writes,

> I believe [. . .] that theistic science contains concepts and information that are richer than what is provided by secular science, as the former guarantees for man not just material and moral sustenance but also full description of the universe. It must be emphasized, however, that we do not want from this theistic science we are talking about – and certainly not the Islamic science in particular – that it produce for us a new scientific method or that it start referring to the Holy Book or to the Prophet's tradition in our physics and chemistry research . . . We certainly are not calling for science to go back to its methods of several centuries ago (something that is neither desirable nor possible); instead we call for the upholding of 'divine vision' in our research outlook, meaning that science be constructed on the basis of assumptions of the existence of a creator and sustainer of the universe and upon a vision that does not reduce the world to matter, does not negate the existence of purpose in the universe, and recognizes the existence of a moral order. All this will help solidify science and make it more positive in its investigations and applications[56].

The Indonesian scholar Zainal Abidin Bagir, who specialises in the history and philosophy of science, has reviewed[57] the question of theistic and Islamic science and pointed out that Alvin Plantinga, a renowned contemporary American philosopher who has specialised in issues at the interface of epistemology, metaphysics, science and religion, has recently argued for a theistic science and called for a careful distinction between the 'factual' part of science and the 'metaphysical' aspects of science (in a manner akin to what Maududi has stated); Plantinga further justifies his call for a theistic science on grounds similar to the above (Alexander, Golshani) arguments, namely that this kind of science will be more 'in the service of a broadly religious vision of the world'.

summary

My personal experience of the public's misunderstanding of the science, both as a basic knowledge and as a methodology, is confirmed by other reports from the Arab-Muslim society[58] and the world at large[59]. I have in this chapter particularly insisted on the idea of falsifiability; in my view, this concept most acutely defines and underscores the nature of the scientific enterprise; unfortunately, it seems to largely escape laymen as well as most of the educated public, particularly in the Muslim world.

As we shall see in the remaining parts of this book, science has always had and continues to have a potent effect on the world view and the general beliefs of a society. For this reason at least, science needs to be handled with care, meaning that its philosophy, methods and limits need to be carefully understood and delineated, lest one quickly falls either on the purely technocratic or the simplistic materialistic view of science. Unfortunately, scientists in general and Muslim ones in particular are widely oblivious to the philosophy of science and its battles, as indeed very few universities require its science and technology students to take a course in the philosophy or the history of science. Towards the end of the book, I will focus greater attention on the educated elite in the Muslim world; I will present results of surveys on their attitudes and general views on science and Islam issues. Let me just say that much work needs to be done in this regard in the future; I shall make some recommendations in this regard in Chapter 10.

This chapter was meant to introduce the reader to the current debates around science (its metaphysical bases, its methods, its limitations) and to show that scientism and naturalism are not automatic outcomes of modern science but rather constructed mindsets, which most (Western) scientists adopt without thinking about the issues much. Perhaps, the brief overview given here of the philosophy of science and of the questions presently being argued can help the reader (Western or Muslim) better gauge the related issues, if not take part in the discussions.

Last but not the least, I have tried to show that one version of the metaphysical cloak that science can be enveloped in – without affecting its methodology and rigour whatsoever – is the so-called theistic science, which can provide a way out of the current multiple crisis of science (ethical, environmental, social etc.). Most importantly, it can constitute a possible common ground for many (if not most) scientists across the world. Some Muslims have espoused it (e.g. Golshani); others have advocated an 'Islamic Science'; but as I shall review in the next chapter, the latter propositions seem to have gone nowhere, and so we will ultimately come back and adopt the theistic science solution, which constitutes a much large common denominator.

4

can one develop an 'Islamic Science'?

'Scientific thought is the common heritage of mankind.'

Abdus Salam

'Ever since children began to learn [. . .] that water is composed of oxygen and hydrogen, in many Islamic countries they came home that evening and stopped saying their prayers. There is no country in the Islamic world which has not been witness in one way or another, to the impact, in fact, of the study of western science upon the ideological system of its youth . . .'[1]

Seyyed Hossein Nasr

science in Islam

'Are those who know equal with those who know not?' (Q 39:9).

'Allah will exalt those of you who believe, and those who have knowledge, in degrees' (Q 58:11).

'The ink of the scholar is more sacred than the blood of the martyr'.

'An hour's contemplation (or study) of nature is better than a year's adoration (of God)'.

These verses and statements, sometimes attributed to the Prophet or his companions or to the general Muslim tradition (and other such Islamic references[2]) clearly show the importance of knowledge and the status of scholars in Islam.

The scholars of the golden era of Islam, particularly the philosophers and scientists, heeded this call and understood the moral and ethical standards that must be upheld in the pursuit of knowledge. Indeed, they related the quest for knowledge 'even in China' directly to that other prophetic injunction: 'State the truth even if it hurts you'. al-Kindi thus wrote: 'Our goal must be to garner the truth from wherever it may come, for nothing is of higher priority to the seeker of truth than truth'. al-Biruni stressed the same moral imperative; in prefacing his substantial work on India, he wrote, 'I have composed this book on the beliefs of the people of India, and although they differ from us in religion, I did not reject what I found to be correct in their knowledge or beliefs'[3]. In another masterpiece of his (*The Chronology of Ancient Nations*), he concluded his work with the following verse: 'That he who lives might live by clear proof, and most surely Allah is Hearing, Knowing' (Q 8:42). He also stressed the necessity of justifying every claim, something that Ibn Sina emphasised by stating the following: 'He who gets used to believing without proof has slipped out of his natural humanness'[4].

This is all fine and great, but the issue is not so straightforward. Indeed, all of the above injunctions and references are about 'knowledge', or *ʿilm*, to use the highly important Arabic term. The problem is that the word *ʿilm* has been understood in so many different ways and has been assigned so many different meanings, including 'knowledge', 'science', 'religious knowledge', and 'knowledge of God' that one finds himself having to struggle to delimit its meaning and usage before going further. Muzaffar Iqbal reminds us that '[a]lmost all reformers translated the Arabic word *ʿilm* (knowledge) as "science" (meaning modern science) and framed their discourse on the necessity to acquire knowledge'[5].

Ziauddin Sardar has argued that the concept of *ʿilm* in Islam, although it is usually rendered merely as 'knowledge', distinguishes itself from the Western concept of knowledge in the moral dimension it carries; he states, 'The concept of *ʿilm* [. . .] integrates the pursuit of knowledge with values, combines factual insight with metaphysical concerns and promotes an outlook of balance and genuine synthesis'[6]. Elsewhere he has asserted that this Qur'anic concept 'originally shaped the main features of Muslim civilization'[7]. Likewise, Munawar Anees (a former collaborator of Sardar) declares the concept of *ʿilm* to be 'one of the most sophisticated, all-comprehensive and profound notions to be found in the Qur'an', one that is 'in its significance [. . .] second only to *tawheed*'[8]. Anees adds that to translate *'ilm* as 'knowledge' is to unjustly limit its many facets, meanings and implications.

However, Sardar himself is aware of the fuzziness and multifaceted aspects of the concept of *'ilm*. Indeed, in one article[9] he tells us that up to 1200 definitions of *'ilm* have been mentioned in the Islamic literature. He then explains that the word *'ilm* eventually came to signify 'science', *'ulum* is now the Arabic word for 'sciences', and *'alim* refers to either 'scholar' or 'scientist'.

The somewhat easy solution that has been adopted by a number of scholars (including Golshani) is to claim that the Qur'an – and thus Islam – does not distinguish between different forms of knowledge, whether natural science, social science, religious knowledge or any other. Indeed, Golshani insists on the global nature of knowledge in Islam, making it a crucial characteristic and a goal in itself. He writes, 'Muslim scholars during the golden era of Islam adopted a global vision and outlook toward the various fields of knowledge; indeed, they considered the scientific branches as continuations of the religious quest . . . Islamic art played a similar role, for it – like the sciences of Muslims – aimed at exhibiting the unicity of God's plan in the universe'[10].

Sardar fully agrees; he writes, 'Polymaths, like al-Biruni, al-Jahiz [and 10 others he mentions by name] and thousands of other scholars are not an exception but the general rule in Muslim civilization. The Islamic civilization of the classical period was remarkable for the number of polymaths it produced. This is seen as a testimony to the homogeneity of Islamic philosophy of science and its emphasis on synthesis, interdisciplinary investigations and multiplicity of methods'[11].

On the other hand, one must remark that Muslim scholars were not all fully embracing all knowledge and sciences. Many adopted a carefully selective stand with regard to those branches that did not seem to always have direct benefit or bearing on the needs of the Muslims. And so we find al-Ghazzali, that major figure in the history of Islamic thought, declaring that 'natural sciences are a mixture of truth and falsehood, correctness and errors' and that mathematics had to be avoided because it was often the preliminary and foundational science to 'erroneous sciences'; indeed, he wrote that 'the study of the science of Euclid and Ptolemy, with the details of calculation and geometry, although he makes the mind and the spirit stronger, we forbid dealing in it for what it leads to; indeed, it is the preliminary to the sciences of the ancients, which contain wrong and harmful creeds'[12]. He goes further in *The Deliverance from Error* (his intellectual testimony) and writes the following[13]:

> The first risk from studying mathematics is that the student is struck by the precision of this science and the imposing power of its proofs. He extends this high esteem to all the philosophical disciplines and attempts to generalize the clarity and firmness had by mathematical proofs. Then, when he hears mathematicians being reproached as heretics, for having negative attitudes or for being scornful of revelation, he rejects the truths which he had admitted previously through pure conformism. If faith were true, he will say to himself, how is it that these mathematical experts have not perceived it? As people say that they are heretics and irreligious, truth must consist in rejecting and denying religious beliefs. How many people have lost their faith because of this simple argument! [. . .]

al-Ghazzali then further expounds his views with arguments of the same kind.

Other conservative scholars have expressed more hostility toward non-religious sciences. Ibn Taymiyyah has written: 'Only the knowledge inherited from the Prophet deserves to be called science; anything else can either be a non-beneficial science or a so-called science, and in both cases if there was any good in it then one would find it in the Prophet's tradition'[14].

Some conservative Muslim scholars, for instance the contemporary researcher Farouq Ahmad al-Dassouqi[15], insist that 'science' must be defined broadly, such as to include (and thus give equal status to) the 'religious sciences' (along with the natural sciences). al-Dassouqi writes, 'Some westernized and secularized thinkers refer to "religious sciences" as "religion" and limit the term "science" to "experimental sciences"; in fact this trend has almost made the practice the de facto definition of the terms, so that when they say "science", it is clear that they mean "experimental science"'[16]. He adds the following: 'And if "science" is the human knowledge that rests on evidence that leads to certainty, then the Islamic religious sciences do rest on such bases, and therefore we are not abusing the truth or being biased to the religion when we denote them as "sciences"'[17].

The twentieth-century fundamentalist writer Anwar al-Jundi sheds heavy doubts and suspicion on, and ultimately outright rejection of, the experimental scientific method itself:

> Islam has placed constraints and reservations on the western experimental science, its principles and applications. Indeed, western science springs forth

from a deceitful arrogance, for it denies the creator of life and of the universe and replaces the divine with nature, and this is western science's biggest mistake. Furthermore, it rests on materialistic philosophy and thus denies all spiritual and moral aspects of man and of life, which leads it to miss important connections between man, science and God. [. . .] Both Darwin's theory and the Big Bang are null and void; indeed, the former has a missing link, and how can a theory that misses an important element be at the forefront of science and be the dominant paradigm for over a century; the latter (the Big Bang) lacks any evidence and is mere speculation, and the Qur'an rejects such speculation, particularly on the topic of the universe's creation.[i]

It is clear that the positions of the orthodox school tend to betray a serious misunderstanding of the methodologies of science. Finally, even a 'rationalist' like Ibn Khaldun, the great pioneer of sociology, has dismissed natural sciences as 'irrelevant to us in our religion and our lives and must therefore be renounced'[18].

Seyyed Hossein Nasr (1933–) is a master, in every sense of the word, including the intellectual and the Sufi meanings. I have attended some of his class lectures, I have heard him give talks to general audiences, I have watched him being interviewed at length, and I have sat down and talked with him one-on-one in his office a few times. He is mesmerising. Nasr has the look and mannerisms of a spiritual sage, the fluency of thought and speech of a skilled orator, and the rich command of a master academician; indeed, I have never seen Nasr use any notes. Most impressive about him is his calm ever-smiling face and his gentle humble dealings.

The first time I visited him in his office (at George Washington University, where he holds the chair of Islamic studies), he offered me his bibliographical index, which then (in the mid-nineties) contained a list of no less than 50 books and 400 articles; he autographed it for me with 'To Professor Nidhal . . .' Some of his books have been translated to a good half-dozen languages,

---

[i] 'I made them not to witness the creation of the heavens and the earth, nor their own creation; nor choose I misleaders for (My) helpers' (Q 18:51).

but – as far as I have been able to check – no more than handful have been translated into Arabic!

And yet I do not subscribe to this philosophy, on the contrary – as will transpire throughout this book.

Seyyed Hossein was born in Iran in 1932, but he was sent to the USA for education at the age of 12; he graduated from high school in 1950 as valedictorian of his class and won the Wyclifte Award for the school's all-round outstanding accomplishments. He then received a scholarship for physics at MIT, thus becoming the first Iranian undergraduate student at the prestigious university. He was soon attracted to the fields of history and philosophy of science, and to metaphysics and philosophy more generally. It is also in that period that he became familiar with the works of Frithjof Schuon, the famous authority on perennial philosophy, who was to become his spiritual and intellectual master.

He was barely in his mid-twenties when he published two seminal and influential books, *An Introduction to Islamic Cosmological Doctrines*, a slightly modified version of his doctoral thesis at Harvard, and *Science and Civilization in Islam*, which rocked the field of history and philosophy of science by insisting that the nature of science in the Islamic civilization was fundamentally different from the modern one and that one had to review the concept of knowledge and civilization before understanding the enterprise of science through history. Indeed, Nasr's first book was his *Introduction to Islamic Cosmological Doctrines*; I read it some 20 years ago, and it blew me away, even though I did not understand much of it (then). I have since read it again and read many reviews of it; it is a classic work, and I will be referring to it extensively in the chapter on cosmology.

After his doctorate, Nasr returned to Iran, where he was successively appointed university professor, dean, academic vice-chancellor and then president of Sharif University, and later chancellor of Melli/Beheshti University. During the 1970s, he was asked by the Iranian Emperess to head the Imperial Iranian Academy of Philosophy, where he was given a chance to revive and implement the principles of the traditional school. That experiment, which saw the participation of illustrious world thinkers of Sufi inclination, like Henri Corbin, Sachiko Murata, William Chittick and Allameh Tabatabaei, ended with the occurrence of the Islamic revolution in Iran. For these reasons, Nasr found himself having to leave Iran for the USA. I remember the sadness on Seyyed Hossein's face when he told me that his greatest loss was the huge library of personal books he had to leave behind.

To understand Nasr, whom I have elsewhere referred to as a 'neo-traditionalist', one must keep in mind that he is in effect a Sufi master. Sufism has been defined by some as both an approach and a special type of knowledge; the Sufi follower seeks to ultimately find God in him, in and by way of his heart. This inner search, the Sufis insist, leads to the discovery of esoteric truths, a whole world of knowledge. The process by which this search is conducted, individually and in groups, is known as the *tariqah* (the way/path/method). The *tariqah* always refers to the Sufi orders, which are schools of Sufi traditions established by Grand Masters (e.g. the Qadiriyyah of Sheikh Abd al-Qadir al-Jilani, the Tijaniyyah of Sheikh Ahmad Tijani etc.) with a particular method for seeking God within the general Sufi tradition. *Tariqah* is only a way, and it is meant to lead to *haqiqah* (Ultimate Truth), which is God, or at least the knowledge of Him.

The Sufi philosophy revolves around the following central principle: Existence is one; all phenomena are partial aspects or reflections of Existence or Truth; our senses are subjective, limited and deceiving tools, particularly when they lead us to see/believe in dualistic distinctions between beings, phenomena and experiences. Furthermore, Sufism emphasises the personal experience of things and makes extensive usage of parables, allegories and metaphors in understanding various aspects of Reality and their relations, similarities and interdependences.

Moreover, many have noted the important fact that Sufism transcends classical theological divisions within Islam (Sunism, Shiism etc.); some have seen this Islamic mysticism as being quite similar in its methods and general philosophies with other spiritual traditions such as Zen Buddhism and Gnosticism.

For the past half-century Nasr has remained a major intellectual figure of Islam, recognised as such both in the West and in the Muslim world. Among other honors he has received, one must note that he was the first (of only two so far) Muslim(s) to have been invited to give the prestigious Gifford Lectures (on science and religion), which have been taking place for over a century and have so far received more than 200 illustrious thinkers, each delivering a series of lectures. The lectures he delivered in 1981 were later published as *Knowledge and the Sacred*, which is considered as one of his most significant books.

Islamic science

The expression 'Islamic Science' was coined[19] by Seyyed Hossein Nasr, though for that he has quite a special meaning and approach in mind. In fact the whole concept of an Islamic science (of whatever flavour) was first introduced by Nasr, back when there were very few, if any, Muslims seriously interested in such topics. Indeed, in this regard Nasr is a pioneer in every way: He was the first Muslim student to major in physics at MIT, the first student ever to establish a Muslim students' association at Harvard (in 1954), and the first Muslim thinker to insist that science within the Islamic world view, both during the golden era of that Arab-Islamic civilization and today, has different defining features than in the western world view.

Nasr has explained[20] his philosophy of Islamic science and its genesis in his mind. He came to realise that most Muslim thinkers who adopted a position vis-à-vis modern science missed completely the crux of the matter. He divides the Muslim thinkers into two main groups:

1. The modernists, who are essentially apologists (at least in this regard); these consider science to be necessary – because it gives power to a nation – and find nothing seriously wrong with western science; in fact they consider it a legacy of the Islamic civilization. In this group Nasr puts a whole series of important Muslim figures, from Jamaluddin al-Afghani and Muhammad Abduh to Muhammad Abdus Salam and his followers.
2. The ethicists, who reject the ethical ills of western science and call for an Islamization of science; Nasr does not name the members of this group either, but it appears that he has in mind people like Ismail Raji al-Faruqi (see later) and Ziauddin Sardar.

Nasr thus believes that most modern/contemporary Muslim thinkers did not identify the objectionable metaphysical and methodological bases of modern science, namely its insistence on materialism and naturalism and denial of any role – or even the very existence – of the spiritual dimension of man and nature. As the quote given at the top of this chapter indicates, Nasr believes that the naturalistic descriptions of nature (water is composed of oxygen and hydrogen) has led to an abandonment of worship and spirituality by the Muslim youth and to a general negative impact of western science on the Muslim society.

Sardar, on the other hand, has divided the Muslim thinkers into the following three categories with regard to their positions vis-à-vis modern science:

(1) The traditionalists, who in his assessment, view science as a *tariqah* (pun intended), literally a method but with some mystical connotation, that leads man to God by contemplation of his creation. Clearly, here Sardar has Nasr's school in mind, though he tends to lump in it al-Faruqi and his group. I do not believe that Ziauddin scorns the role of science in leading

one to the discovery and love of God; after all, this idea is very clearly spelled out in the Qur'an. Rather, I believe Sardar's strong objection to the traditionalist camp is that they ignore the role of science in solving the problems of society, its effects on man and nature, and – in the worst usage of science – its utilisation for colonial (classical and neo) purposes.

(2) The conventional scientists, people like Abdus Salam and his followers, who having been trained mainly in the West or at least in western methods, and having little or no understanding of the philosophy of science and its ills, have convinced themselves that science is neutral, objective and universal, and that our duty as Muslims today is to learn as much of it as possible and use it to contribute both to the production of knowledge and to solving social and developmental problems. (As we saw in the previous chapter, to this Sardar presents a scorching critique of science and lists a whole catalogue of serious flaws in and crimes by science.)

(3) The *I'jaz* (miraculous scientific facts in the Qur'an) school, which he calls Bucaillism after Maurice Bucaille, the French doctor who in the mid-seventies wrote one of the most famous (and dubiously influential) books to appear on the Muslim cultural landscape in the twentieth century, *The Bible, the Qur'an and Science: the Holy Scriptures Examined in the Light of Modern Knowledge*; the book led to the appearance of a whole literature and industry of works purporting to find scientific facts in the Qur'an – and now also in the Prophet's tradition. I will show in the next chapter, which will be totally devoted to this huge and dangerous cultural and religious phenomenon, that Bucaille was neither the first nor the most extreme in this school; by virtue of being French (and though it was never established that he embraced Islam), however, he was given unprecedented attention and ended up giving school, which had existed for decades if not centuries, a new and strong impetus.

So, one may identify the main viewpoints of the following four: Nasr, Sardar, Abdus Salam and *I'jaz* (or 'Bucaillism')[21].

Zainal Abidin Bagir has recently[22] come up with a somewhat different classification of Muslim schools in their stand with regard to modern science: (1) the instrumentalist school, which consists of the conventional scientists (Abdus Salam and company); (2) the Islamic science group, which is made up of different subgroups (Nasr, Sardar, al-Faruqi, Golshani etc.); (3) the *I'jaz* school; and (4) the creationist school, led by Harun Yahya, the Turkish thinker and writer who has in recent times come to dominate the landscape with his books, website and other media materials. (More on him and his campaign later.)

It is clear that the classifications are quite similar and philosophically and culturally only the Islamic science thinkers have anything serious to bring to the debate. Nasr has stressed the general Islamic science approach by calling on the Islamic world to 'master modern science, criticize it in the light of Islamic teachings, create a paradigm drawn from Islamic sources, and develop a new chapter in the history of Islamic science'[23]. Thus Nasr stands apart from other proponents of Islamic science like Sardar and al-Faruqi in his insistence that this new paradigm be 'based upon earlier Islamic tradition whose history and philosophy must be thoroughly resuscitated'[24].

I will thus focus on that Islamic science school, noting however that it consists of very different branches.

Nasr's school (perennial philosophy and science)

The philosophy of nature and man presented by Nasr stems from his belonging to what he calls the Traditional school, of which the 'perennial philosophy' (*Sophia Perennis*) is the core. Indeed, this Tradition is defined quite differently by Nasr and his group; it is not to be confused with the usual traditional (classical, orthodox) brand of Islam that prevails among the general public and the *ulama* (religious scholars of Islam). Nasr's Traditional philosophy, which was publicised and further developed by the twentieth-century thinkers like René Guénon, Frithjof Schuon, Ananda Coomaraswamy and others, is presented as an old[ii] and common perception of nature, Reality, existence, humanity and meaning by various cultures across the ages. These masters argue that such similarities clearly point to the underlying universal principles, and that if one looks carefully, one finds that these form the common ground of most religions. Only cultural conditioning then leads to the apparent differences among human societies in their descriptions of the world. In its Islamic version, this philosophy is put in the framework of *tawheed* (unity/unicity/unification), which in that optic proclaims not just that God is one but also that all existence and all history (of nature and of humanity) stems from that unicity or unity.

[ii] One of the main advocates of this philosophy in the history of Islam is Yahya Suhravardi (see later).

As far as the description of nature and Reality, this philosophy asserts the following main principles:

- Existence is one, and all phenomena are partial reflections of it. *Tawheed* then also implies the unity of truth and knowledge, the unity of creation, the unity of life and the unity of humanity. Consequently, in Nasr's view, the goal of all Islamic sciences, from mathematics to medicine, is to reveal that unity and the 'interrelatedness of all that exists'[25].

- The physical (or phenomenal) world does not constitute the only Reality or existence; another (spiritual) world exists that cannot be accessed by reason or the senses, but rather by other means (such as *tariqah*). Western knowledge/science denies this second reality altogether.

- Humans are a prime example of the existence of these two realities, which are distinct but not independent; the material part of humans is subject to the laws of nature, while the spiritual and intellectual part (the human soul) is free from those constraints and thus can have a life of its own.

- Humans can use their spiritual and intellectual dimensions to seek ultimate truths that transcend the methods of ordinary science. In monotheistic religions, the final goal of this quest is, of course, God; in non-theistic religions (such as Buddhism and Taoism), the Ultimate may be characterised differently, but the essence of a core is basically the same.

One of the consequences of the above principles, as deduced by Nasr and his disciples (Osman Bakar, most notably), is that knowledge (*'ilm*) is of two kinds: an absolute, infallible knowledge of certainty (*yaqin*), which actually comes directly from God, such as what was revealed in the Qur'an, and conjectured knowledge (*zann*), which is obtained by rational methods. Nasr proclaims that the human mind is capable of both, as it is made up of two components: a (rational) partial mind (*'aql juz'i*) and a global, intuitive mind (*'aql kulli*). The former can only perceive external, exoteric or apparent (*zahir*) phenomena, while the other penetrates to the deeper, internal or esoteric (*batin*) aspects of Reality. The second component of the mind is fed and empowered by the Divine Spirit (perhaps through *tariqah* exercises). The contemporary researcher Ibrahim Kalin finds some justification for this external/internal, apparent/real viewpoint in a famous prayer in which Prophet Muhammad asks God to 'show him the realities of things as they are in themselves' (*arini haqaiq al-ashya' kama hiya*). Kalin notes that this prayer 'suggests that the ultimate reality and meaning of things can be attained only through the aid of Divine guidance'[26].

This perceptive intuitive mind can then, according to this philosophy, help obtain information about the world, including scientific truths. As Nasr writes, 'A truly Islamic science cannot but derive ultimately from the intellect which is Divine and not human reason . . . The seat of intellect is the heart rather than the head, and reason is no more than its reflection upon the mental plane'[27]. Kalin explains Nasr's view: 'Whereas reason by its nature analyzes and dissects the world around it into fragments in order to function properly, the intellect synthesizes and integrates what has been fragmented by reason'[28]. He adds, 'Just as the reality of God is not limited to His creation, the reality of the natural world is also not confined to the analysis and classification of natural sciences'[29].

Nasr thus tries to keep both dimensions active, that is, he attempts to fuse the rational and empirical features of science with the intuitive aspect of religious faith, but clearly subjects reason to the 'heart' (the intuitive mind). For him, science can only represent one part of human knowledge, and it must be integrated into 'higher levels of knowledge'[30].

Bakar tries to shed some additional light on this dual concept of mind by making the analogy with the two approaches of Qur'an interpretation that I have introduced in the second chapter: *tafsir*, which is a rather literal explanation of the scripture, based on the meaning of the words and the context of the verse in question; and *ta'wil*, which is an exercise in interpretation of the text, based on additional knowledge, assumptions and global understandings. This is a good analogy, which helps point to the essential problem in the above dualistic approach to knowledge, namely that it guarantees no objectivity whatsoever, as the more important truths are then obtained by individual intuition rather than collective checks (falsifiability, repetition, peer-review etc.).

This is the first novel aspect of Nasr's Islamic science. The other important and controversial aspect is his insistence on bringing back science to its traditional nature. Indeed, Nasr notes that modern science, being a secular enterprise, is an anomaly with regard to human history. He remarks that the Western civilization is the first one to construct a science, a knowledge and description of nature that negates the sacred altogether. He makes a causal link between this fact and the problems that have resulted from science and its applications (technology); indeed, Nasr blames modern science in toto for all the ills that can be found in society, from the onslaught on the environment to the 'debasement' of man. He then valiantly calls for the vindication and revalorisation of traditional science(s), which he defines as sciences/knowledge that puts the Divine, the sacred and man at the centre of all considerations. In this way, René Guénon (another leading thinker of that school) frankly described the transformation of alchemy into chemistry and of astrology into astronomy as a degeneration, i.e. from 'traditional sciences' as sciences of the soul to 'sciences of the material world'.

Nasr uses the term *scientia sacra* (sacred science/knowledge) to describe this conception of knowledge. (Nasr insists on using the Latin expression.) He first defines sacred knowledge as that 'which lies at the heart of every revelation' and 'the center of that circle which encompasses and defines tradition'. *Scientia sacra* is then the application of sacred knowledge to various areas of Reality; it is simply the quest for spiritual perfection. Almost by definition, sacred science encompasses traditional sciences.

Nasr insists that what distinguishes his Traditional philosophy (and thus his Islamic science) from the contemporary Western approach is the holistic, unifying and sacred approach he and his school demand from any system of knowledge. Indeed, what makes modern science, philosophy and life so wrong and broken in his view is, on the one hand, its fragmentation and disorder and, on the other hand, its expulsion of spirituality from the world, nature and the cosmos. He concludes that modern science cannot explore Reality as a whole; it will always be handicapped and defective, unable to reach real truths.

Most importantly, modern science is very quantitative and avoids dealing with concepts like meaning, purpose, value and such. Nasr insists that science is truer and more valuable if it is made qualitative rather than quantitative. In his view, a true science should be able to describe such aspects of the world's phenomena as beauty, harmony, value, purpose and meaning, and these can be found in the traditional sciences but not in the modern ones. He readily sacrifices the ideals of objectivity and universalism for the sake of these qualities, which he regards as much more important and relevant to our knowledge and understanding of the world.

Finally, it must also be stressed that in Nasr's general approach, the way to explore Reality and thus to seek cosmic truths is by way of an individual Sufi experience. Only an inner (*batini*) and intuitive journey can make a seeker (explorer, researcher) see any aspect of Reality. Furthermore, one must be cognisant of the fact that Reality is whole, and it must thus be approached as such. In Nasr's and the Traditional school's world view, the world is a harmonious, ordered, organic totality; everything (natural, physical, cosmic, spiritual and divine) is interrelated; it cannot be approached in any reductive manner, and cannot be separated from the revealed truths. The study of nature is intimately related to the quest for God, and revelation only came to guide in this regard.

Sardar has expressed much sarcasm towards Nasr's philosophy of (traditional and sacred) science, especially during the eighties, when various views and propositions on Islamic science were being formulated and presented. In a paper presented in 1995, he refers to Nasr's program as a 'mystical quest'[31]. In his colourful and bold book *Explorations in Islamic Science*, published in 1989, Sardar refers to Nasr's perennial philosophy of science as a 'magical mystery tour'[32] and to the illustrious scholar as "Nowhere Man."[iii] In that book, Sardar produces a violent attack against Nasr: He starts with the very bases and roots of his philosophy, which he ends up describing as essentially non-Islamic (a mixture of Gnostic Greek traditions, Ismaili doctrine, mystical teachings, Hindu and Zoroastrian beliefs etc.); he goes on to Nasr's brand of Islamic science, which he condemns for its glorification of symbolism, "hidden (*khafiyyah*) or occult (*gharibah*) sciences"[33], as well as the reinstatement of alchemy,[iv] astrology, numerology,[v] angelology and such non-science into some globalistic knowledge system, all under the pretext that Islam promotes a unified scheme of science, where hidden (*batin*) and apparent (*zahir*) meanings and symbols are fused together. This is partly why Sardar insists on distinguishing Sufism from Nasr's religious philosophy, which he refers to as Gnosticism. Sardar thus concludes: 'A study of [. . .] Nasr's oeuvre related to his notion of Islamic science leads to the conclusion that he is a nowhere man, occupying a nowhere land: his discourse is neither about

---

iii In that book, Sardar, with his usual colorful style, used titles and references from pop music songs ('Nowhere Man' and 'Magical Mystery Tour' from the Beatles); furthermore, explaining how he sees Nasr's efforts as completely unrealistic and out of touch with the very nature of science, he refers to Nasr by using a David Bowie song, 'Ground Control to Major Tom,' where the title character is an astronaut who is lost in space aboard a spacecraft that he is unable to command and steer properly.

iv Nasr rehabilitates alchemy by defining it as a science that embraces the cosmos and the soul; the way it works, according to Nasr, is 'processes of giving birth to precious metals and minerals [that] are accelerated by the alchemist through the power of the spirit . . . ' (*Islamic Science: An Illustrated Study*, op. cit., p. 206).

v Numerical values are attached to the letters of the Arabic alphabet through some hidden, sacred, ancient esoteric scheme, which is claimed to be traced back to Ali ibn Abi Talib. I should make clear that 99.99 percent of Muslims of any tradition would reject any such esoteric numerological scheme or philosophy.

Islam, nor about science, but is a purely totalitarian enterprise'[34] (i.e. where one must follow the Gnostic/Sufi master, or order, blindly).

Recently, Sardar seems to have decided to simply ignore Nasr. In his intellectual autobiography, *Desperately Seeking Paradise*, Sardar does not mention the traditionalist even once! And in the edited volume *How Do You Know?*, which consists of 16 of Sardar's most important articles (at least a third of which being devoted to science), Nasr is mentioned exactly twice, and only marginally. One should also note that Nasr dismisses Sardar as a non-scholar, someone who has (in his view) very superficial knowledge of the Islamic tradition (philosophy, theology and science) and one who is therefore not entitled to speak or write on such questions; Nasr believes that Sardar is too quick to add Islamic terms and vague ideas from Islam to modern western thought and to come up with Islamic science, Islamic futures etc.

Still, surprisingly enough, one can find quite a few important points of agreement between the two thinkers. Leif Stenberg[35], in his doctoral thesis on the subject, has pointed to the following common presuppositions in their positions with regard to science: (1) the current crisis of science as it has developed in the West, though our two thinkers use different references for their claims (Kuhn, Feyerabend, Ravetz, for the first; Hegel, Heidegger and Kierkegaard for the second); (2) the myth of the presumed objectivity, neutrality and universality of science, a point on which the two thinkers are in total agreement; and (3) the grounding of Islamic science in Islam's history, though divergences soon appear, since Nasr takes his foundation in the Sufi, Gnostic and unifying traditions, while Sardar claims al-Ghazzali as his classical root[36]. It remains clear, however, that the two programs are fundamentally different; indeed, Nasr's program is deeply traditional and spiritual in nature, while Sardar's is post-modernist in essence, although it is given an Islamic formulation.

Switching to another major Muslim intellectual figure of the seventies and eighties, Sardar devotes many pages to Ismail Raji al-Faruqi, who is widely credited with the development,[vi] the propagation and attempted implementation of the concept of Islamization of knowledge/science. We will not be very surprised to find that Sardar is almost equally dismissive

---

[vi] Several Muslim intellectuals of high calibre have claimed 'paternity' for the concept of 'Islamization of knowledge,' most insistently by the Malaysian scholar Syed Muhammad Naguib al-Attas. See Fethi Hasan Makkawi, 'Hiwarat Islamiyyat al-Ma'rifah: 'ard wa tahleel' ('The Islamization of knowledge dialogues: review and analysis'), *Islamiyyat al-Ma'rifah* 24 (2000), pp. 99–135.

of this programme and that he developed a personal enmity towards the man he admired early on; one should point out, however, that to Sardar's credit, when al-Faruqi was assassinated, Ziauddin devoted a special issue of the cultural magazine (*Inquiry*) he was supervising to the man.

With al-Faruqi, however, we are not quite addressing a conceptual pro-gramme or position with regard to science, but with knowledge, a rather larger outlook. But Sardar states the following:

> It is not always clear [. . .] where the boundaries of the debate on Islamization of knowledge end and discourse on Islamic science begins [. . .]. Ideally, the discourse on Islamic science should be an integral part of the discourse on Islamization of knowledge – science, after all, has shaped contemporary epistemology and philosophy of knowledge and is the main source of our understanding of the material, physical and natural world[37].

Let us thus review al-Faruqi's programme and Sardar's (and others') critique of it.

al-Faruqi and the islamization of knowledge

The Palestinian American scholar al-Faruqi and his wife Lamya (a respected art critic) were murdered in their Pennsylvania residence one night in the month of Ramadan in 1986. Because the two were famous scholars who each was developing a major Muslim theory and programme of knowledge, art and science, and because the assassin had come in combat attire, dressed as a black ninja and carrying a Rambo knife (to use Sardar's depiction of it), the conspiracy theory regarding their murder emerged quickly and seized the Muslim communities (including intellectuals) everywhere. One should also add that only three months earlier, their son, a soldier in the US army, died of food poisoning during his normal training in a New Mexico military barrack. Had al-Faruqi come up with a theory so objectionable to the powers-that-be that he needed to be eliminated? Hardly so, if one listens to Sardar, who continues to show great respect but total disagreement toward the man he refers to as a 'towering intellectual figure'. In fact, Sardar dismisses al-Faruqi's seminal work (or 'manifesto'), *Islamization of Knowledge: General Principles and Workplan* (1982), as 'a pretty mediocre work'[38].

Before I launch into a description and critical review of this programme, I would like to note that it is first a much more general agenda that relates to science.

The development of the concept of Islamization of knowledge in al-Faruqi's mind proceeded in the following way: First and for many years, al-Faruqi was mainly concerned with the roots of the current ills of the Islamic nation (he insisted on using the Arabic/Islamic term *ummah*) and how to address those; then he came to realise that the problem lies essentially within the educational system, the faulty nature of which he largely attributed to the colonial transformation of the Muslim societies; finally, he concluded that the nature of the knowledge that had been inculcated to Muslims for many generations was flawed, as (Western) knowledge was fundamentally secular and thus in primary opposition with the Muslim culture, society and civilization.

As Sardar recounts, al-Faruqi announced his intellectual program for restarting the Islamic civilization, a programme he would shortly thereafter turn into a work plan, at the First Conference on Muslim Education in Mecca in 1977; there he stated, 'The task before us is to recast the whole legacy of human knowledge from the standpoint of Islam. In concrete terms, we must Islamize the disciplines, or better, produce university-level textbook, recasting some twenty disciplines in accordance with the Islamic vision'[39]. Before retelling the dramatic showdown he had with al-Faruqi at that conference, Sardar comments, 'I suspected he was aware that I was the only one at the conference making snide remarks about the new Messiah and his theory of Islamization of knowledge'. If it were not for space considerations, the whole western-movie high-noon showdown between Sardar and al-Faruqi, as retold by Ziauddin, would be worth reproducing here, not only for its humorous and dramatic entertainment value but also for the clear exposition and apposition of the two thinkers' views on the whole concept of Islamization of knowledge (and sciences).

In a moment I will flash some highlights of that encounter, but before that I need to summarise the main ideas that make up this theory or programme, which had excited the Muslim intellectual circles so highly that almost every scholar was working on it, at least from the mid-eighties to the late nineties.

The Islamization program starts from two observations or premises: (1) The failure of the Muslim modern reformers to produce a real civilizational renaissance; and (2) the failure of the post-modernist critics of the western civilization to steer the modern world away from its various disastrous trends, particularly the destruction of religion, and the loss of a sense of meaning and purpose.

Then the proponents of the Islamization theory state clearly that any serious attempt at reawakening the Muslim civilization and setting it on a course that will transform humanity altogether must start from the Qur'an. Taha Jabir al-'Alwani, al-Faruqi's heir and torch carrier, wrote the following: 'The Islamization of knowledge is an attempt to re-introduce the majestic Qur'an to the World [. . .] as the only book that is capable of delivering – not only our Ummah but also – the whole mankind. Solely, the majestic Qur'an has the alternative, universal, epistemological and systematic conception'[40]. But just exactly how are we supposed to proceed from the Qur'an to a reformulation of knowledge methods? al-'Alwani here realises the difficulty and admits the following: 'As for the Qur'anic (transcendental, epistemological, systematic) alternative outlook, it is embedded within the Qur'an itself. Nonetheless, due to their shortcomings within the field of the theory of knowledge, Muslims are unable to uncover such alternative outlook'. Why? Because, he says, 'Whatever we come across of epistemological systems in the Qur'an is usually inhibited by an enormous, cumulative and centuries-old heritage of *tafseer* (exegesis) and other classic disciplines, or *'uloom*, of the Qur'an'. In other words, it is all there, but we won't be able to see it because our minds have been conditioned by the traditional methods of our religious heritage. This means we need to reform our traditional education and reframe our thinking, as then we will be able to extract the proper epistemological principles from the Qur'an – at which point the Islamization of knowledge can proceed. No wonder the program has to be stalled from the start; it's a vicious cycle of finding the proper methodology to reform ourselves, at which point the golden epistemology of the Qur'an will become clear, and the correct pursuit of knowledge will begin. And doesn't that mean, knowledge has never been pursued Islamically? What do we make of the whole Muslim civilization and its sciences, was it then not influenced by the traditional methods of religious heritage?

Leaving such fundamental conceptual contradictions aside, let us continue in our attempts to understand this programme. Before we consider whatever plans its proponents have proposed for its implementation, we may wish to take a look at its stated objectives[41], which can be summarised as follows:

(i) To construct the contemporary Muslim epistemological system essentially from the Qur'an.

(ii) To construct methods of dealing with the Qur'an and the Messenger's tradition as a source of knowledge, thought and civilization.

(iii) To construct methods of dealing with the Muslim classic heritage so as to transcend its eras of imitation and disruption.

(iv) To construct methods of dealing with the modern legacy so as to establish interaction with the modern global thought and civilization and to tackle their crises.

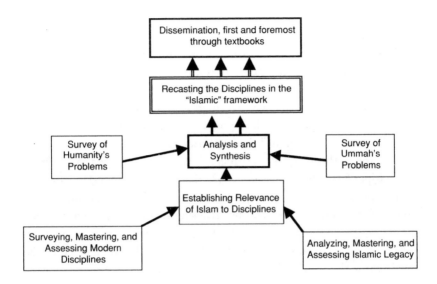

One may or may not be surprised to find these objectives very general and rather vague; after all they are 'objectives'. Indeed, one must rather look at the plans that the leaders of this program have devised and proposed. In his original manifesto, al-Faruqi had listed 12 steps as his work plan, which I can present in graphic form (inspired, redesigned and simplified from an original diagram by Sardar[42]).

al-Faruqi's followers[43], however, have since reduced the original 12 steps, which I have summarised in eight blocks graphically, to five steps:

1. Critically analyze, dissect and master the modern sciences in their highest forms of development.

2. Digest the contributions of the Islamic civilization to knowledge, considering that they stem from a proper understanding of the Qur'an and the Prophet's tradition, all the while assessing the strengths and weaknesses of those contributions regarding their potential relevance to today's Muslim needs and considering the discoveries of science since then.

3. Establish links between specific fields of knowledge and Islamic principles.

4. Perform a 'leap of innovation' in synthesising the fundamentals of the Islamic civilization with the results of the contemporary knowledge in a way that achieves the higher goals of Islam.
5. Steer various aspects of Islamic thought towards achieving the divine (or theistic) model.

Obviously, these steps of the plan remain fuzzy and quite impossible to implement in practice. And indeed, the researchers working in this field seem to be fully aware of the inability of their programme to generate any dynamic and to produce any progress on that front. And so they initially attempted to delineate the methods of their program somewhat more precisely, but later they simply downgraded their ambitions and expectations.

As to the methodology of this program, al-'Alwani has tried to construct it in a paper[44] that he published in 1987. He adopts Fakhr al-Din al-Razi's double approach for the acquisition of knowledge: 'the experiencing of emotion, and the exercise of reason'. But realising that this approach can open the door to all kinds of knowledge, al-'Alwani attempts to put some requirements for their validation; he says that the results thus obtained must be consistent with the following criteria:

(a) Human nature.
(b) The natural laws of the universe.
(c) Islamic teachings: principles and injunctions.
(d) Islamic values: moral and aesthetic.

How results can be consistent with human nature, the natural laws of the universe and Islamic teaching and values is anyone's guess at this point. So our thinker gives it one last valiant attempt, as he writes the following[45]:

(a) Whatever knowledge can be proven to be scientific fact may legitimately be accepted as Islamic.
(b) All knowledge must be fixed in the overall framework of the scheme of things...
(c) Anything found contrary to the universal principles of Islam must be rejected. This is where the efforts of Muslim social scientists will contribute to those of Muslim scientists and technologists in the establishment of an integral Islamic society by means of which man may fulfill his mission on earth.

Some of the criticism levelled at the Islamization of knowledge programme stemmed from ideological agendas; indeed, liberals, secularists and
anti-Islamists saw in it a stealthy attempt to Islamize society by changing the
educational upbringing (conditioning) of new generations[46]. Furthermore,
it has been argued[47] that the Islamization of knowledge program tends to get
lost in metaphysical considerations, e.g. the principles of unity, vicegerency,
matter–spirit balance etc. Ali Harb, for example, decries the absence of any
spirit of innovation in the Islamization programme. Moreover, he says, it
willingly gives up any quest for new knowledge to the West (and other cultures) and thus keeps the Muslim mind in a defensive mode[48]. Sardar fears
that this program will try to paint an Islamic colour on western sculptures;
in that case, one would be bringing Islam under the West's fold by attempting to Islamize fields of knowledge that had originally been produced in the
western mold[49]. Likewise, the idea of assessing and sifting knowledge by
means of the 'Islamic filter' has been ridiculed as 'aesthetic surgery'[50] and
'flat-earthism'[51].

Other critics have pointed out the fact that none of the scholars of Islam's
golden era ever attempted to Islamize the knowledge that had been received from the previous civilizations; instead, they all proceeded to learn,
master, digest and critique it, not from the point of view of its accordance with Islam but rather from its accordance to the truth as dictated
by methods of discovery. al-Marzouqi, for instance, goes further[52] and insists that certain features of today's scientific method, such as peer review
and collective acceptance, can be found in the Islamic tradition in the famous process of *ijma'* (consensus), so dear to the Islamic jurists, who have
often used it to deny the validity of individual efforts of interpretation
(*ijtihad*).

Noting the huge opposition that this program elicited from the start, and
realising their own inability to generate much positive production along
their agenda, the proponents of the Islamization of knowledge soon tried to
both justify their failure and settle on a more modest and fairly easily feasible
program. The main justification was, as usual, blaming the world and the
general anti-Muslim climate. al-'Alwani thus explains,

> 'the Qur'an is somewhat implicated [in the conspiracy against Islam] as it is
> associated with extremism, fanaticism and terrorism, while its adherents are
> subject to aggression, bullying and world-wide campaign of demonization.
> Furthermore [. . .] the exclusion of certain passages of the Qur'an from

education, media, preaching and public guidance so as to divest it of its potency and efficacy and to render the Muslim reading of the Qur'an partial. As a result, none of the following, the systematic approach, the rubrics of construction or the methodological rules of the Qur'an could be explored . . . '[53]

At the same time, trying to come up with a 'realistic' agenda, Ibrahim Rajab has proposed the following new plan[54]:

1. While the Islamization program is, in principle, concerned with all fields of knowledge, it should be clear that it is more directly relevant to social scientists; let then the efforts focus on social sciences in the first phase.
2. Let the social scientists attempt to reform their own fields from within, while the traditional religious scholars should also start reforming their field from within. Let each group try to update its methods and research topics; in particular, each should revisit its field's adherence to 'deduction from fundamental texts' (religious scholarship) and to 'empirical induction' (social and nature science). Once each field has been properly reformed, then a merging of the two will become possible and natural, as the distinction between the two areas will clearly look artificial and fuzzy.
3. Produce university textbooks in various fields, ensuring that any obvious contradictions with the Islamic creed and with Islam's conception of Man and the universe are removed, and sprinkles of Islamic conceptions are added wherever appropriate. It should be understood, however (says Rajab), that such efforts will be presented only as individual propositions, with no claim that they stem from a commonly accepted methodology or that they carry any scientific credibility, for they will not have been subjected to peer reviewing. Still, Rajab adds, this should not diminish the importance and value of such efforts, which are highly needed and necessary at this juncture.
4. In a subsequent phase, new theories will have to be produced, stemming from the Islamic conception of Man and the universe, theories for which the scientific validity will have been checked on the ground.

Now, as a way of dramatically highlighting the disagreements over this school's conceptual program, let us go back to that OK Corral scene of confrontation between al-Faruqi and Sardar, which – it must be recalled – occurred right after the birth announcement of the Islamization of knowledge theory; as I have pointed out, Sardar seems to have nuanced his views of the whole project somewhat, although he has remained adamant that it is fundamentally flawed in its methodology.

Starting an intellectual and verbal gunfight, Sardar began by pointing out the fact that the Islamization programme will inevitably lead to suppressions and censorships of any knowledge that is deemed non-Islamic, thus creating a caste of scholars who will subjectively filter out knowledge as they see fit. al-Faruqi rejected this criticism, insisting on the need to project Islam and its world view on disciplines such as economics, sociology, psychology, political science, anthropology etc. With a metaphor that Sardar found highly naïve, al-Faruqi explained that Islam and these disciplines are like knife and fork that we use to eat with; in the West, the knife is held with the right hand, the fork with the left; in Islam one is encouraged to eat (and perform all acts) with the right hand; therefore Islam and the sciences must find a new way to cooperate for the benefit of humans. Sardar then counters that eating with a knife and a fork is itself a product of a particular society and lifestyle, which itself is related to a particular setting (say eating at a table, as opposed to eating on a carpet) and environment (in certain parts of the world, eating with chopsticks is more appropriate because it's cheaper and easier to maintain). Likewise, the various sciences, which we must realise, come from specific environments and historical developments. What Muslims must do is not Islamize the present disciplines and thus produce an Islamic sociology or an Islamic cosmology, but rather realise that anthropology is altogether a western production originally aimed at understanding others (read: primitive societies). Muslims must rethink the sciences and knowledge from their own philosophical and metaphysical foundations and produce knowledge in new terms and categories that are organically consistent with their whole world view and useful to the progress of humanity. In Sardar's view, as I will explain shortly, Islamic science does not mean Islamizing science as it is known today, and certainly not producing science for Muslims, but rather rethinking science and knowledge in the way Islam, with its global human outlook, has taught, for the benefit of humanity as a whole.

The showdown ended up in a stalemate, as al-Faruqi, not as naïve as anyone would think or easily dismiss as such, retorted to Sardar: 'You want us to reinvent the wheel?', which is what Sardar's programme would appear to be. Ziauddin, as quick and witty as anyone could be in such high-intellect debates, simply replied: 'I am only pointing out that different vehicles need different kinds of wheels. You can't land a plane on bicycle wheels . . . Civilizations are like vehicles. They need appropriate wheels – their own disciplines – to move forward . . . '[55]

I should note, however, that Sardar did in the meantime soften his rhetoric somewhat with regard to the Islamization of knowledge programme, which he ends up considering as 'a formidable, but not an impossible task'[56]. He added, 'In their respective ways, both al-Attas and al-Faruqi have shown that Muslim scholars are gearing up to meet the challenge'.

One last issue needs to be discussed briefly with regard to the Islamization of knowledge/science theory: Do its proponents really intend it to apply to natural sciences? In 1987, a three-day workshop was organised in Washington, DC, around the theme 'Islamization of attitude and practice in science and technology', with the declared intention to 'focus on problems and challenges of science with respect to the ideology, personality, education and environment of a Muslim scientist in the contemporary world' and the aim of 'Islamizing all contemporary fields'[57]. Indeed, the workshop directed today's Muslim scientists to 'bring the spirit of sciences back in accordance with the spirit of Islam'. But do these advocates really target the physical scientists by such directives? Yes, indeed, a quick look at the workshop's themes and talk titles confirms this: Islamic perspective on knowledge engineering; use of Islamic beliefs in mathematics and computer science education; a blueprint for the Islamization of attitude and practice in earth sciences with special emphasis on groundwater hydrology; Islamization of attitude and practice in embryology etc. In the introductory paper that Taha Jabir al-'Alwani delivered to the workshop, which he praised as 'a landmark in the series of seminars on the Islamization of Knowledge', he stressed that 'while the social sciences are perceived as the true focal point for the Islamization process, there can be no doubt about the relevance of the natural and applied sciences as fertile ground for the same process'[58]. He decried the reluctance of the mainstream Islamic circles to address the Islamization of the natural sciences and technology, noting for instance that the Algerian international conference on Islamic thought series (which had been a yearly gathering of tens of leading Muslim thinkers for two decades) had rejected a proposal to include the Islamization of science and technology as a subject for debate. He justified his insistence to include the natural and applied sciences within the 'legitimate ambit of the Islamization process' on the following two grounds: (1) the division between social and natural sciences is, in his opinion, an 'arbitrary and utilitarian division'; and (2) physics has shown in the twentieth century that the observer has an important 'influence [. . .] on events', which – in his view – forces us to reconsider the whole concept of objectivity in the natural sciences. He concludes, 'We have every reason to suppose that the Muslim scientist who approaches his work from an Islamic point of view will indeed produce solutions with an Islamic tinge'[59]. One is tempted to ask al-'Alwani about the equations which will get 'solutions with an Islamic tinge', but it is quite clear that his understanding of the twentieth-century physics is lacking.

Sardar and the Ijmali school

In formulating their Islamic science philosophy, Sardar and his group (known as the Ijmalis[vii]) base themselves on two fundamental ideas: (1) Modern science is flawed and dangerous (see the long discussion in the previous chapter), in both its metaphysical bases and its technological applications (environment, ethics etc.); and (2) Islam makes strong exhortations for the pursuit of science, but considers ethics, moral values and harmony between humanity and nature a sine qua non condition in any such quest. Sardar considers science as 'a form of worship which has a spiritual and social function'[60]. He also writes: 'We therefore need to develop mechanisms by which Islamic science, as is dictated by the notion of *'ilm*, is moved to the centre of Muslim cultural, social and economic life. In other words, Islamic science, as a pursuit of objective knowledge and as *'ibadah* [worship], occupies the same place in Muslim everyday concerns as prayer, fasting and other forms of worship'. He adds, a bit more realistically, 'However, given the current status of Muslim societies, this is a tall order'[61].

In *Islamic Science: The Way Ahead*, Sardar explains the foundations upon which the Ijmalis' version of science rests: 'Attempts to rediscover Islamic science must begin by a rejection of both the axioms about nature, universe, time and humanity as well as the goals and direction of western science and the methodology which has made meaningless reductionism, objectification of nature and torture of animals'[62]. As we saw in that dramatic confrontation between al-Faruqi and Sardar, the latter calls for the development of 'new wheels', that is, organically developed scientific disciplines within the Islamic world view.

[vii] The term 'Ijmali' was chosen by Sardar's group as describing their philosophy well, for it 'captures the substance of synthesis with the style of aesthetics' (Sardar: *Explorations in Islamic Science*, p. 112). Indeed, while the word 'Ijmali' in Arabic simply means 'globalistic' or 'synthetic', it is also derived from 'jml', the root word for beauty.

Munawar Anees, one of the prominent members of the Ijmali group, has described what their version of Islamic science *cannot* be[63]:

1. *Islamized science*, for its epistemology and methodology are the products of Islamic world view that is irreducible to the parochial western world view.
2. *Reductive*, because the absolute macroparadigm of *tawheed* links all knowledge in an organic unity.
3. *Anachronistic*, because it is equipped with future-consciousness, that is, mediated through means and ends of science.
4. *Methodologically dominant*, as it allows an absolute free flowering of method with the universal norms of Islam.
5. *Fragmented*, for it promotes polymathy in contrast to narrow disciplinary specialisation.
6. *Unjust*, because its epistemology and methodology stand for distributive justice with an exacting societal context.
7. *Parochial*, because the immutable values of Islamic science are the mirror images of the values of Islam.
8. *Socially irrelevant*, for it is 'subjectively objective' in thrashing out the social context of scientific work.
9. *Bucaillistic*, as [this] is a logical fallacy.
10. *Cultish*, for it does not make an epistemic endorsement of occult, astrology, mysticism and the like.

This beautiful wish list indirectly attacks the other schools (al-Faruqi, Nasr and Abdus Salam) by declaring that the Ijmalis' Islamic science will not carry such ills and flaws, while promising to deliver the kind of science that has never been seen by humanity (polymathy, distributive justice, immutable values, future consciousness, organic unity).

And just how is this greatness supposed to be achieved? Sardar and the Ijmalis have identified the following ten fundamental Islamic concepts that constitute 'the framework within which scientific inquiry should be carried out'[64]:

- *Tawheed*, which is projected as an underlying unity of the natural world.
- *Khilafa* (human trusteeship of the earth's resources), which then forces upon man a sense of responsibility and accountability for his acts with regard to nature (research and application).

- *'Ibadah* (worship), which can take many forms, including the discovery, contemplation and further development of God's creation.
- *'ilm* (knowledge), which is both a source and an outcome of *'ibadah*, understood as an active quest for God.
- *Halal* (lawful or, for the Ijmalis, 'worthy of praise') versus *haram* (prohibited or 'worthy of blame'); these two opposite values are considered as means of assessing the knowledge and science being sought, obtained and applied in this program.
- *'Adl* (justice) versus *dhulm* (tyranny); these two opposite values are also standards of assessment for the applications of the knowledge derived by the above *'ibadah*.
- *Istislah* (public interest or, more accurately, benefit or welfare) versus *dhiya'* (waste); another axis for positioning and measuring the potential applications of the above *'ilm*.

These general statements and brief explanations of what the Ijmalis identified as the fundamental Islamic concepts or framework do not impress hardline scientists like Pervez Hoodbhoy (who belongs to the conformist science-is-universal-and-objective school led by Abdus Salam). Hoodbhoy in his polemical book *Islam and Science: Religious Orthodoxy and the Battle for Rationality*, published in 1991, quickly reviews the above fundamental concepts and then dismisses the whole programme with a slight of hand: 'If the reader wants more than platitudes, he will be disappointed'[65]. This is not a fair judgment of the Ijmali program and of its philosophical bases. Indeed, Sardar and his colleagues have tried hard (harder than any of the other schools) to make explicit the applicable aspects of their program. Indeed, Leif Stenberg has summarised[66] a few attempts that the Ijmalis have made in producing specific Islamic science disciplines. For example, Munawwar Anees has tried to formulate an Islamic vision of biology, Hussein Mehmet Ateshin has written a paper entitled 'Urbanization and the environment: an Islamic perspective', and Merryl Wyn Davies has made a valiant effort at coming up with an Islamic anthropology[67]. In his assessment of these efforts, Stenberg writes the following: '[Anees] mostly states what he is against but does not actually set up an Islamic alternative. For Anees it appears to be difficult to construct a specific Islamic biology out of the critique of contemporary biology'[68]; 'Anees appears to formulate an Islamic ethics of biology rather than creating the outlines of an Islamic discipline of biology'[69]; 'Davies, however, does point out some guidelines for a methodological approach and gives examples of possible questions that the anthropologist may ask in the field. The balance between theory and practice makes her book one of the most elaborate attempts to shape an Islamic discipline of science'[70].

In *Desperately Seeking Paradise*, Sardar defends the Ijmalis' program (years after the original fervour has dissipated and the group had dispersed):

The notion of Islamic science we developed would transform Muslim countries into knowledge-based societies. It would encourage research fine-tuned to solving local problems. Diarrhea and dysentery in Pakistan, flood control in Bangladesh, tackling schitosomiasis (bilharzia) in Egypt and the Sudan would have replaced the international agenda dominated by the predilections of western science and blindly adopted by Muslim nations.

He goes on:

And I thought indigenous knowledge too, would receive a tremendous boost. Traditional medicine, healthcare systems, agriculture and water management would be researched, built upon and improved in collaboration with a new outlook on science. I imagined certain big philosophical issues would be addressed: What happens to modern science if its basic metaphysical assumptions about nature, time, the universe, logic and the nature of humanity are replaced with those of Islam? What if nature, for example, is seen not as a resource to be exploited but instead as a trust to be nursed and nourished? What would then replace vivisection as the basic methodology of biology? How would science itself change when we consider human values, moral and ethical principles to be integral to the process of doing science?[71]

Bitterness and a sense of failure fill Sardar and his friends. Regarding vivisection, I should mention that Adam Wishart, in his very informative and poignant book *One in Three*, points out[72] that 'most advances in cancer have thus far relied partly on animal experimentation' and that because 'anti-vivisectionists are making it increasingly difficult, for instance, to carry out animal research in Britain', advances in cancer research may be significantly slowed down.

I should end this review of Sardar's programme by mentioning that Sardar concluded his short review of the Islamic philosophy of science, published in 1997, as follows: 'It is too early to say whether either of these movements [Nasr's sacred-science philosophy and this Ijmali vision of science] will bear any real fruit'[73]. Sardar is at least aware that the ambitious programme of 'creating new (science) wheels' from within the Islamic world view hasn't progressed very far. Andrew Jamison has remarked:

> The programme of Islamic science appears to have increased in rhetoric but lost something in practical achievement and focus. Indeed, in this respect the attempt to develop an Islamic science seems to be repeating much of the same process that the attempt to develop a 'science for the people' went through in the early 1970s. In both cases, a critical identification of problems leads to an overly ambitious formulation of an alternative that has proved impossible to realize in practice. While the alternative becomes ever more extreme and absolute in terms of rhetoric, it thus fails to solve the particular problems that were initially attributed to Western science[74].

Muhammad Abdus Salam was a genius. And for a genius to end up receiving the Nobel Prize is a no big surprise. But in this case, the feat was extraordinary. Indeed, Abdus Salam was the first Muslim to ever receive that most prestigious award, especially in the sciences[viii]. Born in 1926, he came from a small village of the Punjab region of what was then India and today Pakistan. In fact he received most of his education in that poor part of the world; yet he showed his genius from very early on, as when he obtained the highest marks ever recorded for the matriculation examination at the University of the Punjab at the record age of 14. He then earned a Master's degree from that university and a scholarship to Cambridge University, where he received the Smith's Prize for the most outstanding pre-doctoral contribution to physics, and went on to obtain his PhD in theoretical physics. By the time his thesis (in quantum electrodynamics) was published (in 1951), it had already gained him an international reputation. He was then only 25 and was only beginning.

[viii] Only one other Muslim has been awarded the Nobel Prize in the sciences: Ahmed Zewail in Chemistry in 1999. Other Muslims who have received the Prize are Naguib Mahfouz in 1989 (Literature); and Anwar Sadat in 1978, Shirin Ebadi in 2003, Mohamed El-Baradei in 2005, and Muhammad Yunus in 2006 (Peace).

After a few years back in Pakistan, he realised that he would not be able to continue the kind of research he had the capacity for, and so he returned to Cambridge. But he continued to contribute to his country's scientific and technological development, at least in terms of policy; he thus participated in starting Pakistan's nuclear technology program as well as its space and atmosphere agency Suparco.

His major contributions to science, to the Third World and to humanity in general came in the sixties. It was then that he – independently from two Americans (Sheldon Glashow and Steven Weinberg) who would in 1979 share the Nobel Prize with him for the same achievement – developed the electroweak theory of particle physics, a mathematical and conceptual synthesis of the electromagnetic and weak interactions (or forces), a major step in the physicists' long quest to unify the four fundamental forces of nature (gravity, electromagnetism, the weak nuclear force and the strong nuclear force). The Salam–Glashow–Weinberg theory was experimentally confirmed in the early eighties by a team of several hundred researchers at Conseil Européen pour la Recherche Nucléaire (CERN, now renamed Organisation Européenne pour la Recherche Nucléaire) in Geneva with the discovery of the W and Z particles. The theory today constitutes the central part of the 'standard model' of high-energy physics.

In addition to his seminal and long-lasting contributions to physics, Salam was driven by two other major preoccupations: (1) the need to help scientists in the Third World achieve their potentials and contribute fully to humanity; and (2) the need to harmonise faith in God with the methods of modern science. He thus spent the second half of his life devoting most of his time and energy to creating the conditions for helping Third World scientists. So first, in 1964, he founded the International Centre for Theoretical Physics[ix] (renamed after him upon his death in 1996) in Trieste in Northeastern Italy. Salam also founded the Third World Academy of Sciences (TWAS) in 1983 (also in Trieste), which aims at facilitating contact and cooperation for excellent research between Third World scientists and organizations.

He used his share of the Nobel Prize entirely for the benefit of physicists from developing countries and did not spend a penny of it on himself or his family; likewise for the money he received from the Atoms for Peace Award.

Additionally, Salam was always deeply imbued with his Muslim faith. Salam belonged to the Ahmadiyya Muslim community; some have claimed

---

[ix] To give an idea about the scale of achievement of ICTP, one may simply note the following statistics (from the ICTP website): since 1964 the Centre has received about 100,000 visits from scientists worldwide; in just 2005, it received some 6,000 visitors from 122 countries.

that because of this, i.e. belonging to a non-orthodox branch of Islam, the Pakistani government never gave him the regard he should have received for all his extraordinary contributions. One of his deeply held beliefs was that there was no contradiction whatsoever between his faith (and faiths of believers in the Divine in general) and the discoveries that science allows us to make about nature and the universe. In his Nobel Prize banquet address, Salam quoted the following verses from the Qur'an: 'Who hath created seven heavens in harmony. Thou seest not, in the creation of the All-Merciful any imperfection, Return thy gaze, seest thou any fissure? Then Return thy gaze, again and again, it will come back to thee dazzled, aweary' (Q 67:3–4). He then added, 'This in effect is the faith of all physicists: the deeper we seek, the more is our wonder excited, the more is the dazzlement for our gaze'[75]. He also often stressed that the Qur'an encourages us, sometimes even commands us, to reflect on the order of nature, and that is what science now is capable of doing, in a way never before possible; our generation, he has remarked, has been privileged with the gift of discovering the intricacies of God's design in the universe.

Salam, however, has not written extensively on the philosophical issues surrounding the attempts to harmonise faith with modern science. One of the few texts he has written to describe his vision of science, religion and humanity today has been repoduced as a chapter in the *Selected Essays of Abdus Salam*, volume published in 1987 under the title *Ideals and Realities*[76].

I had the honour and pleasure of meeting Abdus Salam and spending a few days with him in 1991 when he was invited by the president of the University of Blida (Algeria), where I was then a young professor. The university president, Nouredine Zettili, an Algerian theoretical physicist who was closely familiar with Salam's work, invited the illustrious scientist to visit our university and give a series of talks. Salam accepted, and I was asked to act as the translator. Salam addressed the scientific council at the university, met with the minister of higher education and other high officials, and gave two long talks at the university, one on the standard model of particle physics (including, of course, his central contributions) and the other on science and the Muslim world. The second talk, as it turned out, was essentially the content of the chapter from *Ideals and Realities* I have mentioned above. The audience was mesmerised by this humble man, who easily manipulates Einstein's equations, quotes from the Qur'an, cites Newton, Maxwell and Faraday, pauses when the call for prayer is heard in the middle of the seminar, resumes to relate a story from al-Biruni's life and so on. No one had ever seen a man at the highest levels of the scientific establishment in so perfect a harmony between his rigorous methods and knowledge, his faith and his cultural identity and history.

Abdus Salam and Hoodbhoy: 'science is objective and universal'

Muhammad Abdus Salam, Nobel Prize winner in physics (see sidebar) has been labelled 'a conventionalist scientist', for he believes that science is universal, and only its application is affected by various cultural factors. He does not believe that there are any serious metaphysical problems inherent in modern science that should warrant any reconstruction. In his typical flair, Sardar described this position as the 'business as usual'[77] school.

Abdus Salam starts with the Qur'an and underscores the insistent call made to the believers to observe, contemplate, reflect, understand and learn from nature. As he was fond of repeating, in contrast to a few dozen verses of legalistic nature, the Qur'an comprises hundreds of verses that exhort the believers to study nature by using the most valuable gift given to humans, their minds (reason).

Salam then moves on to show that the Qur'an and the Islamic spirit of study and rational reflection was the source of that extraordinary civilizational development, including the flourishing of the sciences in the Islamic golden era. This is a point of serious contention among historians of science, where most of the western experts give primary importance to the Greek influence, while some of the Muslim traditionalists, including S. H. Nasr and Muzaffar Iqbal, downplay the importance of the Hellenistic influence after the translation movement and instead give more credence to the internal factors. This is not Salam's concern; what he means to show is that the Islamic civilization did develop a genuine scientific tradition, especially after the year 1000 CE (or thereabouts), when the methods became experimental and the science became modern. Salam highlights the works and methods of Ibn al-Haytham, no doubt a pioneer of optics in the modern sense, and al-Biruni, whom he considers 'as modern and as unmedieval in outlook as Galileo six centuries later'[78]. He cites al-Biruni to show how clear he was on the sharp difference that existed between his method and those of Aristotle, who personifies the ancient methods. He concludes by citing two historians of science, Briffault and Sarton, who support his viewpoint (and that of others, like Sardar) that the science of the Muslims, at least after 1000 CE, was modern, in that it was based on experimentation. Salam also cites similar statements by Sarton, but without reference: 'The main, as well as the least obvious, achievement of the Middle Ages was the creation of the experimental spirit and this was primarily due to the Muslims down to the 12th century'.

Up to this point, there is no disagreement between Salam and Sardar. Nasr, as we have seen, chooses to highlight the traditional science of the medieval era, in total contrast to the modern aspects that Salam and Sardar identify there.

But then Salam makes a seriously different assessment. In trying to insist on the universality of science and its methods, Salam asks: 'Was the science of the Middle Ages really "Islamic" science?' He answers: 'The story of famous Muslim scientists of the Middle Ages such as al-Kindi, al-Farabi, Ibn al-Haitham and Ibn Sina shows that, aside from being Muslims, there seems to have been nothing Islamic about them or their achievements. On the contrary, their lives were distinctly unIslamic'. He adds that for Ibn al-Haitham, 'Truth was only that which was presented as material for the sense perception. No wonder that he was generally regarded as a heretic, and has been almost totally forgotten in the Muslim world'[79].

A we will see shortly, a follower of Abdus Salam's view of science as universal and objective, the aforementioned Pervez Hoodbhoy, who wrote an anti-orthodoxy book on the subject with a foreword by Salam, goes further and picks five of the greatest thinkers of that golden era (al-Kindi, al-Razi, Ibn Sina, Ibn Rushd and Ibn Khaldun) and calls them, with a degree of provocation, 'heretics', in the sense that they either held unorthodox beliefs or did not allow their faiths to influence their works. This claim can only be considered to be correct if by that one means that religious beliefs did not affect these thinkers' research programmes and methodologies; interpretations and world views are, of course, always subject to one's philosophy and faith.

What is important at this point is to note that Salam considers modern science to be more than compatible with Islam; indeed, Muslim scientists essentially produced its experimental methodology, and that modern science, wherever it is developed, comes out with the same features.

So Salam strongly rejects the idea of an Islamic science. He refers to oriental cultures, which

> are not seduced by diversionary slogans of 'Japanese' or 'Chinese' or 'Indian' science. They recognize that though the emphasis in the choice of disciplines on which to research may differ from society to society, the laws, the traditions and the modalities of science are universal. They do not feel that the acquiring of 'western' science and technology will destroy their own cultural traditions: they do not insult their own traditions by believing that these are so weak[80].

Finally, he quotes a hadith to support his philosophy: 'Wisdom/knowledge is the believer's quest, wherever it may be found, he is entitled to it'.

Then how are faith and modern science to mesh? Salam basically finds some limits to science and identifies questions that are, in his view, beyond the grasp of the scientific realm. As he puts it, 'Science has achieved its success by restricting itself to a certain type of inquiry'. So, for example, the doctrine of creation from nothing is metaphysical in his view, and so here science stops and gives way to religious considerations. He says, 'My own faith [is] predicated by the timeless spiritual message of Islam, on matters on which physics is silent and will remain so'[81].

Sardar has criticised Salam's philosophy; he calls the great scientist and his followers 'conventional positivists', which is hardly a fair characterization of at least Salam's position and approach. Sardar then pushes the criticism too far and ascribes personal motives to the stand of the conventionalists; he considers that 'they have both an image and a myth to maintain'. Perhaps, he was incensed by the more radical position taken by Hoodbhoy, as Sardar adds, 'However, positivist defence of the purity of western science has now turned into scientific fundamentalism, as illustrated by the semi-literate diatribe of Pervez Hoodbhoy'[82].

There is no question that Hoodbhoy is a secularist[83], and he is totally free to adopt any such general philosophy. But one must acknowledge the fact that he has written extensively[84] on the questions of science, society, metaphysical foundations, connections to faith and religion etc., but that his writings are almost always eloquent, well-informed and oftentimes moderate.

In his book of 1991, Hoodbhoy starts out strongly defending the secular nature of science; on page 2 he writes, 'Science is a secular pursuit, and it is impossible for it to be otherwise. . . . Scientists are free to be as religious as they please, but science recognizes no laws outside its own'[85]. His intention is not to deny religion any role or value, but rather to clearly separate the two spheres of relevance. In his concluding chapter he writes the following:

> Science can be used to enhance the moral values of life because it insists upon searching for the truth. At its deepest level, it does create a feeling of reverence because an advance of knowledge brings us face to face with the mystery of our own being. But while recognizing that religion and science are complementary and not contradictory to each other, a clear demarcation between the spheres of the spiritual and the worldly is necessary . . . While science must be vigorously pursued both for development and for

enlightenment of the mind, one must be clear that science is not a replacement for religion and that it does not constitute a code of morality[86].

And so one is not surprised that the chapter titled 'Can there be an Islamic science?' starts with, 'The answer to this question, in my opinion, is simple. No, there cannot be an Islamic science of the physical world, and attempts to create one represent wasted effort'. He goes on to give three arguments to support this view[87]:

1. 'Islamic science does not exist. All efforts to make an Islamic science have failed'.
2. 'Specifying a set of moral and theological principles . . . ' [note that he doesn't say 'metaphysical', as he should] 'does not permit one to build a new science from scratch'. The example he gives here is that Salam was a believer and Weinberg an avowed atheist, and yet they constructed the same theory in particle physics.
3. 'There has never existed, and still does not exist, a definition of Islamic science which is acceptable to all Muslims'.

None of these objections are very convincing. Still, in his conclusion he calls upon the Islamic science advocates to basically cease and desist:

A truce needs to be declared in the continuing opposition to modern science as an epistemological enterprise, although debate on its utilitarian goals must continue and even be sharpened . . . While science must be vigorously pursued both for development and for enlightenment of the mind, one must be clear that science is not a replacement for religion and that it does not constitute a code of morality[88].

The most serious criticism levelled at the 'conventionalist' school was, not surprisingly, made by Sardar. With his typical flamboyance, and making use of lyrics from a Beatles song[89], he describes Salam and his group (with Hoodbhoy, Kettani etc.) as 'all those lonely people'[90]. Why do I say this is the most serious criticism levelled at Salam's position? Because, as Leif Stenberg explains, by lonely Sardar means 'without guidance', so that this school's science remains (in Sardar's view) without values, ethics or norms. Clearly, this is one of the core concerns of the Ijmalis and one that the conventionalists hardly address or regard as significant enough to warrant a reshaping of science's philosophy or a redress of its course.

summary and discussion

We have seen that Islam places great emphasis on the concept of *'ilm*, but we have also found that it carries many meanings, which has led contemporary Muslim thinkers to adopt very different positions with respect to science in general (nature of the enterprise, place and role within society etc.). One should therefore not be surprised to find so much disagreement and dispute over what can constitute an Islamic science. Proposals for addressing 'Islam's quantum question' have been diverse, to say the least.

Several schools of thought can be identified in the cultural Muslim landscape with regard to science: from the highly mystical school (Nasr's Gnostic, traditional and sacred approach to knowledge and science) to the secular, universalist, conventionalist (Salam and Hoodbhoy's insistence that science is universal and objective, there can be no Islamic or Chinese or Japanese science). I also reviewed the Islamization of knowledge/science school of al-Faruqi and al-'Alwani, which we have found to be conceptually and methodologically flawed and which – no surprise – has not moved forward even one inch in the past three decades since its inception. And I have postponed dealing with the Bucaillist science-in-the-Qur'an school to the next chapter. That leaves us with Sardar's (and the Ijmalis') school of Islamic science, which claims to be potentially universal even though it is clearly inspired from Islamic principles. The Ijmalis' model, although it too has made only modest progress in moving beyond the general principles and into the realm of serious formulations of new disciplines, has shown some promise; it is still not solidly grounded and conceptually robust, but it is neither 'spaced out' nor 'naïve'.

One important criticism that can be addressed to Sardar in particular is his exaggeration in critiquing modern science, especially what he describes as violent methods and unethical standards. Sardar has many times used the technique of vivisection as an example of modern science's violent and immoral methods. Finding this example so many times in his writings, one begins to suspect that he cannot find any other examples to strengthen his case; indeed, the only other similar example I have found in his texts is the statement, 'How can we combat the intrinsic violence – so vivid in vivisection and experimental theoretical physics – of modern science?'[91] In other words, he considers experimental high-energy physics, which uses beams of subatomic particles that are accelerated to high speeds and energies and collided with targets of particles, to be a violent field of modern science. I don't know of any physicist who has ever expressed ethical concerns about using beam-collision methods in studying subatomic particles and interactions. And though one might agree that vivisection is immoral (in principle), I am not convinced that it is so widespread and such a primary method in biology to warrant making it a flagship issue in characterising modern science and its presumed unethical and immoral ways.

Indeed, sometimes Sardar tends to be categorical and excessive in his judgments. I do not disagree with him that the scientific method is far from objective and scientists, mostly unconsciously, affect their research with their preconceived ideas. But I would not go so far as to say, 'bias-free observation is a myth'[92]. And I would not subscribe to the view that 'Scientists often modify their observation with their own ideas and prejudices, the values and norms of their society. Not just observation but experimentation also cannot be undertaken in a cultural vacuum . . .'[93] In fact, the cultural effect can be (and are) averaged out by having the experiment and observation repeated in many different settings.

Similarly, the following statements can only be described as extreme:

> That science is inescapably linked to repression and domination is no accident: it is a direct product of a rationalistic epistemology, just as Newton, Darwin, Freud, B. F. Skinner and Edward Wilson are the products of the same epistemology. Social Darwinism, sociobiology, stockpiles of nuclear weapons that could destroy the earth several times over, are a logical outcome of the epistemology of instrumental rationality[94].

I can understand – though not necessarily agree with – Sardar's criticism of Darwin, Freud and E. O. Wilson, but the jumps from Newton to nuclear weapons and from Darwin to social Darwinism are truly gigantic and gross; moreover, to simply lump all those ills and abhorrent results of modern technology and present them as 'logical outcomes of the epistemology of instrumental rationality' is unreasonable. I know Sardar knows better; he must have gotten carried away with this one.

Now let me discuss the following issue with Ziauddin. Assume we have implemented an ethical, moral, God-fearing and socially beneficial type of science, what the Ijmalis regard as Islamic science. I am a Muslim physicist; suppose my university or research group decides to investigate new interactions (say something similar to nuclear forces). Then, suppose we start uncovering new forces; one of my team members remarks that this can lead to the usage of new forms of energy, while another researcher points out that it could also be used as a bomb. What do we do at this point? Do we continue our research on the basis of its potential benefit to our society and to humanity at large? Or do we stop because it may be misused? Is Science 'praiseworthy' or 'blameworthy' at this point? I believe we must be careful to distinguish science from technology, which is always a mixture of science, policy, social environment and circumstances.

One other critique (or at least interrogation) to the Ijmali conception of science is how they envision society to steer science, how 'every member' of society can be made to participate in the debates on science. Sardar says: 'Islamic science [is made] accountable not just to God, but also to society'[95]. In a more recent book, he talks about the democratisation of science, where science is not handed over to untrained personnel, but 'rather it brings science out of the laboratory and into the public debate where all can take part in discussing its social, political and cultural ramifications'[96]. Furthermore, he (approvingly) quotes Funtowicz and Ravetz, who writes that 'there must be an *extended peer community*, and they will use *extended facts*, which include even anecdotal evidence and statistics gathered by a community'[97] (emphases in the text). I doubt that we will find any practicing scientists who will subscribe to any such free-for-all scientific undertaking, which redefines facts as 'extended', and where data are collected without proper methodology, anecdotal evidence is used so freely, and inexperienced and unknowledgeable members of the community are invited to debate the social, political and cultural ramifications of the scientific enterprise. The discussions are complicated and stressful enough as they presently are among the select few (out of the educated elite) who appreciate the issues enough to take part in the debates.

At other places Sardar has made pretty much the same point by calling for the Islamic concept of *ijma'* to be broadened and for *ijtihad* (effort for investigation and innovation) to be turned into a right of practice for every single lay Muslim[98]. I am all for reopening the gates of *ijtihad*, but not for everyone. No educated Muslim, let alone any of the scholars, will accept such a level-playing field where everyone has (equal) rights to participate in the debate. *Ijma'* is equivalent to peer review; it functions the same way in both the Islamic tradition and the scientific tradition; let us keep it that way, use it properly, and not wreck one of the good practices that have been arrived at independently by these two civilizations.

Finally, I would like to note that all these works on Islamic science originated in the West. None of the ideas presented here, with the small exception of Golshani, who has not contributed any major idea even though he adopts and expounds a reasonable and reasoned position, came from the Muslim world. Furthermore, very little of these works has made it to the Arabic language, which is a measure of the negligible effect this whole discourse has had on the Muslim nation. This in no way reduces the value of such endeavors, but it shows how much we still have to do in our attempts to make the Muslim society adopt science (in some form) that will serve well.

5

I'jaz: modern science in the Qur'an?

'If it were upon the Prophet to explain the natural and astronomical sciences, that would be the end of the activity of human senses
and intellect, and that would spoil human freedom . . . The
Prophet advised people succinctly to use their senses and intellect on whatever improves their welfare, broadens their knowledge, and in the end advances their souls . . . Therefore the doors
for these sciences are intellect and experimentation, not tradition and religious sciences.'[1]

Muhammad Abduh (1849–1905)

the speed of light and the age of the universe from Qur'anic verses

About a decade ago, a young, bright Egyptian astronomer told me of a paper
he had recently read; he said it was the most impressive piece his mind had
encountered. I thought he was referring to a real science paper, and though
his field of specialty was quite different from mine, I was very curious to
know what the paper was about and why it had impressed him so much. He
then stunned me by saying that it was a calculation of the speed of light by
an Egyptian physicist from nothing but a few Qur'anic verses. I exclaimed:

'You mean he derived the value of c (the speed of light) by combining a few verses like algebra on a few equations? That's impossible!' I laughed. But he calmly replied that he would bring the paper and let me judge by myself. In fact, he was eager to see my reaction upon reading the piece.

I only got to see and read the paper a few years later, when it was posted on the Web[2] and circulated by email. The author, Mansour Hassab-Elnaby, who has since passed away, was then a professor of physics at Ain Shams University in Cairo. The paper was titled 'A New Astronomical Qur'anic Method for the Determination of the Greatest Speed C'. Originally written in Arabic, it has since been translated into (bad) English, French and Spanish (the versions I could find on the Web), and probably other languages too.

The full scrutiny and exposition of the paper and its many serious flaws are given in Appendix B. For, although the methodological faults are striking, and even a non-scientifically inclined critical thinker will readily identify them, the full dissection may be useful for various reasons. I here only wish to bring up the main serious transgressions that were committed by the author and the lessons that must be learned from this dangerous precedent, especially since it has now led others to follow the suit (more on that shortly).

The calculation is based on one Qur'anic verse, essentially; in the author's paper, it is rendered as follows: 'God rules the cosmic affair from the heavens to the earth. Then this affair travels to Him (i.e. through the whole universe) in one day, where the measure is one thousand years of your reckoning' (Q 32:5). For comparison and useful implications, here is how a typical translation (that of Yusuf Ali) renders the verse: 'He rules (all) affairs from the heavens to the earth: in the end will (all affairs) go up to Him, on a Day, the space whereof will be (as) a thousand years of your reckoning'. It is worth noting that the 'Day' being referred to here is the Day of Judgment, not any generic day.

The author then combines this verse with some astronomical and geometric quantities to (artificially) arrive at a value of c (the speed of light) that is very close to the one determined experimentally (i.e. 299 792.46 km/s).

The main conceptual and methodological errors made by Hassab-Elnaby are as follows:

(1) The Qur'anic verse talked only about times, and no distance is mentioned; obtaining a speed from the verse is therefore necessarily invalid. (See Appendix B for an explanation of how the author artificially introduces a distance.)

(2) The author is seen to manipulate physical quantities in just the way to get the right result, sometimes fixing the equations with the artificial inclusion of the cosine of an angle or an average value of a quantity (such as the moon–earth distance, which varies throughout the year and over the ages).

(3) The Qur'an refers to 'a thousand years of your reckoning', that is to synodic (lunar phase) months, not sidereal (lunar orbit) months, which Hassab-Elnaby uses.

(4) The author overwhelms the reader with technical physics terms and concepts, including a contrived and erroneous interpretation of special relativity. If his method were actually robust, he should have explained it as simply and as clearly as possible.

(5) If one can calculate the distance travelled by 'the affair' to reach God in one 'day' ('1,000 years'), it would necessarily imply that (a) God is located at a certain distance from us; (b) information takes quite some time to reach Him.

(6) Last but not the least, several writers have pointed out that the idea of a day being like a thousand years also appears in the New Testament ('With the Lord a day is like a thousand years', (2 Peter 3:8)) and, more importantly, the word 'thousand' in ancient cultures simply meant 'a very large number' or 'much larger than what humans are used to in their everyday experience'.

Why should one be concerned with a paper that is so seriously flawed? Aren't there many such ludicrous claims in the web literature? Indeed there are, but this piece is important first in the way the Qur'an is now being approached by Muslims, and second in the impact it has had on the Muslim culture. In addition to being so widely disseminated, this idea has also been reproduced, in worse form, in one of Zaghloul al-Naggar's books[3], a 600-page volume with 200 full-colour images that has had at least four editions, where Professor al-Naggar[i] devotes one page[4] to a derivation of the speed of light, which starts with the equation:

Speed of the cosmic affair = 1,000 years of what we reckon/
a day of what we reckon.

[i] Professor al-Naggar is presented at the start of the book with a mini-CV of credentials: PhD from the University of Wales, professorship at several universities, including some academic stays at UCLA, publication of over 150 papers and 20 books, supervision of 45 theses, etc.

However, even any student could quickly see that this equation is fundamentally wrong: the speed can never be the ratio of two time periods. And when the equation gives 334,056.8 km/s as the result, al-Naggar tries to get the right result (300,000 km/h) by using Hassab-Elnaby's 'suggestion' to multiply by 'the month factor' and the 'year factor', which he defines as, respectively, 'sidereal month/synodic month, i.e. 27.32/29.53 = 0.9252' and '309 lunar years/300 solar years', which somehow he claims to be equal to 0.97087. Gone is the cosine of the angle that had artificially been introduced by Hassab-Elnaby; instead we get some new numbers that are again multiplied abracadabra-like to just yield the expected (known) value.

al-Naggar's different version of the derivation is very significant, for it confirms that numbers are just used when they combine well to yield the final result, not because they make sense in some coherent conceptual framework.

Sadly, this is not the end of the story. This approach seems to have developed into a real 'school', for now there are followers who apply the same methodology to other such problems.

In July 2007, I received a paper titled 'A New Method for Estimating the Ages of the Earth and of the Universe'[5] by Prof. Kamel Ben Salem; note the similarity of the title with that of Hassab-Elnaby's 'seminal' paper. In this work, Ben Salem follows exactly the same methodology that I have decried above. He starts with a Qur'anic verse ('To Him ascend the angels and the Spirit in a day the measure of which is fifty thousand years' (Q 70:4)), which will be central to his thesis and which he combines with other verses ('Verily a Day with your Lord is like a thousand years of your reckoning' (Q 22:47, Q 32:5)). In the process, he too throws in expressions like 'Relativity of time in the Qur'an'. He begins his derivation by stating that 'one of the companions of the Prophet realised that the verse of the 50,000 years gives the period which separates the start of the creation from its end (the Day of Judgment)' and starts multiplying:

1 day = 50,000 years = 50,000 × 365.256363051 'terrestrial days' = 18,262,818.152 'terrestrial days'
and each 'terrestrial day' is worth 1,000 years,
so the period under calculation is equal to 18,262,818,152 years of 'what we reckon'.

Our author then cites a few more verses (Q 50:38, Q 41:9–10) so as to infer that the total time decreed for the universe (from creation to the

Day of Judgment) consists of eight 'days' (or periods P), of which the past cosmological time represents 6 periods P, the last two of which give the age of the earth. From this it is a trivial step to obtain

the age of the earth, $T_E = 2\,P = (18,262,818,152/8) \times 2 =$
4,565,704,538 years, i.e. 4.5657 billion years;
the age of the universe, $T_U = 6\,P = 6 \times (18,262,818,152/8) =$
13,697,113,614 years, i.e. 13.7 billion years;
the remaining time between now and the Day of Judgment = 4.5657
billion years, but this Ben Salem avoids pointing out, because this
is considered by Muslims to be exclusively divine knowledge, so he
simply refers to it as the 'time remaining for the solar system'.

What are we to make of such a 'derivation'? Nothing much, except to note that Ben Salem must have noticed that the latest value of the age of the universe (13.7 billion years), obtained from the data of the recent WMAP mission, seems to correspond almost exactly to three times the well-known age of the earth (4.54 billion years) as given by the US Geological Survey. (Note that this 3:1 ratio is only approximately true in our era and will no longer hold in say 100 million years.) Then any Muslim would have recalled that this 3:1 ratio is what can be inferred in some Qur'anic verses that talk about the creation of the heavens and the earth in four days, two days etc. Similarly, it is easy to notice that $1/4$ (or $2/8$) of $50,000 \times 365$ gives 4.5 million, so if one multiplies by 1,000, one gets 4.5 billion years, which would be the age of the earth. The rest of the digits in the derivation, which are really meaningless, are included for impressive purposes.

Now, lest anyone be too impressed with the above numerical calculation, let me point out that at the time when Ben Salem was constructing his 'new method', which is fully based on some Qur'anic verses, Zaghloul al-Naggar and Marwan al-Taftanazi were publishing books, which, from the same verses, were inferring and insisting that the earth must have been created before the rest of the cosmos. Let me quote al-Naggar (the professor of geology) on this point:

Verse 2:29 decrees that Allah had created for us all that is on Earth before fashioning the initial smoky heaven into seven heavens. Verses 41:9–12 state that all the elements necessary for life on Earth, in fact the primordial earth itself had already been created before the separation of the initial smoky heaven into seven heavens. So can Muslim astronomers, astrophysicists, and geologists, starting from these Qur'anic verses, revise the current theories and scientific calculations to establish these facts as *I'jaz*? This, in times when the specialists are unanimous that heavy elements could only be produced in the universe inside stars. [. . .] This would strengthen the believers in their faith, and invite the non-Muslims in these times when people have been seduced by science and its data to the point of going astray[6].

Similarly, Marwan al-Taftanazi (another *I'jaz* expert[ii]) wrote on the subject:

These verses (Q 41:9–11)[7] decree an established cosmic fact [. . .] that the earth after the initial 'separation' was created first, then the heaven was shaped and constructed from the primordial 'smoke'. [. . .] And those among exegetes who, out of a strong urge to conform to scientific theories, tried to put the creation of the heavens before that of Earth have committed a serious mistake[8].

Even the highly respected Algerian thinker Malek Bennabi (1905–1973), whose first and the most important book *The Qur'anic Phenomenon* was devoted to an exploration of both the Qur'an's characteristics and the Prophet's experience of it, got seduced by the *I'jaz* siren calls and devoted a whole chapter of that book to the 'scientific miraculous content' of several verses. The most astonishing of his proclamations is his explanation[9] of the famous 'Verse of Light' (Q 24:35, previously cited), where (for him) the niche is the projector, the lamp is the filament and the glass is the bulb. Unfortunately, this was not his only slip-up; he had a few other such stunning interpretations.

a growing cultural phenomenon

This Qur'anic *I'jaz 'ilmiy* ('miraculous scientific content') theory has exploded and expanded to quickly occupy large parts of the cultural landscape of the Islamic world (particularly the Arab part) over the last few decades, so much so that a whole industry of 'scientific content in the Qur'an' has sprung

[ii] According to the brief biographical note at the back of his first major book (*al-I'jaz al-Qur'aniy fi Daw' al-Iktishaf al-Ilmiy al-Hadith* – Qur'anic *I'jaz* in Light of the Modern Scientific Discoveries, Beirut: Dar al-Ma'rifa, 2006), Marwan al-Taftanazi is an Imam, in his thirties with two master's degrees and a PhD. He has recently published a three-volume (hardcover) work on *I'jaz*; he regularly appears on Radio and TV in the United Arab Emirates.

forth. A quick Web search for such literature will show books with titles such as *Subatomic World in the Qur'an*[10] and book chapters such as 'Science and Sunnah: the genetic code'[11], 'The grand unification theory (GUT): its prediction in al-Qur'an', and 'Islam and the second law of thermodynamics'[12]. Articles have been written to show, for instance, that the Qur'an foretold the invention of the telephone, fax, email[13], radio, telegraph, television[14], lasers[15], pulsars[16] and black holes[17].

In his intellectual autobiography *Desperately Seeking Paradise*, when Sardar defends his Islamic science thesis to one of his friends, the latter, highly skeptical of the whole concept, retorts: 'Your idea of Islamic science has been hijacked by fundamentalists and mystics. The fundamentalists are looking for scientific miracles in the Qur'an. Everything from relativity, quantum mechanics, big bang theory to the entire field of embryology and much of modern geology has been "discovered" in the Qur'an'[18]. (By 'mystics', Sardar's interlocutor was referring to Nasr's school.)

I must also stress that most of the advocates of this method are highly educated people,[iii] and despite the objectionable nature of most of their propositions, they are sincere, if badly in error, in trying to make what they think are genuine assertions.

Indeed, one of the most striking ideas that one encounters in all of the Islamic literature is the claim that the Qur'an contains 'all knowledge', sometimes with the emphatic addition 'of the ancients and the moderns', This is usually justified by reference to the Qur'anic verse (6:38): 'We have neglected (or ignored) nothing in the Book'. The modernists (reformers), however, interpret this verse to simply mean that the Qur'an contains general principles of all matters that are important for human beings to know. But another trend has appeared in the twentieth century claiming that this principle of 'Qur'anic knowledge completeness' must also be extended to include modern science, so that everything true that humans have discovered and will discover can somehow be found in the Qur'anic text, if explored properly.

There are two versions of this trend: (a) the 'scientific exegesis' (*Tafsir 'Ilmiy*) school, which stipulates that modern scientific knowledge must be utilised, along with other tools, to better comprehend some passages of the

---

[iii] Zaghloul al-Naggar is an Egyptian retired university professor of geology (see earlier note); Bechir Torki was a Tunisian university professor of physics; Abdelmajid al-Zandani is a Yemeni religious and political figure, who received education in pharmacology; Kamel Ben Salem is a professor of data analysis in the department of computer sciences, University of Tunis.

Qur'an that could not be properly interpreted in earlier times; and (b) the school of the 'scientific miraculous content of the Qur'an' (*I'jaz 'Ilmiy*), which claims that many verses of the Qur'an, if read and interpreted 'scientifically', express in semi-explicit ways scientific truths that were discovered only recently, and therefore the Qur'an is scientifically miraculous and points to a divine origin.

In recent times, this bold school has received not only popular but also official support and funding. Indeed a 'Commission for Scientific Miracles of the Qur'an and Sunnah' has been established in Mecca, under the auspices of the World Muslim League; to date it has published dozens of books on such topics as 'Qur'anic miracles in geology' and organized at least nine international conferences in various countries. The most recent example of the popularity and official support for this 'school' is the attribution of the Dubai International Holy Qur'an Award of the 'Islamic Personality of 2006'[19] to Zaghloul al-Naggar[20], the biggest star of this phenomenon.

Before I scrutinise this *I'jaz* theory, I wish to first examine the other, more moderate approach, which is commonly referred to as 'scientific exegesis' and advocates using scientific knowledge in one's interpretations of various passages of the Qur'an.

Tafsir 'Ilmiy – scientific exegesis

Unlike al-Taftanazi, al-Naggar (like many other writers) makes a clear distinction between the *I'jaz* approach and its objectives and that of *Tafsir 'Ilmiy*. He gives the following definition of each[21]:

> By 'scientific *I'jaz*' we mean the provision of proofs that the Qur'an did precede modern science in referring to one of the facts of the cosmos or by providing an explanation for a natural phenomenon many centuries before the acquired sciences were able to; 'scientific exegesis', on the other hand, is a human attempt to reach a good understanding of some Qur'anic verse(s); if this attempt is successful then the interpreter receives two awards (one for trying and one for succeeding – as per the Islamic tradition), and if he fails he receives one.

This school of Qur'anic interpretation is intimately linked to the reform movement, which emerged around the middle of the nineteenth century with leaders like Sir Syed Ahmad Khan, Jamal Eddine al-Afghani and Muhammad

Abduh, all of whom were seriously concerned both with 'proving' the modern potential of Islam vis-à-vis the West and producing a vibrant and genuine revival movement among Muslims, one built on modern attitudes of reason and progress. (Some authors[22] claim that the great classical twelfth-century Qur'an commentator Fakhr al-Din al-Razi must be considered as a precursor of this school, for he extensively used the scientific knowledge of his time to illuminate his exegetical efforts.) Khan and the reformers insisted that both the Qur'an and nature must be regarded as covenants between God and humans, and hence instead of ever being in contradiction, they must mirror each other in harmony.

In his (unfinished) ambitious exegesis of the Qur'an, Abduh used science to reinterpret some concepts and events,[iv] but more importantly gave science the final word on the meaning of any verse that dealt with natural phenomena. Still, none of these reformers could be considered as pioneers of the scientific exegesis trend, for this was not their main goal; in fact, they tried to harmonise the Qur'an with science in order to show the modernity inherent in Islam.

According to the Islamologist Rotraud Wieland[23], the scientific exegesis approach was started by Muhammad al-Iskandarani, a physician who around 1880 wrote two books which purported to 'uncover the luminous Qur'anic secrets about heavenly and terrestrial bodies, the animals, the plants and the metallic substances'. He was followed by others, with bolder agendas, particularly Tantawi Jawhari, who in 1923 produced nothing less than a full Qur'anic encyclopedia of scientific subjects, complete with pictures and tables, trying to show that the Qur'an contained many 'jewels' (*jawahir* in his book's title) of knowledge.

Earlier, however, illustrious thinkers had adopted this approach to the Qur'an; from the classical era I must name the literary genius and polymath al-Jahiz (ca. 781–869), the Andalusian theologian and jurist Ibn Hazm (994–1064) and al-Ghazzali (1058–1111), and from the modern times al-Kawakibi (1849–1905).

[iv] For example, the supernatural creatures known as jinns became 'possibly' microbes, the birds which shot the Avicennian invader of Mecca the year Prophet Muhammad was born could have been swarms of flies carrying diseases and infecting the army, and so on. Needless to say, the conservative and orthodox writers and commentators have strongly objected to this kind of rationalistic, 'scientific' interpretation of the Qur'an (e.g., al-Taftanazi: op. cit., p. 122).

Recent advocates of the scientific exegesis approach include the famous French surgeon Maurice Bucaille (though he sometimes falls into the 'scientific miraculousness of the Qur'an' trend[v]), Waheed ad-Deen Khan, Muhammad Jamal-Eddine al-Fendi, Ahmed Mustafa al-Maraghi, Muhammad Metwalli al-Sha'rawi, Mustansir Mir, Jalees Rehman[24], Abd-al-'Aleem Abdul-Rahman Khudr and even Mohamed Talbi (to some extent). In a recent article on the subject, Rehman[25] cites the following as examples of 'attempts to explain Qur'anic verses in the light of modern science': 'explanations of the flood in Prophet Noah's time as a melting of ice caps', 'diseases associated with the consumptions of pork and alcohol', etc. Perhaps, realising the feebleness of such programs, he adds, 'Many [. . .] authors [of such attempts] have the best intentions and often believe that showing correlations between the Qur'an and modern science produces Islamization of science'. He admits, 'One danger of such attempts to correlate modern science with the Qur'an is that it makes a linkage between the perennial wisdom and truth of the Qur'an with the transient ideas of modern science'[26].

Let me give a few examples of how this exegetical approach proceeds. Typically, a verse relating to a natural phenomenon is considered; for instance: 'It is He Who has let free the two bodies of flowing water: One palatable and sweet, and the other salt and bitter; yet has He made a barrier between them, a partition that is forbidden to be passed' (Q 25:53), or 'And the sun runs on to a term/resting-place determined for it' (Q 36:38). Performing scientific exegesis then consists of mustering all the scientific knowledge at one's disposal and trying to draw the most reasonable explanation (or at least one that is consistent with the scientific knowledge of the time). Thus, Waheed ad-Deen Khan will draw upon the concept of 'surface tension' in waters of different composition to explain[27] the first verse cited here; Muhammad Abduh will use the idea of undersea volcanic activity to interpret the second verse; and Ahmed Mustafa al-Maraghi will use modern astronomical knowledge to try to describe the motion of the sun referred to in Q 36:38.

---

[v] See for instance how he starts out enlightening with his scientific knowledge the verse 'Have We not made the earth as a wide expanse, And the mountains as pegs?' (Q 78:6–7), writing: 'The pegs mentioned here are those used to hold a tent to the ground (going to the roots of the Arabic term). Modern geologists describe folds in the ground, allowing landscape relief to take hold with variable sizes going from kilometers to dozens of kilometers. From this folding results a stability of the terrestrial crust'. But then he veers into the 'miraculousness' trend as he adds: 'I don't believe that anyone had such notions of geology fourteen centuries ago'.

One particularly important classical Muslim scholar who adopted that approach many centuries ago is the great exegete Ar-Razi, whose masterpiece *Mafatih al-ghayb* or *Kitab at-tafsir al-kabir* (*The Keys to the Unknown* or *The Great Commentary*) made extensive usage of the sciences at his disposal then, to the point of sometimes losing the reader in some astronomical or other discussion, straying far from the initial issue brought up by a given verse. He himself realised his exaggerations and attempted to justify his approach by preempting the criticism he expected. He wrote the following:

> And perhaps some ignorant and idiotic people will come and say: you have overloaded this interpretation of Allah's Book with Astronomy, contrary to the norms? To this poor mind we say: if you did contemplate the Book of Allah deeply and correctly enough you would have realized the wrongness of your objection; indeed, the defense of our treatment comes from two angles: first Allah Himself has filled his Book with examples from the heavens and the earth to emphasize the concepts of knowledge, power and wisdom (of His), and He has repeated that in many verses and chapters, and secondly He has stated that 'Assuredly the creation of the heavens and the earth is a greater (matter) than the creation of men: Yet most men understand not' (40:57), thereby underlying the importance of understanding heavenly phenomena . . . [28]

Many scholars have expressed objections to the whole approach. Following were the most important opponents to this school: In classical times al-Shatibi (d. 1388), the Andalusian Muslim jurist, and in the twentieth century Rashid Reda, Mahmud Shaltout and Sayyid Qutb. As to the objections themselves, Wielandt[29] summarises them as (a) often assigning untenable meanings to some of the Qur'anic vocabulary; (b) downplaying the occasions of revelation (*asbab al-nuzul*) and the textual context of the verses under consideration; and (c) disconnecting the verses from the social and cultural context in which they were revealed. The contemporary writer Sami Ahmed al-Musili[30] adds two more critical remarks: (1) In scientific exegesis, the perfect Qur'an is made subject to the imperfect knowledge (science) of humans; and (2) this approach is elitist by nature and is not accessible to all Muslims.

None of the above objectives are really serious problems in my view; indeed, the first three disregard the idea that the Qur'an must not be culturally bound to the seventh-century Arabia, that it must be relevant to all people, provided they make an intellectual effort to make it so, and secondly all these objections ignore the main idea I have advocated in this book, namely that the Qur'an carries a multiplicity of meanings and can, therefore, be illuminated by prior knowledge, and so by reflection it can enlighten any reader, scientifically or literally inclined, rationally or spiritually minded.

In a more recent article, Mustansir Mir[31] considers the viability of the scientific exegesis project. Attempting to defend the approach in a reasonable way, he points out that this type of exegesis should first be considered as one approach to the Qur'an just like other (linguistic and theological) approaches have prospered before, and secondly that this 'scientific' approach arose 'in response to real and concrete needs [. . .] Today the dominance of science and the scientific worldview would seem to encourage, even necessitate, the cultivation of *tafsir 'ilmi* (scientific exegesis)'. He justifies his guarded support for this approach on the following basis:

> From a linguistic standpoint, it is quite possible for a word, phrase or statement to have more than one layer of meaning, such that one layer would make sense to one audience in one age and another layer of meaning would, without negating the first, be meaningful to another audience in a subsequent age.

He offers the following example:

> The word *yasbahun* (swim or float) in the verse 'And He is the One Who created the night and the day, and the sun and the moon—each 'swimming' in an orbit' (Q 21:33) made good sense to seventh-century Arabs observing natural phenomena with the naked eye; it is equally meaningful to us in light of today's scientific findings [i.e. celestial mechanics].

Mir is, however, very critical of many of the amateurish and ill-informed 'scientific interpretations' that have appeared everywhere, particularly on the Web. He concludes that 'no credible scientific exegesis of the Qur'an has so far been produced' and 'like Sufism, *tafsir 'ilmi* may have to wait for its Ghazzali'. I agree.

I'jaz 'Ilmiy – a brief historical review

Authors have placed the origin of this trend as far back as the Prophet himself, through some of his statements (hadiths), or his companions, through some of their declarations. The Prophet is reported to have stated that the Qur'an 'has no end to its marvels'[32] and has advised people to 'explore the meanings of the Qur'an and to search for its mysteries'[33]. Similarly, one of the primary companions of the Prophet, Abdallah ibn Mas'ud, is reported to have said: 'Whoever wants the knowledge of the ancients and of the moderns should study the Qur'an extensively'. I should add that others have found justification for this miraculous Qur'anic scientific content in the hadiths (and other classical discourses) that make the Qur'an 'the Prophet's miracle'[34].

Other writers consider the serious ignition of the phenomenon to have come from the great classical theologians and jurists al-Ghazzali (d. 1111) and al-Suyuti (d. 1505), who, perhaps, in a hyperbolic flight of rhetoric and impressed by the verses 'and We have revealed the Book to you as an exposition of all things' (Q 16:89) and 'We have not neglected anything in the Book' (Q 6:38), exclaimed that the Qur'an contains all thinkable knowledge. al-Ghazzali defended the 'exegesis by personal opinion' (*tafsir bi-r-ra'y*) and pointed out that the above statements of the Prophet and of Abdallah ibn Mas'ud clearly imply that one must go beyond any superficial reading of the Book; in his masterpiece *Ihya' 'Ulum ad-Deen* (*The Revival of the Religious Sciences*) he wrote, 'In the Qur'an are symbols and indications which can only be understood by the specialists, so could a simplistic rendering of its apparent meanings do?'[35] He went further in his next book *Jawahir al-Qur'an* (*Pearls of the Qur'an*): 'Haven't you heard that the Qur'an is the ocean from which branch out all the knowledge of the ancients and the moderns like rivers branch out from coasts of the ocean?'[36] And in a similar analogy, he referred to the exploration of the Qur'an as finding shells, then either limiting oneself to admiring and polishing them or opening them and seeking to find the pearls inside.

But as Ahmad Dallal[37] points out, both of the above verses (Q 16:89 and Q 6:38) when read in their contexts more likely refer to knowledge about the afterlife; moreover, he insists that 'despite their claims, neither al-Ghazzali nor al-Suyuti proceeds to correlate the Qur'anic text to science, in a systematic interpretative exercise'. Dallal goes on to show that no such trend ever appeared in the golden age of Islam, even when science was at its peak.

It is often claimed that the recent explosion of *I'jaz* was started by Maurice Bucaille, the French surgeon who in 1976 published a book titled *La Bible, le Coran et la Science: Les écritures saintes examinées à la lumière des connaissances modernes* (*The Bible, the Qur'an and Science: The Holy Scriptures Examined in the Light of Modern Knowledge*), in which he stated that the Qur'an not only does not contain any statement that can be contradicted by the most recent scientific knowledge but moreover contains some references to facts that could not be known to anyone 14 centuries ago. The book has since had at least 14 French editions and has been translated to probably every language that humans read today. It was an instant sensation in the Muslim world, and Dr. Bucaille became a cultural icon. There is no doubt in the minds of all observers that this was largely due to his being French and a medical doctor; and the fact that he always remained elusive about whether he embraced Islam only added to his mystique.

In his book with Mohamed Talbi (*Réflexions sur le Coran*[38]), he relates how he presented his findings to the Académie Nationale de Médecine in a lecture he delivered shortly after the publication of his book:

> In my book I gave numerous examples on a great variety of subjects without ever finding in the Qur'anic text the slightest statement that would contradict modern science. A scenario of the creation of the world, different from the Bible's but in total conformity with the general modern ideas on the formation of the universe, some considerations on celestial objects, their motions and their evolutions, all in harmony of current notions, the prediction of space exploration, some reflections on the water cycle in nature and on the earth's landscape which were confirmed many centuries later, all these observations, troubling for anyone who considers them objectively, shifted my work to a level which places the question in much larger dimensions. But the question that one must ask remains: are we not here in the presence of facts that seriously challenge our natural propensity to try to explain everything by materialistic considerations; indeed, the existence in the Qur'an of these scientific statements appears to be a challenge for humanity[39].

As we can clearly see, Bucaille moves between a *Tafsir 'Ilmiy* approach and a bona fide subscription to *I'jaz 'Ilmiy*. In the book with Talbi he devotes much space to the defense of his opinions on the question of science in the Qur'an.

A similar case to Bucaille's is that of Professor Keith Moore, who in 1986 published an article titled 'A scientist's interpretation of references to embryology in the Qur'an'[40] in *The Journal of the Islamic Medical Association*. Moore was then professor of anatomy and associate dean of basic sciences in the Faculty of Medicine of the University of Toronto. Moore also soon ended up a celebrity in the Muslim world (much less than Bucaille, though) for what he had proclaimed. Indeed, reviewing the Qur'anic verses[41] which describe (succinctly) the phases of development of the embryo, he reaches the following conclusion:

> The interpretation of the verses in the Qur'an referring to human development would not have been possible in the 7th century A.D., or even a hundred years ago. We can interpret them now because the science of modern Embryology affords us new understanding. Undoubtedly there are other verses in the Qur'an related to human development that will be understood in the future as our knowledge increases[42].

One must note, however, that in all of his analysis, Moore refers to only one specific translation of the verses under consideration, that of Abdelmajid az-Zandani (a strong proponent of *I'jaz*)! One understands then why Moore made such an impressed reading and drew such enthusiastic conclusions.

But let us read those verses with an open mind and from a neutral perspective: 'Then We made the sperm into a clot of congealed blood; then of that clot We made a (foetus) lump; then we made out of that lump bones and clothed the bones with flesh; then we developed out of it another creature. So blessed be Allah, the best to create' (Q 23:14); 'He makes you, in the wombs of your mothers, in stages, one after another, in three veils of darkness' (Q 39:6); and a few others (e.g. Q 22:5), which are similar to these two. The reasonable conclusion that one reaches is that these descriptions are both general and correct enough to be understood at different levels by people of different knowledge across the ages. One is hard-pressed to find evidence for the claims of Az-Zandani and the *I'jaz* proponents, namely the existence of 'facts' that have been discovered only recently. It is not debasing to the Qur'an to say that much of our knowledge of nature (human body, celestial motion etc.) was known in the past but often not understood – this is where modern science comes in picture. I am convinced that this is one of those cases.

These two examples and approaches to *I'jaz* by Bucaille and Moore appear to me as a good introduction to the methodology of *I'jaz* with its many flaws. It quickly becomes clear that the proponents of this approach let themselves gradually slide from a position of using the scientific knowledge at one's disposal to better understand some Qur'anic verses dealing with natural phenomena, a method suggested and practiced by Abduh and other careful thinkers, to impressed readings and fanciful interpretations of verses that are claimed to refer to space exploration, radio, relativity, black holes and the speed of light.

methodology of *I'jaz* – in theory

Zaghloul al-Naggar has presented his general programme of trying to prove miraculous cases of the Qur'an preceding various modern scientific discoveries[43] with some of its statements. First, he insists that Qur'anic formulations are often couched in such a language as to make people of various eras understand them differently, but without contradiction. And so with the expansion of the sphere of knowledge, scientific and other, our understanding of various Qur'anic verses can be of a different nature. I agree with him, up to this point.

But he quickly makes a big leap by claiming that the Qur'an contains much scientific information; he finds 'justification' for his conviction in the following verses: 'For every message there is a term, and you will come to know [it]' (Q 6:67), 'We shall show them Our signs on the horizons and within themselves until it will be manifest to them that it is the Truth. Is it not enough that your Lord does witness all things?' (Q 41:53), and 'This is no less than a Message to (all) the Worlds. And ye shall certainly know the truth of it (all) after a while' (Q 38:87–88).

Now he is ready to set guidelines[44] for the *I'jaz* enterprise; he lists the following ten principles:

1. Understanding the Qur'anic text well, according to the Arabic language's rules of meaning.
2. Taking into account the old 'Qur'anic sciences' (occasions of revelation, the verses that were made obsolete, the hadiths, which explain certain verses etc.; see Chapter 2).
3. Assembling the various verses which pertain to a common topic before attempting to propose any new interpretation.
4. Avoiding exaggerated interpretations, not 'twisting the neck' of some verses to make them accord to scientific results.
5. Staying away from issues of the 'unseen' (*ghayb*, exclusively divine knowledge).
6. Keeping to one's scientific expertise when addressing verses dealing with specific topics.

7. Upholding accuracy and intellectual honesty when dealing with divine statements.

8. Making use of established scientific facts only, not uncertain theories and conjectures, except in one case: verses and prophetic statements which deal with the creation and extinction of the cosmos, life and humans; indeed, these topics do not lend themselves to direct observation and determination, thus human sciences in these fields can only remain at the conjecture level, and Muslims can – by using statements in the Holy Book or in the Sunnah (Prophet's tradition) – help elevate some of them (one theory or another) to the rank of established truth.

9. Distinguishing between *Tafsir 'Ilmiy* and *I'jaz 'Ilmiy*, both in the Qur'an and in the Sunnah. In *Tafsir 'Ilmiy*, in some cases the human sciences have not reached definitive results yet, and it is then acceptable to employ the dominant scientific theory to explain the cosmic verses or hadiths, even if the theory turns out later to be incorrect, for then the error is upon the interpreter, not the holy text. In *I'jaz 'Ilmiy*, however, one must only use fully established facts, for one cannot claim Qur'anic or Prophetic miraculous precedence with statements that may turn out to be totally wrong later.

10. Respecting the efforts of previous scholars in all related matters.

Examining the above guidelines quickly shows that most of them are rather trivial in nature and fall squarely under normal academic method-ological rules that we teach and hold our students to. But as we'll see throughout this chapter, the problem is that the practitioners of the *I'jaz* field hardly, if ever, stick to such proclaimed methodology.

I must note, however, that the guidelines 8 and 9 above do present something new and controversial. To some extent guideline 9 can be accepted under the guises of 'human intellectual endeavor' and 'encouraging *Ijtihad*' (intellectual effort aiming at producing new views that may be more useful than the old beaten ones). I am certain, however, that few orthodox scholars will accept to let 'experts' project their scientific understanding on various holy statements. Guideline 8, on the other hand, is a stunning proposition: Professor al-Naggar is willing to let – and is even encouraging – researchers to use Qur'anic verses to 'elevate' one model (say the closed universe) over another (the open universe), as he (erroneously) does!

Now, one important issue to bring up from this theory is the primacy of the Qur'an over scientific knowledge versus the independence of science in its methods and results. Two views can be found in this regard among the proponents of this theory: (a) One should not rush to try to find references in the Qur'an for scientific discoveries, for the latter are temporary and limited, and the former is eternal and absolute; or (b) one can find definitive truths in science, and only those should be sought in the Qur'an.

An example of the first viewpoint is the article titled 'Scientific I'jaz, regulations and limits'[45], written by Fahd A. al-Yahya (a leading member of the international commission on the scientific miraculousness of the Qur'an and the Sunnah). In this article the author warns those who, for example, try to find the space conquest in the Qur'an, insisting on

> the doubts and rejections by scientists – including one American – of the NASA claim of manned trips to the Moon [. . .] while Muslims know for certain that the Moon was split during the life of the Prophet (that is only centuries ago), yet none of the scientific theories are able to prove or explain that but still claim knowledge of events going back thousands of light-years [sic] . . . .

On the second viewpoint, one may cite Abdellah al-Mosleh (the secretary general of the aforementioned Commission), who, in his article 'Regulations of research in the field of the scientific miraculousness of the Qur'an and the Sunnah'[46], lists the following criteria for the 'validity' of any concordance to be found between scientific results and Qur'anic/Prophetic verses/statements:

1. Guarantee that a particular scientific discovery has been established as a 'permanent and durable' fact by the specialists.

2. Exactness of the meaning of the text (Qur'anic or Prophetic) concerning the given scientific fact, without any overdue effort of interpretation of the text, while still showing that knowledge of that fact was impossible during the Prophet's times.

3. The proof of the concordance between the above two points must be accomplished via the following steps:

   (a) Proof of the existence of the scientific fact under consideration in the religious text.

   (b) Proof, in the case of the Sunnah, that the Prophetic statement under consideration is absolutely authentic (established) and suffers no doubt in its having been uttered by the Prophet.

   (c) Proof of the scientific 'validity' of that fact.

(d) Proof that knowledge of that fact came only recently and was impossible during the times of the revelation.

(e) Proof of the concordance between the scientific fact and the religious statement.

It will not be difficult for us, with our review of science, its methods and philosophy (Chapter 3), as well as of the Qur'an and the principles of its interpretation (Chapter 2), to realise that the establishment of the above 'criteria' is objectively impossible. Let me give here just one example. Would these *I'jaz* proponents consider gravity as an established fact of nature and Newton's theory (law of attraction between masses) an established theory? Wouldn't Einstein's theory (general relativity), which explains gravity as a curvature of space and which superseded Newton's, be described as an established one? Is Einstein's theory the final word on the subject? At what point do we proclaim an idea to be an established 'fact' or 'theory'?

Other leading proponents of this theory (al-Naggar, al-Taftanazi etc.) have also proposed guidelines and criteria for validation of this type of 'research'. They are largely similar, so I would like to move to an interesting idea presented by another leader of the *I'jaz* school, Abdulmajid Az-Zindani, who has attempted to produce a theoretical framework[47] consisting of the following two methods: (1) a cosmic truth, already expressed in the Qur'an but not understood by Muslims is discovered by science; this then puts an end to the multiplicity of interpretations of the verse or passage, unless there is a contradiction between the scientific conclusion and the obvious meaning of the verse; and (2) Muslim scientists take 'leads' from Qur'anic verses and pursue scientific research until new discoveries are made that corroborate the text. Note that Az-Zindani, not surprisingly for a conservative leader, gives the Qur'an precedence and veto power over scientific discoveries and truths.

For fairness, however, I would note that such a theory, with its many startling claims and ill-founded methodology, is not exclusive to the Muslim world. Observers[48] have pointed out that similar defensive and apologetic discourse has existed among some Christian fundamentalists, creationist circles[49], as well as some Hindu fundamentalists[50].

A light of hope shone from Mecca in December 2007. In a very interesting article[51] published in the *Islamic Studies* journal of the Umm al-Qura University (in Mecca), Saud bin Abdelaziz al-'Arifi presented a rare critique of the *I'jaz* theory from within the religious scholarly community. First and foremost, he proceeded to attack its foundational methodology, but also mentioned a few examples from the *I'jaz* corpus to highlight factual errors as well as failures to follow the claimed methodology. Let me briefly summarise the main ideas in that important article:

- There is a fundamental and huge difference between the Qur'an's approach, which argues for God's existence, creative power and sustenance of the universe and the *I'jaz* theory, whose goal is to prove the divine origin of the Book, which is tantamount to arguing for the prophethood of Muhammad.
- There is a large conceptual quantum leap between presenting an 'argument' or some (suggestive) evidence, which is the Qur'an's way, and attempting to 'prove' the divine origin (of the world or the Book); the most that the *I'jaz* people should shoot for is to challenge anyone to find Qur'anic verses that show clear contradictions with scientific discoveries or established knowledge;
- Many of the 'scientific discoveries' that the *I'jaz* proponents claim and find in the Qur'an were in fact known to ancient physicians, philosophers and naturalists; what is new is the depth and precision in our knowledge and understanding of them.

Now, I disagree with al-'Arifi on only one (important) point: He believes that no 'modern scientific meanings' can exist in any Qur'anic verses simply because at least the Prophet and his companions had full knowledge and understanding of all the various meanings that every verse carried. As I have mentioned, I believe that the Qur'an presents multiple levels of reading, so that many meanings can be found in verses, depending on one's education and era; one may then still read some scientific fact of nature in a verse without turning it into a miraculous (definitive) claim of precedence. And needless to say, believing that the reading of the Qur'an is an open venture is much more progressive than the claim that the Prophet and his companions understood it all and that any new meaning they did not see is simply not there.

Still, I strongly applaud al-'Arifi's brave, cogent and potent effort at both sapping the flawed methodology of the *I'jaz* program and decrying the extent to which this theory has become a staple of today's Islamic culture (the author bemoans the huge popular and financial support given to it nowadays and notes that a 'silent majority' of religious scholars do not subscribe to it but refrain from any criticism for fear of being seen as attacking 'firm ideas').

methodology of *I'jaz* – in practice

Let me now review how the *I'jaz* enterprise is actually carried out; I will use examples from books by some of the leaders in the field. The main serious methodological transgressions committed by al-Naggar and al-Taftanazi are as follows:

1. Insisting that science is an *ideology*, thus encouraging people to discard any result, and certainly any theory, which does not conform to 'our' view of nature and the cosmos. al-Naggar writes:

   > There are as many theories related to creation as there are different ideological types: believers, atheists, polytheists, agnostics, happy folks, miserable and stressed people, straight or deviant ones etc. The Muslim, however, has a light from his Lord, which he obtains from [reading] a Qur'anic verse or a Prophetic statement, and that light shall help him discriminate among the many confusing theories and elevate the right one to the status of established truth, not because the acquired sciences will have claimed that but because he will have gotten a sign from the holy texts[52].

   He tries to justify his claim by referring to modern cosmology: 'And from a starting point of denying creation, modern astronomers push for the idea of an open universe, that is one which will expand to infinity; however, the estimates of the missing masses in the calculations of the balance of the observed universe confirm that the universe is closed'[53]. Unfortunately, the author is wrong on the current cosmological paradigm and, of course, on his claims that models and scientific results are the product of ideological stands.

2. Claiming that science's inability to give definitive answers on a given topic is tantamount to an admission of a divine origin of that particular phenomenon:

   a. 'Experimental sciences confirm that there is in living organisms a secret which nature is a mystery to us, for we know the constituents of the cell, and the material composition of man's body, and still science is unable to build a single cell'[54].

   b. 'A simple look at any region of this universe, along with everything and everyone it carries, will confirm its need for the supervision and safekeeping on the part of its mighty creator at all times of its existence; for the universe is full of dangers, and Earth is surrounded by many from all sides'[55].

3. Decreeing that certain areas of knowledge shall remain conjectural no matter how much material evidence they may produce:

> God in His mercy has left for us in the records of the sky and of Earth's rocks concrete evidence that may help us – with our limited capabilities – to figure out the process of creation; such a model, however, shall remain of theoretical and conjectural nature and can never reach the status of truth, for scientific truths must come under man's senses and perception[56] [whatever that means].

al-Naggar then applies this approach and decrees that the universe must be closed (on the basis of Q 21:104) no matter what the (real) cosmological research says.

4. Having a very simplistic if not a seriously mistaken understanding of science, especially physics:

   a. The slowing down of the universe's acceleration [which is the opposite of what we know now] implies that gravity will, at a time which only Allah can know, overcome the expansion, leading to a reassembly of all matter and energy in one high-density object, which will then explode like the first big bang, so that a new earth and new heavens shall appear, conforming to the Qur'anic verses Q 21:104 and Q 14:48 ('One day the earth will be changed to a different earth, and so will be the heavens, and (men) will be marshalled forth, before Allah, the One, the Almighty')[57].

   b. al-Naggar confuses 'dark matter' with 'primordial gas' ('smoke')[58]; he further thinks that the Cosmic Background Explorer (COBE) satellite, which made accurate measurements of the cosmic microwave background radiation (a relic of the Big Bang) in the nineties actually observed the 'primordial smoke'[59].

5. Despite a long list of references to scientific works given at the end of al-Naggar's book, one must note that (a) only one book (out of about 60) is less than 10 years old, and that one is 'Powers of Ten, a flipbook'; (b) those works are hardly ever really cited and used by al-Naggar; the listing

aims at comforting the reader and convincing him that a large collection of western (and, if possible, recent) references have been used.

6. Most of al-Naggar's (and al-Taftanazi's) works consist of the following strategy: pick a 'cosmic verse' (a Qur'anic statement describing some natural phenomenon), present a dozen pages of scientific information that can be found in any encyclopedia, and end by proclaiming that it is truly miraculous (*I'jaz*) that the verse had foretold all these scientific facts; here are a few examples:

a. 'So verily I swear by the planets/stars, That run and hide...' (Q 81:15–16). These verses are claimed to refer to black holes, and one is given 20 pages of text with 20 colour images[60], before the conclusion is drawn that the Qur'an foretold the existence of black holes.

b. 'Praise be to Allah, Who created the heavens and the earth, and made the darkness and the light' (Q 6:1). This produces 13 pages of exegesis from various classical references, followed by a detailed eight-page description of the Big Bang scenario, where each phase is described: the age of quarks and gluons, the age of nucleons and antinucleons, the age of nucleosynthesis, the age of ions, the age of atoms and the age of stars and galaxies. This is then followed by a description of 'darknesses': the primordial darkness of the universe; the current darkness of the cosmos; the darkness in the depths of the seas and oceans; the darkness of the wombs; and the darkness of the tombs. How all this is inferred from that starting verse is truly beyond me.

c. 'And the sun runs on to a term/resting-place determined for it . . . Each (Sun and Moon) swims along in (its own) orbit' (Q 36:38; 36:40). In these verses, al-Taftanazi gives us seven pages of detailed information on the sun (mass, temperature, composition etc.) and its motions (rotation around its axis, revolution around the centre of the Milky Way) and then concludes that there is

> amazing scientific concordance with the Qur'anic verses about the true, not apparent motion of the Sun in space, about the orbital revolution of the Sun around our galaxy; this kind of miraculous presentation, in how the various motions of the Sun have been described, is one that must astound those who have minds and carry knowledge/science, and it must lead the objective analysts to admit to the glory of this great book[61].

d. 'And the Moon, We have ordained for it mansions/stages till it becomes again as an old dry palm branch' (Q 36:39). This verse, according to al-Taftanazi, 'definitely implies that the Moon revolves around the Earth, and this is a fact that no one knew at the time that the Qur'an was revealed and after that for centuries'[62] (which is totally wrong as any quick review of astronomy's history will reveal).

e. '[I swear by] the Moon in her fullness; That ye shall journey on from plane/stage to plane/stage' (Q 84:18–19). Here, as one can already guess, it is the trips to the moon that are (claimed to be) referred to[63].

And so it goes in the field of *I'jaz*. It is always the exact same 'methodology'. al-Shatibi, the aforementioned fourteenth-century theologian and jurist had anticipated such trespassings and decried them: 'Many people overstep all bounds and make undue claims about the Qur'an when they assign to it all types of knowledge of the past and the present such as natural sciences, mathematics and logic. It is totally impermissible to ascribe to the Qur'an what it does not call for as it is not right to deny it what it calls for'[64]. Likewise, the thirteenth–fourteenth-century Islamic scholar and historian al-Dhahabi decried such 'lunatic innovations'[65].

I'jaz in the hadith

With no huge surprise, we find a new subgenre of *I'jaz* literature, now addressing the Prophet's statements (hadiths), also claimed to be scientifically miraculous. There is no point in challenging that in full, for one finds the same methodological flaws as above – and worse; still, I need to make the readers aware of the extent of the factual and methodological erring in such works.

al-Naggar has published a two-volume book titled *The Scientific I'jaz in the Prophetic Tradition (Sunna)*[66]. Within a year of their publication (2004), the two volumes had gone through seven and five editions, respectively, a stunning testimony to the popularity of such books in the Arab world. After an introduction justifying the need for such 'scholarly treatments', al-Naggar goes over 28 and 25 hadiths, respectively. Over five pages on average, he presents the different versions of a given hadith, then relates it to the scientific knowledge it presumably contains, and invariably concludes that it is 'so ahead of humanity's knowledge that it must be considered yet another testimony to the divine prophethood of Muhammad'.

Before I (briefly) give a few typical examples, let me here point out the foundational problems with the methodology: (1) While the author strives in his introduction to convince the reader that the hadiths he considers are historically genuine and that their wording is certain, he knows and remarks from time to time that there are thousands of 'hadiths' that the Islamic scholars have labelled 'false' or 'weak', and that there are, for most genuine ones, several versions; (2) he does not ever – not once – give a single reference to the huge scientific claims he makes. Here are a few examples from his work:

- 'The water of Zamzam (the well in Mecca) is a remedy to whatever one drinks it for'; al-Naggar gives a list of its (very heavy) mineral contents, then goes on to state that such mineral water is good for the treatment of various ills (stomach acidity, indigestion, vascular problems etc.). No supporting evidence or scientific reference is given to any of this.

- 'The Day of Judgment shall not come until the Sun rises from the west'; the author explains that the earth has been slowing down in its rotation (around its axis), and therefore one day it will stop and start rotating backward! He hastens to add, however, that this does not mean one can compute the date of the Day of the Judgment, for that will occur 'without regard to cosmic laws, calculations or the Earth slowing down'! One is left trying to make sense of this logic. Two things become clear here: (1) Prof. al-Naggar does not understand tidal breaking, which results in a rotation–revolution lock, not a total halt, and certainly no reversal of the rotation; (2) he did not bother to conduct any systematic scientific investigation of this.

- 'The lunar months (by which Islamic occasions, such as Ramadan, are set) are either 29 or 30 days long'; here the author claims that the Prophet was presenting information that became available only in the past two centuries (which is totally wrong, as anyone with the slightest knowledge of Astronomy and its history will know) and that this statement could only be made if one understood the revolutions of the moon around the earth and of the latter around the sun, as well as the rotations of the moon and the earth around their axes – an incorrect claim in toto.

- 'Allah created Adam in his same shape, making him 60 cubits tall'. Our author here tells us that this hadith is another proof that the Darwin's theory is false and that there is a fossil proof that creatures have gotten smaller over the ages.

for a reasonable and credible muslim discourse on the Qur'an and science

After all these incredible statements and erratic methodologies, the question that we must first ask is: Why is this *I'jaz* theory so extraordinarily popular? It is not easy to formulate any convincing answer to this question, for it relates to several sociological and historical factors. Indeed, that question raises several others: What kind of understanding does the Muslim world have of science today? What level of critical thinking and analysis do we find in the Muslim society? How eager are the Muslim peoples today to turn their general defeat in all fields into a position of precedence and superiority and to convince themselves that their religion and civilization are indeed true – in the absolute sense – and superior?

Without giving any clear answer to our question, the contemporary author Mustafa Abu Sway provides an element of understanding; he says[67]: 'Rather than attaining science and maintaining its proper status within the Islamic worldview, it seems that the "scientific interpretation" provided a comforting cushion. The rest of the world can do science, and we, Muslims, can discover it anew in the Qur'an!' The Sudanese writer Bustami Mohamed Khir emphasises the special relation and attitude adopted by Muslims with regard to science more generally: 'The Muslim experience with science is in many ways distinct [from that of the West]. Instead of being in defense of Islam against the offense of modern science, Muslims attempted to use science as new evidence to support the truths of the Qur'an'[68].

As I have shown, the Qur'anic *I'jaz* theory rests on erroneous principles, two of which should be noted and emphasised: (1) The interpretation of Qur'anic passages can be univocal and definitive, thus allowing for a comparison with specific scientific results and statements; (2) science is simple and clear; it contains definitive facts that can easily be distinguished from 'theories'. I have also insisted that this theory is the product of a confusion, which took place, first gradually but today almost completely, between the legitimate and commendable attempt to infuse the scriptural exegesis enterprise with newly acquired human knowledge (which will, of course, include modern scientific discoveries and insights) and the principle that the scientific results, laws, discoveries and inventions, from the most general to the specific and arcane, can be found in the Qur'an and even in the Hadith, only if we made the effort to search and re-explain the verses in a 'scientific' way. The *I'jaz* theory is a snowball that started out small and white but then rolled and collected dirt (ludicrous claims built on flawed methodologies); it is also a mass of dirty ice that easily melts under the intense light of objective and methodical scrutiny.

Furthermore, I consider the *I'jaz* approach as perilous because it claims that one can identify scientific 'facts' and compare them with 'clear Qur'anic statements', which shows clear misunderstanding of the nature of science (Poincaré said 'science is not a bunch of facts just like a house is not a bunch of bricks'). Moreover, it distorts in young Muslim minds the very nature of science and the approach we should have towards the Qur'an.

But if at the start this theory was born from an interesting and valuable idea (making use of science in trying to better understand some passages of the Qur'an), can it be salvaged, cleaned up and redirected – at least for the general public – by reformulating it in a way that serves a new view and reading of the Qur'an for a new discourse on the relation between Islam (the foundational texts plus the human heritage built upon them) and science? I think this is possible, and here I will attempt a quick brush of such an approach.

I believe that one must 'simply' replace the above erroneous principles with two, more correct ones: (1) the Qur'anic text allows multiple and multilevel readings; (2) science and its philosophy must be fully comprehended before anyone puts forward any interpretation of religious texts that appears to have some possible relation with science. The second principle is clear: One must study science, particularly its methodology and philosophy, very well and not use any superficial understanding of it. In my view the first principle is even more important for the Muslim world and its religious development: It can potentially alleviate various ills, from this *I'jaz* theory to fundamentalism (the claim of monopoly of truth by some leaders, scholars or schools).

This principle of multilevel readings of the Qur'an is not new; we have presented aspects of this in Chapter 2; it goes back to the classical era of Islam, particularly to Averroes, who had built large parts of his thesis of harmony between philosophy and religion on the principle that the reading of the Qur'an and the interpretation of the Law allow at least two levels: a general, basic one, which is accessible to the whole public, and a more profound and sophisticated one that only erudite and elite minds can attempt. (For example, in the Prologue of this book, I have mentioned this suggestion that the famous 'Verse of Divine Light' could be understood in various ways by people with different levels of sophistication.)

Now before I explain how this approach applies to the Qur'an–science field, I must clarify one issue: the idea of multilevel readings of Qur'anic verses encompasses, but is not limited to, the metaphorical interpretation that must be performed in certain cases. It is widely accepted that some scriptural statements cannot be taken very literally, for instance, the anthropomorphic descriptions of God, although some conservative commentators have refrained from acknowledging even that much. We shall see in the upcoming chapters that even verses that pertain to cosmic and natural phenomena, including those of the creation of the earth, the heavens, life and humans, will require some degree of interpretation.

Still, the idea of multiplicity of readings goes beyond this metaphorical approach. It is intimately connected to another principle of the Qur'an's scope in addressing people: people vary widely in their abilities to understand various ideas, especially when they range from the cosmic to the spiritual and social; consequently, the language used must accommodate various minds, intelligences, learning styles and eras. And that is why the Qur'an by necessity uses phrases, statements and stories, which can be understood in different ways; these diverse interpretations are not necessarily all 'correct', but their objective is not to be 'accurate' but rather to convince the reader/listener.

The question of correctness and accuracy is an important and sensitive issue, for some people will object to the implicit implication that some Qur'anic statements may not be 'absolutely true'. To address this point, let me first explain that 'correctness' and 'accuracy' can only be achieved in an asymptotic way; our quest for understanding and our ability to achieve it only tend towards the truth; we may never reach it completely, and so at any time we would have only achieved a certain degree of certainty of our 'correct' understanding of a given verse. But the important thing is that our comprehension continuously improves with time.

Let me give a few examples to illustrate these ideas. The Qur'an says of Noah that he remained among his people 'a thousand years save fifty' (Q 29:14) before the flood engulfed them. In previous times, people could take Noah's stay to have literally lasted for 950 years, although even classical exegetes often merely said 'a long time' (e.g. Ibn Kathir's *Tafsir*). More recent scholarship (e.g. Muhammad Asad's exegesis[69]) has pointed out that the figure of 950 is also given in the Bible and that the Qur'an uses 'this element of the Biblical legend' (Asad's words) to stress (to Prophet Muhammad and to the early Muslims) the fact that the long stay of a messenger among his people never guaranteed its acceptance. Moreover, it should be recalled that the term 'thousand' only meant 'very large/long' in ancient times.

Another important question that falls within this metaphorical/multi-level reading of Qur'anic verses is that of miracles. As we shall discuss in Chapter 10, a large number of 'physical' miracles of the previous prophets, e.g. Moses and Jesus, can today be understood on more naturalistic bases; for example, Jesus's healing of the blind may be imputed to the placebo effect (induced by the mind), while Moses's parting of the red sea could somewhat easily be explained by sudden meteorological conditions. Now what is important to note here is that this still leaves a wide margin of understanding, including the miraculous divine intervention, as one can still claim that the placebo (mind's) effect was triggered by divine intervention at the quantum level, while the meteorological conditions, which led to the wind blowing the red sea waters apart, were made possible by divine intervention at the microscopic level, which then led to a 'butterfly effect' (non-linear, chaos-type) magnification to large scales. One may or may not like or be convinced by such explanation, but I think this is a good illustration of the multilevel approach to some Qur'anic verses. Still, there are other miracles that cannot be understood naturalistically; for instance, the magic transformation of a stick into a snake, and here one may try another level by insisting on a metaphorical interpretation; I should add in support of this approach, that in the story of Moses the Qur'an first explains that 'it appeared to Moses' (and to others) that rods and ropes had been turned into snakes, but that this was merely a magicians' trick, and so God instructed him to do the same trickery (not any miracle) in order to defeat the magicians.

The same approach could be adopted with the Qur'anic verses that seem to imply some miraculous events from the life of Muhammad, particularly the assertion that during the battle of Badr 'one thousand angels in succession' (Q 8:9) were sent by God to support the believers (who were outnumbered three to one). This is another good example of multilevel understanding: the classical commentators (e.g. Ibn Kathir[70]) rely on a story that a companion came to the Prophet and told him that some of unbelievers seem to have died from strange hits on their heads and was told by Muhammad that 'it was from the reinforcements from the third heaven', but more modern interpreters (e.g. Muhammad Asad) emphasise the metaphorical aspect: 'As regards the promise of aid through thousands of angels [. . . the] spiritual nature of this angelic aid is clearly expressed by the words, "and God ordained this only as a glad tiding"'[71].

Finally, another important example is the famous story of *al-Isra'* of the Prophet, his night journey from Mecca to Jerusalem, his Ascension to Heaven and his return (in one night) to Mecca, for this story is not only often taken literally but sometimes used to compute distances to heavens etc. Here too, a classical exegete like Ibn Kathir will relate a dozen stories, including one from the Patriarch of Jerusalem, who presumably found evidence of 'a prophet who prayed last night in our sanctuary' and of 'an animal having been tethered' at the corner of the sanctuary, whereas Asad makes a 'mystical' reading of the event, to which he devotes a whole appendix[72] of his commentary of the Qur'an.

Now how does this multilevel hermeneutical approach apply to the Qur'an–science discourse? As Campanini explains[73]:

> Scripture contains neither explicit scientific statements nor proper laws of nature, but the truths the Scripture reveals in a simple language, comprehensible to common people, are the same basic truths that science and philosophy express in sophisticated and polished demonstrative language. It would be absurd to look for a scientific demonstration of the duration of the universe in the Qur'an that is whether the universe is created or [eternal], finite or infinite.

Let us then apply this multilevel hermeneutical approach to some of the *I'jaz* writers' favorite examples. al-Naggar is fond of quoting the verse 'We sent down Iron, in which is (material for) mighty war' (Q 57:25) and claims that iron is produced in stars, which is true, but that is also the case for almost all elements in nature other than the first four or five elements in the periodic table, and thus – al-Naggar says – iron 'came down'; moreover, he says, this stellar-evolution information only became known 50 years ago, so this is clearly *I'jaz*. However, meteorites, which in antiquity were often found in the desert and sometimes seen to fall from the sky, were known to be iron-rich (or metal-rich), and in fact that is the material that was always used to make swords. So iron 'coming down' can easily be understood (as it has usually been in the past) as the stuff in the meteorites. Today, some may prefer al-Naggar's interpretation, which would square fine with my multilevel reading, but note that iron was part of the material that existed at the time of the formation of the solar system; that iron did not 'come down' to earth (after its formation). And that is why I prefer the meteoritic interpretation – and I don't see any *I'jaz*.

Another oft-cited verse is that of the sky as protection: 'And We have made the heavens as a canopy well guarded (saqfan mahfudhan)' (Q 21:32). This used to be understood differently along the meaning of Qur'an verses 72:8 and 37:7, i.e. protection from devils; today many understand it to mean the atmosphere, the ozone, the magnetosphere etc.; in a century or two, who knows how it will be understood. But if reading it with an open mind and from a neutral perspective, then it is very difficult to see any scientific content there.

Taking these principles and applications (or examples) as the bases of one's approach to the Qur'an and science, it becomes readily evident that one cannot insist on a particular reading and explanation of specific verses, much less declare them to already contain scientific truths. A scientific exegesis of the Qur'an, however, could constitute one possible reading of the Text, informed and enlightened by knowledge gathered through science by 'men of understanding', a reading that is one among others, and an interpretation that only a high degree of knowledge of science could permit. Such an exegesis, no longer an *I'jaz*, then becomes nothing more nor less than a collection of individual, personal interpretations, some of which will be more convincing (to people) than others.

A large number of Qur'anic verses (of 'allegorical' nature, as stated in Q 3:7, the 'key phrase of all Qur'anic key phrases') then lend themselves to this approach. Among such verses one can cite those which have so impressed the proponents of the *I'jaz* theory, e.g. those dealing with the cosmos (Q 41:11 etc.) or those dealing with the development of the human embryo (Q 23:13–14, Q 39:6 etc.). The seven heavens could then be understood as the orbits of the seven planets of the ancients, the many galaxies of what we know of today's cosmos, the multi-verse theory if one is inclined towards it, or metaphysical 'worlds' beyond our universe. Such readings and interpretations would be for the individual to make, depending on his/her knowledge and insight; some would appear more 'correct' than others to sophisticated minds, but in no case would any of them be declared as miraculous statements containing scientific knowledge that preceded human discoveries by many centuries.

To sum up, I would like to emphasise the extent to which we have found the whole *I'jaz* theory to be flawed in the methodological principles it sets for itself in theory, in the methods it uses in practice and in the many examples I have given from the leading authors in the genre. I have tried to make a constructive distinction between *Tafsir 'Ilmiy*, the attempt to make use of scientific knowledge to bring new understandings to some of the many verses that describe nature, and *I'jaz 'Ilmiy*, which claims that some verses contain clear scientific facts that were discovered only recently; this I have shown to be a utopian idea and programme both in theory and in practice. Finally, I have proposed a new paradigm which can potentially erase all these conceptual flaws by bringing a new approach to our reading of the Qur'an, what I have called 'multiple, multi-level' readings, whereby various meanings are found depending on one's education and era; one may then still read some fact of nature in a verse without turning it into a miraculous claim of precedence.

summary and conclusions of part i

The first two major ideas that I have tried to emphasise in the previous five chapters, for western readers in particular, are the concept of God in Islam (and how that relates to science) and the extraordinary place and influence that the Qur'an occupies in the lives and the minds of Muslims. That strong emphasis aimed at several objectives: (a) to explain why the Muslim discourse on science and religion (and other social and political issues) tends to often be filled with references to the Qur'an; (b) to explain how the emphasis on the Qur'an led to a dangerous distortion and derailment of the discourse into a 'scientific miraculousness of the Qur'an'; (c) to show that for a credible and reasonable discourse on science and Islam and to have a chance to be well received by the public as well as the elite, it must at least ensure the Qur'anic acceptability (or non-objection), if not full compatibility, to the ideas being put forward; (d) to stress the fact that the Qur'an can be read and interpreted, in at least one of the multiple approaches, rationally, and it, more generally, can present us with a rational world view.

Indeed, one must first note that the Sacred Book repeatedly draws one's attention to the 'general predictability of the world's physical phenomena'; for example: 'The sun and the moon follow courses (exactly) computed; and the stars and the trees both prostrate in adoration; and the Firmament has He raised high, and He has set up the balance' (Q 55:5–7). That is why the Qur'an continually encourages people to observe, reflect, and search.

The second important idea that transpired from our investigation is that despite the Qur'an's many injunctions to humans to observe and reflect on the phenomena of nature and their relation to the Creator, one finds some difficulty in relating the concept of science (in the modern sense) to the Qur'anic discourse. Indeed, the concept of science in the modern sense cannot easily be found in the Qur'an or even in most of the classical Muslim heritage; rather, the concept of knowledge is developed. This distinction, more or less subtle, eludes a great number of Muslim thinkers and educators; indeed, the word *'ilm* is today routinely used for science, although it is quite certain that it originally stood for knowledge, in a much larger sense. This has led to strong disagreements between the traditionalists and the reformers regarding the possibility (or not) of building a case of Qur'anic basis for science, the latter being given several possible definitions, ranging from 'sacred science' to 'Islamic science', and even 'scientific *I'jaz*'.

The position that I have advocated first is the rejection of all extreme positions. Clearly the 'scientific knowledge' ('scientific miraculousness') in (of) the Qur'an is to be rejected for the variety of reasons that I have exposed in Chapter 5. Instead, I have emphasised and promoted a multiplicity of readings (with multilayered nuances and hints) of most, if not all, of the Qur'an, an approach that allows for an intelligent enlightenment of one's interpretation of various Qur'anic verses, using various tools, including scientific knowledge, at one's disposal. I have argued that this approach meshes well with that of some of the most intelligent scholars of Islam, from Ibn Rushd (Averroes) to Mohamed Talbi (see earlier quotes that strongly support this stance).

We have also seen how Mohamad Shahrour has broken with traditional rules of interpretation, not only constructing a very original and radical new approach to the Holy Book but, moreover, opening the gates of interpretation to anyone who has the intellectual capacity to do so, including non-Muslims and non-Arabic speakers. Furthermore, he views the integration of modern knowledge into our reading of the Qur'an as one way to expand some of the Book's potentials as well as to help unite the given society and harmonise its knowledge with God's truths.

To summarise, while the Qur'an cannot be turned into an encyclopedia of any sort, least of all of science, one must keep in mind the fact that if the Qur'an is to be taken seriously and respectfully, one must uphold the Rushdian (Averroes's) principle of no-possible-conflict (between the word of God and the work of God) and his hermeneutical prescription. In practice this principle can be turned into a no-objection or no-opposition approach, whereby one can convince the Muslim public of a given idea (say the theory of biological evolution), not by proving that it can be found in the Qur'an but rather by showing that at least one intelligent reading and interpretation of various passages of the Holy Book is fully consistent with that theory.

In parallel to the attempt to develop a reasonable approach to the Qur'an (its place with respect to science and human knowledge and its interpretation or hermeneutics), I have laboured to establish a clear understanding of science and its philosophy. From my experience with students and professors in the Arab/Muslim world, I have inferred a strong need to emphasise the various aspects of the scientific enterprise, particularly the criterion of falsifiability. Science is not a mechanical process; it starts with principles, which are by nature unproven and often unprovable, it relies on individual intuitions and personal inclinations as much as collective consensus, and follows periods of paradigms and moments of 'revolution'. The concept of metaphysical basis of science has also now become standard in the discourse. With that understanding, it becomes clear that the materialistic nature of modern science is a metaphysical choice that has been made but is not part and parcel of the scientific enterprise. Indeed, past cultures (including of course the Islamic civilization in its classical 'golden' era) did conduct plenty of science, a large amount of which was rigorous and systematic, but it was largely theistic in its foundations and outlook. Attempting to develop science with a theistic cloak is not necessarily going to destroy its pillars and its beautiful high constructs and achievements, if one does it with a full understanding of its various aspects, such as to distinguish the metaphysical facets from the methodological parts. That is what I have tried to show and argue in Chapter 3.

In another chapter (Chapter 4) devoted to science, I examined the attempts that have been made and proposed to construct an Islamic science, a dual ('quantum') combination of Islamic principles with modern science's methods and results. Several schools of thought were identified in this regard: from the highly mystical school (Nasr's sacred approach to knowledge and science) to the universalist and conventionalist one (Abdus Salam) and the secular school (Hoodbhoy). I also briefly reviewed the Islamization of knowledge/science program of al-Faruqi and al-'Alwani, but found it to be rather flawed. We were then left with Sardar's (the Ijmalis') school of Islamic science, which showed some promise one or two decades ago but then lost momentum – as the members of its group dispersed. Despite declaring this school closest to my position, I found much criticism to level at the ideas of Sardar, particularly its exaggerated critique of modern science, its overemphasis on ethics and the utilitarian function of science, and especially its idea of 'democratizing' science, that is, allowing every member of society to participate in the scientific process (e.g. collecting data) and the debates on science. In the end, perhaps the most reasonable and moderate position concerning the repositioning of science and Islam is that of Golshani, who advocates theistic science, one that is fully conscious of God, upholds His objectives (as stated in the Qur'an and other Holy books), but remains as rigorous as possible. That proposal, if not quite detailed and clear in the way it can be carried, has the merit of uniting Muslim efforts with those of theistic philosophers from other cultures.

Equipped with all these principles and the general understanding of Islam and science, we are now ready to address major topics (cosmology, design, anthropic principle, and evolution) in full and others (miracles and divine action) in preliminary fashion.

part ii

Islam and contemporary science issues

6

Islam and cosmology

'It is not without reason that Hawking said: "It is difficult to discuss the beginning of the universe without mentioning the concept of God". Fred Hoyle wrote: "I have always thought it curious that, while most scientists claim to eschew religion, it actually dominates their thoughts more than it does the clergy". Perhaps when C. S. Lewis warned that "a young man who wishes to remain a sound atheist can't be too careful of his reading", the reading to be strictly avoided should include science books.'

Kitty Ferguson[1]

'Islamic cosmological beliefs are rooted in the Qur'an itself, which deals extensively with this issue. [. . .] The Qur'anic perspective on the creation of the physical cosmos can be summarized as follows: the cosmos was created by God for a purpose. After creating the cosmos and all that it contains, God did not leave it to itself; in fact, the entire created order is perpetually sustained by God; without this sustenance it could not exist.'

Muzaffar Iqbal[2]

cosmology in today's muslim culture

When I visited Ziauddin Sardar in his London suburb home in the spring of 2007, one of the questions I asked him was: what is 'Islamic cosmology'? Without thinking too much about it, he replied: 'Islamic cosmology is what science tells us with certainty about the universe, for that is the result of what God asked us to discover'. I did not necessarily disagree with him, but I wanted to challenge him a bit and probe further, so I retorted: 'But in your writings on science, you have often insisted that if science were constructed on non-western foundations, for example if time were cyclical and not linear, or if logic were four-fold instead of binary (yes/no), our descriptions of nature would be vastly different'. He then compromised a little: 'Any foundations can give us some kind of science, but the results must conform to the observed reality'. We were then in full agreement.

Cosmology is an interesting branch of science, perhaps the only one where thinkers often expound their 'views' freely, building on religious and philosophical principles at least as much as physics and astronomy. This is probably due to the fact that for a long time, that is, until the past few decades, cosmology had very little solid data and was the most speculative branch of science. This freewheeling treatment of cosmology was/is also due to the fact that holy scriptures have all expounded a religious view and description of the cosmos, its creation, its content, its purpose and sometimes its end.

The amount of cosmological data has increased exponentially in recent years, to the point where (to give just one quick example) our knowledge of the age of the universe (since its creation) has improved from the '10 to 20 billion years' estimate that was commonly given to students a decade ago to 13.7 billion years (plus or minus 0.2 billion years) that we now teach, since the Wilkinson Microwave Anisotropy Probe (WMAP) results of 2003. Still, most if not all the books on cosmology that have been written by Arab authors in recent years present cosmology not as a branch of astronomy but practically as a branch of Qur'anic exegesis. Let me mention a few titles:

- *The Qur'an for Astronomy and Earth Exploration from Space*, S. Waqar Ahmed Husaini, 1999 (3rd edition)
- *The Universe and Its Secrets in the Holy Qur'an*, Prof. Hamid M. al-Naimiy, 2000
- *How Creation Was Started*, Muhammad bin Ali al-Mujahid, 2001

- *Astronomical Sciences in the Holy Qur'an*, Ibrahim Hilmy al-Ghury, 2002
- *The Universe, A Traditional and Modern Look*, A. Abdelamir al-Mumin, 2006
- *The Universe and the Holy Qur'an*, Mohammed-Ali Hassan al-Hilly, 2006
- *The Structure of the Universe and the Destiny of Man – A Critique of the Big Bang Theory; Amazing Facts in the Cosmic and Religious Sciences*, Hichem Taleb, 2006.

The typical approach that authors of such books adopt is the following: a Qur'anic verse of 'cosmic content' is picked (e.g. 'Allah is the One who created seven heavens, and of the earth the like of them; the decree continues to descend among them, that you may know that Allah has power over all things and that Allah indeed encompasses all things in (His) knowledge' (Q 65:12), then the author starts to present his cosmological views, which are usually a mixture of his knowledge of 'facts' from modern science, his understanding of the Qur'anic verse from linguistic and exegetical explorations, and his 'ideological' views (for example, the author may prefer a static universe to an evolving one, or vice-versa).

I will give shortly a few examples of this Qur'anic cosmology trend, but I want to first stress that while this is an equally flawed approach, it is quite different from that of *I'jaz*. In that approach, authors were interested in proving the precedence of the Qur'anic 'scientific' content over the results of modern science. In the present trend, writers are simply convinced that one must start not only from the Qur'an but also fully construct a knowledge of the universe from the interpretation of some verses, no matter how few or many, and how general or specific, they may be.

Here are a few examples from the above-mentioned literature:

- Starting from several Qur'anic verses (e.g. 2:29, 11:7, 23:17, 31:10, 67:3 etc.), al-Hilly constructs a scenario for creation whereby 'before God created the universe, His throne was upon the water, i.e. water vapor in space, for at that time there were no ethereal heavens, and when He created them gradually one by one, He let His throne be borne over the ethereal heavens'[3]. He also tells us that '"heavens" here means the gaseous layers'[4] and that '[God] fashioned seven gaseous layers from one layer of smoke'[5]. al-Hilly pays no regard to modern scientific knowledge; in fact at some point he tells us that '"created the earth in two days" means in two thousand years, for one day of God's equals one thousand of our years' (adopting a literal reading of Qur'anic verse 22:47). I actually just stopped reading when I came across statements like: 'the cause of gravitation is the heat' and 'the movement of the particles in the core of the earth causes the heat and the gravitation, and this leads to the rotation of the earth around its axis'[6]. (Recall that this book was published in 2006!)

- Hichem Taleb's book is another marvel of the genre. Hard-covered, with 700+ pages and dozens of images and diagrams, it not only looks impressive but can also fool any reader if it were not for the word 'critique' (of the Big Bang theory) in its subtitle, which sounds the alarm. The moment one reads the introduction and realises that the author is a journalist with no scientific background, one begins to imagine what's in store. Here are a few examples:

1. The Big Bang event, which the author (and every other writer on the topic) reads in the verse Q 21:30 ('Do not the Unbelievers see that the heavens and the earth were joined together before we clove them asunder? And We made from water every living thing. Will they then not believe?') is explained as splitting of the earth from the heavens, at which point rain (water) starts pouring and vegetation starts growing on the earth. The scenario then becomes amazing when winds are mixed up with cosmic 'smoke' etc[7].

2. The six 'days' of creation, which all Muslim writers have come to interpret as 'epochs', are here computed in various ways to yield respectively[8] 6,000, 12,000, 50,000, 155,520 years etc.

   The 'seven heavens' are given a specific and detailed description[9] that starts from the hadiths, which relate the Prophet's 'ascension' (*mi'raj*) to the heavens after his famous 'night journey' (*isra'*) from Mecca to Jerusalem. It is important to note that he builds on a religious text (hadiths) and from what may very well be a mystical, spiritual event to construct a cosmological picture of the universe, a mixture of physical parts and non-physical aspects. And so readers should not be surprised when the author, in the midst of his 'scientific' descriptions, declares Mecca to be the *physical* centre of the world[10]. (I have personally heard university professors at scientific conferences claim that Mecca is the magnetic centre of the world, and there is a TV interview of an Egyptian researcher claiming this on the web[11]).

- In *The Universe, a Traditional and Modern Look*, A. Abdelamir al-Mumin also starts from the Qur'an (e.g. Q 32:5, 21:30, 41:11) but quickly moves to scientific sources (mainly Stephen Hawking's popular book *A Brief History of Time*) to describe the birth of the universe in a way which concords with the verses he cites. And though we do not encounter any astounding pronouncements as given in the previous books, one quickly notes the author's limited knowledge of modern cosmology both in his usage of terms like 'cosmic egg' (by which he means the old 'primordial atom') and in the sources he draws from. al-Mumin also, like many other Muslim writers, wrongly believes that current cosmological data (the existence of much dark matter) points to our universe being 'closed'[12].

Perhaps the astute and rigorous reader will object to the above examples on grounds that they were not written by academics or scientists, and so it is no surprise that such lack of up to date scientific knowledge and methodology would be found so easily. That is correct; however, I wanted to give an idea of the kind of 'science literature' that gets published in the Arab/Muslim world nowadays; indeed, the books I have mentioned above were all published in the past few years, and I have not biased the discussion by selecting only the awful ones.

Still, one should discourse with the academics who are expected to deliver accurate information and expound a rigorous methodology. For that purpose I would like to devote several paragraphs to the aforementioned book by Hamid M. al-Naimiy, who holds a PhD in astronomy from the University of Manchester (UK) and is currently the dean of the college of arts and sciences at the University of Sharjah (UAE) after having been chairman of the Institute of Astronomy and Space Sciences at Al al-Bayt University (Jordan) and director of the Center for Astronomy and Space Science Research in Iraq.

In his book *The Universe and its Secrets in the Holy Qur'an*[13] (published in 2000), Professor al-Naimiy tries to be as cautious as possible not to make any scientifically erroneous statements, while at the same time keeping to the widely held orthodox religious views (for example, the earth was created first[14], as may appear from first sight in the Qur'an). Unfortunately, his attempts to achieve some harmony between the two views fall far short from that goal. Here are a few brief examples of his work:

- Referring to the Qur'anic verse 'Verily, when He intends a thing, His command is to say to it "be", and it is!' (36:82), al-Naimiy writes the following:

> Clearly the concept of 'thing' goes beyond any appearance; that is, the concept of 'be' comprises the photon, the boson, the gluon, the graviton, and U, where U refers to the unknown in a global way (in cases of requirement) and in partial ways (in cases of open choices), so U is what science has not yet uncovered, in matters of energy (forms and units) or motion and speed, though space-time is inherently among the known and the available . . . [15]

- On the Qur'anic verse 27:40 (the story of Solomon and the throne of Sheba's queen), the author writes: 'And that is how the greatest 'transport' was achieved, with the highest speed, to which no applications of Macro-Psycho-Kinesis[16] or what is known as parapsychology can come close, whether in the modern world or in the whole history of humanity, because the speed here is tied to the mass of the queen's throne, by way of Einstein's laws'[17].

- Referring to the Qur'anic verse 21:104 (the folding of the heavens at the end of time), he writes the following:

> And from a purely philosophical viewpoint, this fact is absolutely true and satisfies the third law known as 'Law of Excluded Middle'[18], that is no part of it can be belittled and nothing can come ahead of it in level or in quality, and nothing can lead to it; it is a fact that stands on its own and is produced as the Text stipulates by an order from its Lord, that is by the letter N [We] in; *Natwi'* [We scroll]. . . [19].

- al-Naimiy writes the following two math-like statements (first in Arabic, then in English):
'Universe ⊂ All-Mighty (Divine Will) [the symbol ⊂ indicates inclusion]
R [radius of] Universe ⊂ R of All-Mighty (Divine Will)'[20].

On the author's methodology, we must note two important characteristics of his approach: (1) He starts from the Qur'an in constructing his cosmology and often refers to classical exegeses before presenting his views; (2) he subscribes to the *I'jaz* theory and even goes beyond it to claim and cite Qur'anic 'numerical miracles', some of which he credits himself for their 'discovery'. Indeed, in explaining his epistemology, al-Naimiy tells us very clearly that 'attempting to reach truths by reason does not always succeed. [. . .] however, the indications obtained from the Mighty Qur'anic Text are the best guidance for research and inference *even in the field of physical science,* so Glory be to God'[21] (emphasis added). As to his adoption of *I'jaz*, one can cite his claim[22] that the verse 'You see no incongruity in the creation of the Beneficent Allah; so look again, can you see any disorder?' (Q 67:3) is a direct reference to the Cosmological Principle (that the universe is homogeneous and isotropic).

I can also briefly mention examples of the author's subscription to the numerical *I'jaz* (to which he devotes many pages): (a) He finds miracle[23] in the equality between the two ratios: solar year over lunar (synodic) month and lunar (sidereal) month over 'the difference between synodic month and sidereal month'; (b) he finds it remarkable[24] that the speed of light (300,000 km/s), the speed of the sun with respect to the centre of the galaxy (300 km/s) and the speed of the earth around the sun (30 km/s), are multiples of the numeral 3 (and tens); however, anyone can check that, for example, the sun's speed in the galaxy is about 220 km/s. Furthermore, one may as why it would be so impressive if these quantities shared the numeral 3? al-Naimiy regards this as an important relation because the word Allah contains only three (different) letters (A, L, H)![25] In fact, the author determines numerical quantities of certain words or expressions in the Qur'an from the values that the ancients sometimes assigned to the alphabets (A = 1, B = 2, J = 3, D = 4, H = 5 etc.) and declares them miraculous; for instance, *kullu shay'* (every thing) would have a value of 360, which corresponds to the number of degrees of a complete circle, so this means that the Qur'an was referring to God 'encompassing' (*muhit*) everything in a circular way.

But as we are discussing cosmology in this chapter, the most striking example I must bring up from al-Naimiy's book is his treatment of the structure of the universe: the seven heavens, the seven earths and God's throne. He begins by quoting hadiths, which presumably give the distance between the earth and the heaven(s) as the equivalent of 500 years of human walk. And though he does note that classical Muslim scholars have declared at least some of those hadiths to be 'weak' (unreliable), he proceeds to compute the distances to various heavens and earths. He estimates the distance between the earth and the lowest heaven (which is the same as the distance between two successive heavens) at 58.4 billion km (note that this is but a fraction of the distance from the earth to the nearest star and totally negligible compared to the sizes of the galaxy and of the universe at large). Then al-Naimiy recalls that he did not take into account the 'thickness' of each heaven, so he adjusts his estimate to find the distance from the earth to God's throne, which is just above the seventh heaven[26]: 7.665×10^{12} km! But he realises that we have observations to the farthest galaxies, which put them at 10^{23} km, so he constructs for us the following diagram[27] of the universe:

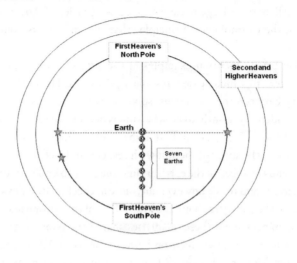

Then he notes that this would make the universe rather flat, because the vertical dimension (to the Divine throne) is billions of times shorter than the horizontal dimension (to the galaxies). So he modifies his picture to the following one[28], where he imposes the observational value of 10^{23} km as the size of the universe (thus discarding the hadiths he started with).

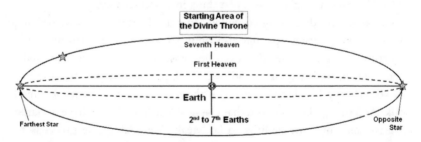

And this leads him to the final question: Are the seven heavens (or at least the six 'external' ones) inside or outside the universe? He concludes that it could be either way, and we do not have definite knowledge to decide on the matter.

Similarly, al-Naggar cites two verses referring to seven heavens one above another[29] and simply concludes: 'It becomes clear from these two verses that the seven heavens are superposed in concentric manner around a common center'[30], which he explains is the centre of the earth. In fact, he fuses this with his interpretation of a verse on 'seven earths' (Q 65:12) to construct

(graphically and textually) a picture of the earth with seven layers surrounded by seven heavenly 'shells'. We are evidently back to Aristotelian–Ptolemaic cosmology (see the diagram of standard medieval cosmology), except that Aristotle's four elements have been replaced by seven earthly layers. How far back we have gone!

general principles of Islamic cosmology

There are strong disagreements on when (roughly) full Islamic conceptions of the universe started to appear and what fueled their formulation. Some scholars have argued that a Qur'anic cosmology was formulated very early and was in fact perhaps the earliest science to emerge within Islam; others have insisted that it was the Muslim philosophers (*falasifah*) who, after having absorbed the Greek heritage, attempted to formulate a solid (rational and astronomically based) conception of the cosmos in a way that would be compatible with Islam.

According to both Seyyed H. Nasr and Muzaffar Iqbal, the cosmological science first emerged in the realm of the Islamic revelation's description of the cosmos[31]. Iqbal refers to the 'Radiant Cosmography' (*al-hay'a as-saniya*)[i] as a science that appeared from the earliest days of the Prophet and his Companions; it was derived from Qur'anic verses and constructed

---

[i] Iqbal refers to a book by that title written by the classical Muslim scholar and exegete as-Suyuti (d. 1505), which 'summarizes eight centuries of scholarship on this topic'.

by way of exegetical exercises. He struggles to show that this approach to cosmology originated not only early and from the pure sources of Islam but also remained an important 'theory', which 'became the counterweight to the Aristotelian cosmology that came into Islamic thought in the translation movement'[32] (he makes it a point to insist that the philosophers' cosmology was external and Aristotelian, but not Islamic). Iqbal justifies his high view of this radiant cosmography by pointing to the many Qur'anic verses that contain 'cosmological data' and to some hadiths that deal with the creation events (from water vapour etc.). The Qur'an does contain many references to cosmic objects, perhaps even a cosmography; indeed, it speaks of seven heavens and earths; it also refers to a Divine chair (*kursi*) and throne (*'arsh*), as well as to a cosmic mountain (*qaf*), not to mention planets and stars. In many verses, one finds a clear connection between cosmic and metaphysical symbols pertaining to God[33] (as in the verse of the *kursi*[34] and that of the Light[35]). Indeed, the Ikhwan as-Safa group, whose cosmology we shall review later in detail, associate the Chair with the firmament of stars, while the Throne is nothing else but the ninth sphere/heaven that separates the realm of the Divine from the cosmos of matter.

I shall devote a long section to the cosmologies constructed by the philosophers (and other thinkers from outside the orthodox mould); let me here first summarise the main cosmological principles that one can infer from the Qur'an[36]:

- The universe was created by God, who has absolute and exclusive power of creation; His act of creation was one of bounty and mercy.
- The universe was created with a purpose.
- It is continuously sustained by God.
- The cosmos is characterised by wholeness, order and harmony between all its elements and events; Iqbal calls this the 'Theory of Balance', which he ascribes to Jabir ibn Hayyan (ninth-century Muslim alchemist).
- Cosmic time and chronologies in the Qur'an are qualitative; 'days', for example, are not to be given specific lengths.

As I have stressed in Chapter 2, one of the surprises that greet the reader who delves into the Qur'an for the first time is the large number of references to nature and physical phenomena. Indeed, many commentators have revelled in the fact that compared to some 250 verses of religious injunctions, one finds over 750 verses (about one-eighth of the Book) encouraging the reader to observe and reflect upon nature (God's creation) and to make the best utilisation of one's mind to understand it. Some writers (e.g. Iqbal) have pointed to this striking emphasis as the core reason for the development of various sciences in Islam.

Another important remark I should make about cosmology in the Qur'an is the fact that the specific word by which the Arabic language refers to the universe (*kawn*) is never encountered in the Qur'an. Instead, somewhat equivalent – but subtly different – words are used: *khalq* (creation), e.g. 'It is Allah Who has created the heavens and the earth, and all between them, in six Days, and is firmly established on the Throne . . . ' (Q 32:4), *as-sama* (*wal ardh*) (heaven and earth), e.g. 'We have built the Firmament with might: And We indeed Have power to expand (the creation? the firmament?)' (Q 51:47), or *as-samawat* (the heavens), e.g. 'Do not the Unbelievers see that the heavens and the earth were joined together (as one Unit of Creation), before We clove them asunder?' (Q 21:30). Note, incidentally, that the words *sama* and *samawat* occur some 300 times in the Qur'an. The concept of 'universe' is, in fact, inferred from the Qur'an essentially from the following important statement which is made over and over: 'Verily, when He intends a thing, His Command is "be", and it is!' (Q 36:82), where the 'order' (the word 'Be') in its Arabic form (*kun*) is very closely related to the word-concept of universe, *kawn*, which, in fact, is synonymous to 'being'.

The Qur'an, like the Bible and other religious/divine Books, raises and addresses other cosmic questions: Why is all this creation around, and what is the finality of it all? The answers given, however, are often of religious or metaphysical nature.

creation of the universe in the Qur'an

Muslims and their religious scholars, particularly the Qur'an's interpreters, today take it for granted that the Qur'an refers to the creation of the universe as an *ex-nihilo* act (by God). In doing so, they base themselves on the latest scientific knowledge in order to reconstruct the story of the creation from the verses that deal with the topic, verses that are in fact scattered in the Book, unlike the Bible's Genesis chapter. The Qur'anic story can be summarised by the following verses (note their distant places in the Qur'an):

> Your Guardian Lord is Allah, Who created the heavens and the earth in six
> Days, then He settled Himself on the Throne; he draweth the night as a veil
> o'er the day, each seeking the other in rapid succession; and the sun, the
> moon, and the stars, (all) are subservient by His command. Verily, His are
> the Creation and the Command; Blessed Be Allah, the Cherisher and
> Sustainer of the Worlds! (Q 7:54).

Say: is it that ye deny Him Who created the earth in two Days? And do ye
join equals with Him? He is the Lord of (all) the Worlds. He set on the (earth)
mountains standing firm, high above it, and bestowed blessings on the earth,
and measured therein its sustenance in four Days, alike for (all) who ask.
Then He turned to the sky, and it had been (as) smoke: He said to it and to
the earth: 'Come ye together, Willingly or unwillingly'. They said: 'We come
(together), in willing obedience'. So He completed them as seven firmaments
in two Days, and He assigned to each heaven its duty and command. And We
adorned the lower heaven with lights, and (provided it) with guard. Such is
the Decree of (Him) The Exalted in Might, full of knowledge (Q 41:9–12).

It is He Who hath created for you all things that are on earth; then He turned
to the heaven and made them into seven firmaments. And of all things He
hath perfect knowledge (Q 2:29).

Do not the Unbelievers see that the heavens and the earth were joined
together, before We clove them asunder? (Q 21:30).

The mere fact that the universe in the Qur'an was created in six 'days'
seems to imply a beginning. However, as some of the great Islamic scholars
of medieval times (Ibn Rushd, Ibn Sina and many others) showed, under
the heavy influence of Greek thought, one can certainly construct an Islamic cosmology that rests on the assumption or belief of an infinitely old
creation.

It is perhaps not difficult to realise, simply by reading the above sample of
'cosmological' Qur'anic verses that the problem resides in how to interpret
them. One of the most telling examples about this difficulty to interpret
the Qur'anic verses is the fact that for a very long time Muslim scholars
have debated whether the Qur'an states that earth was created before or
afterwards the heavens. Two seemingly contradictory verses can be found
in this regard: Q 2:29 (given above) and 'Are ye the harder to create, or is
the heaven that He built? He raised the height thereof and ordered it; And
He made dark the night thereof, and He brought forth the morn thereof.
And after that He spread the earth' (Q 79:27–30).

(As we saw in Chapter 5, al-Naggar and al-Taftanazi still insist that the
earth came first.)

The Qur'an can also be queried for another sub-topic of cosmology, that of
eschatology. It is indeed of interest to investigate the Qur'an's verses about
the end of times, even though one quickly finds that verses to be heavily

cloaked in metaphysics. Several verses have been brought up and discussed in this context:

> The Day when We shall roll up the heavens as a recorder rolleth up a written scroll. As We began the first creation, We shall repeat it. (It is) a promise (binding) upon Us. Lo! We are to perform it (Q 21:104).

> O mankind! Fear your Lord! For the convulsion (earthquake) of the Hour (of Judgment) will be a terrible thing! (Q 22:1).

One may, of course, raise many scientific objections to the idea that one could have a 'global' earthquake that would destroy the earth and humanity in a short moment. Doesn't one need a catastrophic cause of astronomical origin, perhaps a large comet or some other large(r) celestial object crashing on the earth? But then, doesn't man (now) have the means to detect such an object and thus predict the event many years in advance and at least allow a few groups of humans to escape (temporarily in the present time and perhaps permanently as soon as we can establish colonies on the moon or other planets)? On this argument, the Qur'an would seem to rule out this astronomical catastrophe scenario (for instance in Q 7:187[37]).

Secondly, can one imagine a judgment day that would destroy earth and humans and leave the rest of the universe intact? Would there be different judgment days for different species in the universe (assuming we are not alone)?

However, as pointed out earlier, Islamic scholars and Qur'an interpreters are often heavily influenced by the knowledge of their times, and so we now find references to the Big Crunch (the universe collapsing back on itself into a singularity, assuming its mass-energy density is larger than the critical value known to theoretical cosmologists) in relation to the Qur'anic verse 21:104 (given above). Note that this interpretation is subject to a similar counterargument, namely the fact that this would not occur before tens of billions of year (when the Qur'an promises a sudden Day of Judgment).

These examples clearly show why one could not deduce a clear cosmology from the Qur'an and why most scholars resorted (in medieval times) to the Greek models of the world and (in recent times) to the scientific theories.

the traditionalist cosmology

Seyyed Hossein Nasr and his followers insist that cosmology must not be regarded as a pure scientific discipline that deals with the physical aspects of the universe. Nasr insists that most, if not all, previous civilizations saw the latter as a unified realm where physical and non-physical objects and realities coexisted in a holistic structure. He writes: 'ancient cosmologies [...] are sciences whose central object is to show the unicity of all that exists. [...] This unicity is particularly important in Islam where the idea of Unicity (*al-Tawhid*) overshadows all others and remains at every level of Islamic civilization the most basic principle upon which all else depends'[38]. Elsewhere he has stressed that

> the meaning of the term cosmology in the Islamic, or other traditional, contexts differs profoundly from the meaning given to it in the context of modern science. [...] Traditional cosmologies [including the Islamic visions] deal with cosmic reality in its totality including the intelligible or angelic, [and] the metaphysical principles [pertaining] to the cosmic realm[39].

Nasr, who has studied classical (medieval) Islamic cosmology in great detail (see his seminal book *Introduction to Islamic Cosmological Doctrines*, written in 1963) comes to the interesting, perhaps surprising, conclusion that despite the differences in perspective and emphasis, the conceptions of nature found with most Muslim writers 'seem like so many exegeses of the same cosmic text'[40]. One immediately recognises this analogy; indeed the Qur'an repeatedly refers to nature as a book which contains not only *ayahs* (verses) like the Book itself but in fact can be considered as its macrocosmic counterpart.

The cosmos, Nasr insists, is not (in the Islamic conception) the entity containing everything (as it is defined in modern physical science), but rather only a small parcel of a greater cosmic ensemble, which contains it[41]. As such, the study of the cosmos or nature is not an end in itself but a means to an end, the goal being 'to gain knowledge of the Creator whose wisdom is reflected in His creation in such a way that the study of this reflected wisdom leads to the knowledge of the Creator himself'[42].

William Chittick, who fully subscribes to Nasr's school of (Islamic) perennial philosophy, presents the main characteristics of the Islamic traditionalist cosmology as (a) 'The cosmos is a grand hierarchy in which every level of reality is present simultaneously, without regard to temporal succession';

(b) 'This hierarchical cosmos is divided into a basic world, one visible and one invisible | . . . The invisible realm is closer to God and more real than the visible world'; (c) 'Human beings are unique in the cosmos. . . . As a result, everything found in the external universe is found, in essence and reality, in the primordial selfhood known as *fitra*'[43].

---

two worlds – an important distinction rarely appreciated

Let me raise an important point which seems to be often overlooked or not substantially appreciated by readers of the Qur'an and believers of the Islamic world view. The Qur'an makes an important methodological distinction between the 'observable (i.e. physical) world' (*'alam al-shahadah*) and the 'unseen or inaccessible (i.e. metaphysical) world' (*'alam al-ghayb*). The former, defined as containing all material beings and things, is therefore the universe, as scientists define it. The latter, defined as all matters relating to God, angels, paradise, hell, the soul (however one may define it) etc., can only be discussed on purely religious bases, usually inferred from religious texts (holy scriptures, prophetic statements and explanations etc.).

The distinction between these two worlds may appear as an obvious one, but when the topic is cosmology, which is discussed in the realm of a religious text or religious scholars' interpretations, one often finds crossings between the two and inferences that rational thinkers will immediately object to and dismiss. The traditional Islamic cosmology/cosmogony, however, rejects not only such a separation but also makes a point of unifying the two realms.

Several examples will be given in the next section, but two of the most recent ones of this 'unified' viewpoint are Muzaffar Iqbal's exposition of cosmology in the Islamic culture and William Chittick's attempt to present an alternative to the current 'western' approach to cosmology and knowledge (more generally). Insisting that what is meant by cosmology today is entirely different from what was traditionally meant in the Islamic tradition, Iqbal declares that 'the Qur'anic cosmos is not merely physical, made up of stars, planets and other physical entities; it also encompasses a spiritual cosmos populated by nonphysical entities'[44]. I would not object to such a description if the two 'cosmoses' were treated as separate realms, but in Iqbal's, Nasr's and Chittick's doctrines, there is no separation. Indeed, Chittick insists that 'there is no place for the stark dualisms that characterize so much of modern thought'[45]. In his view even rocks are not 'matter'; he sees the cosmos and the soul as intertwined in an 'organismic' relation[46] within an 'anthropocosmic' vision that does not distinguish between object and subject[47].

medieval Islamic cosmological doctrines

As the early Muslim nation came in direct contact with other civilizations and fully absorbed their heritage, intellectual life and thought matured, and thinkers, who expounded highly sophisticated doctrines and conceptions that took full account of ancient philosophical traditions, emerged. Thus appeared *falsafah* (Islamic philosophy), *kalam* (Islamic theology), *tasawwuf* (mysticism, Sufism) and the esoteric doctrines of *fayd* (emanation), spanning a very wide range of tendencies and influences, not to mention geographical areas and historical epochs.

The spectrum of Islamic cosmologies is summarised by Nasr in the following words:

> In the writings of the Ikhwan as-Safa we encounter a wealth of neo-Pythagorean and hermetic symbols . . . With al-Biruni we meet one of the most eminent of Muslim mathematicians and astronomers and at the same time the independent scholar and historian in search of the knowledge of other lands and epochs. As for Ibn Sina, he was undoubtedly the greatest of the Muslim peripatetics and in his writings one can discover the peripatetic philosophy of nature in its most lucid and thorough form . . . [with] many elements in common with the views of the Sufis on nature and on the role of cosmology in the journey of the Gnostic toward illumination[48].

The main Islamic cosmological theories can be divided along philosophical and theological tendencies: the Hellenistic philosophers, the Ikhwan as-Safa group, the independent scholars/scientists and the Sufis.

Ibn Sina, Averroes and the philosophers (falasifah)

Many Muslim thinkers, philosophers in particular, considered themselves free to read the Qur'an without any established *a priori* cosmological or religious conception. Then they relied heavily on ideas from other ancient traditions, particularly Greek philosophy. Let me briefly review the main cosmological views of the *falasifah*.

al-Kindi (800–73) adhered to the doctrine of creation *ex nihilo*, though most of the later Muslim philosophers came to reject it. al-Kindi did, however, adopt many of the old cosmological 'proofs' of the existence of God, beginning with the Prime Mover argument, as there had to be in his view an unmoved, unchangeable, perfect, indestructible mover[49]. This led him to the conclusion that it was God's power and prerogative to start a universe from nothing.

al-Farabi (870–950), however, stayed true to Aristotle: He did not believe that God had 'suddenly' decided to create the world, as that would have involved the eternal and static God in unseemly change[50]. al-Farabi, more heavily influenced by the Greeks, adopted their 'chain of being', which proceeds down from the One in ten successive emanations or 'intellects', each of which generates one of the Ptolemaic spheres: the outer heavens, the sphere of the fixed stars, the spheres of Saturn, Jupiter, Mars, the Sun, Venus, Mercury and the Moon.

Ibn Sina (Avicenna, 981–1037) is considered as one of the greatest and most refined of the Muslim thinkers. A prodigy, he mastered all the sciences of his time, from medicine to astronomy and philosophy, starting from Aristotle's metaphysics and graduating with al-Farabi's works, which he tried to harmonise more fully with Islam. His total acceptance of Revelation as the highest level of knowledge, contrary to the philosophers' adoration of reason, made him lean gradually towards Sufism without, however, neglecting the intellect. He too attempted to produce rational proofs of the existence of God, strongly influenced by Aristotle's thinking.

Ibn Sina divided the process of creation into two synchronous operations: self-awareness and self-discovery of the Divine. He further distinguished between four kinds of creations:

- *Ibdath*: Creation of nature's creatures, both temporary and eternal.
- *Ibda'*: Creation – without intermediary – of eternal non-decaying creatures;
- *Khalq*: Creation through other agents.
- *Takween* (formation): Creation through worldly, decaying and temporary agents.

He also classified creatures into the following four categories:

- Angels, which are creatures that are both moved by God and made to move others
- Cosmic souls, which act as the intermediaries in the moving actions
- Heavenly bodies
- Earthly objects

His concept of the cosmos was similar to that of Greeks, except that he considered nine spheres instead of the usual eight (one for the earth and seven for the planets), because one is needed (in his scheme) to separate the domain of the Divine from the physical spaces of the cosmos; the ninth sphere was supposed to be completely empty. Ibn Sina further imagined that each sphere must be controlled by one intellect commanding a group of angels or souls, the whole collection of intellects, angels and souls being under the prime intellect's command. The total fusion of the physical and metaphysical worlds seemed to him not only plausible but also perfect.

Ibn Rushd, although heavily influenced by Aristotle, remained true to the spirit – but not the letter – of the Islamic/Qur'anic teachings. He subscribed to the doctrine of the eternal cosmos, but insisted that the universe is and has remained in constant change or evolution, as God continues His creation. In his view, the Creator is the true Old/Eternal Being; He imparts change to His creation through different media (intellects, angels and souls), as had been fully developed by Ibn Sina and the Ikhwan. Ibn Rushd, however, refused the doctrine of emanation with its consequences, the top-down theory of creation as presented by Ibn Sina, although the concept itself goes back to Plato. On the other hand, he upheld the Aristotelian theory of final causes, which states that the reason for any phenomenon or event must be found in the goal that any being is attempting to reach; this implied an upward evolution rather than a top-down process in nature.

Averroes introduced an interesting new twist to the old concept of unity of the cosmos; he considered it from the viewpoint of change (or lack thereof), concluding that as content and form cannot be dissociated, and both are eternal, the cosmos must be static, even if it evolves. (Modern cosmologists will recognise in this viewpoint the stationary universe theory with its various features, as was proposed by Hoyle and others in the mid-twentieth century.)

Ikhwan as-Safa

The Ikhwan as-Safa (literally the Brethren of Purity) was an esoteric society that appeared in Basra (Iraq) in the tenth century (known as the 'Shii century') and is widely assumed to be an offshoot of Ismailism. Its members dedicated themselves to the pursuit of science, particularly natural and cosmic topics, as well as to political action. Their *rasa'il* (epistles or letters) became an encyclopedia of philosophies and sciences of their times, and their popularity made them spread far and wide. The Ikhwan represented a complex school of thought that has been classified sometimes as rationalists, as peripatetics and other times as 'pre-Sufis'.

In their cosmology, they subscribed to the Platonic doctrine of emanation rather than to the apparent Qur'anic doctrine of creation *ex nihilo*[51]. The Ikhwan believed in the organic conception of the cosmos, with the added Sufi shroud, which makes all of the existence a single being. Furthermore, the Ikhwan subscribed to a numerological system for the universe, starting from 1, which to them represented 'existence' or 'being', and ending with 0, which represented infinity or the divine essence. Moreover, they set up a formal scheme describing the cosmos, from 1 to 9, as in their view 10 represented a return to 0. This scheme can be summarised as follows:

1. The Creator (Being), which is one, simple, eternal and constant.
2. The Intellect, which is of two types: pre-existing and acquired.
3. The Spirit, which is of three types: vegetal, animal and intellectual.
4. Matter, which is of four types: artificial, physical, cosmic and primordial.
5. Nature, which is divided into five categories: the four natural elements (earth, water, air and fire) plus the heavens' element (ether).
6. The Body or object, which can take six possible directions: up, down, forward, backward, right and left.
7. The Seven Spheres, which contain seven planets.
8. The Eight elements: the four elements being associated with the four physical characteristics (hot, cold, humid and dry).
9. The Earthly bodies: mineral, vegetal, animal, each of which is of three types.

Note that each class of this cosmic scheme contains a number of elements equal to its place on the scale, which further emphasises the importance of numerology in the Ikhwan's cosmology.

In parallel, they describe the process of creation along the following lines:

- Creation of the Intellect, because the Intellect is simple and perfect, and because it must play the role of 'veil' between the divine essence and the world to be created;
- Creation of the Cosmic Spirit, which is the cosmic agent that produces creatures in nature by the will of God;
- Creation of the primordial matter, from which all creation will be constituted.

The Ikhwan, however, insisted that the whole process of creation, although divided into sequential steps, takes place instantly. This creation, moreover, is an 'emanation' process from the Divine. Finally, they reject the eternally old nature of the cosmos; indeed, they subscribe to the Augustinian solution that time and space were all created with nature and matter.

al-Biruni (973–1051)

Al-Biruni is the paramount representative of the Muslim scientific (methodical) school. One of the most competent scientists of the Middle age, he approached the study of nature as a devout Muslim who saw the world as the handiwork of God and considered the observation and study of nature as a religious duty[52].

Closely upholding the Qur'an, Biruni totally subscribed to the doctrine of creation *ex nihilo*; however, he concluded that knowledge of the date of creation is impossible, because God did not mention anything on this in His Book and has therefore decided to retain this knowledge with Himself. Moreover, estimates of that date by ancient civilizations differed widely, some of them being totally unrealistic in his view. Finally, he emphasised that the concept of time as perceived by humans must necessarily be very different from the one as seen by the Creator (Verily a day with your Lord is as a thousand years of what you reckon. Q 22:47).

Biruni also stated that his observations and investigations in the workings of nature led him to conclude that the laws of nature must remain unchanged through time. However, he accepted the ancient belief of the cyclic nature of history, with a gradual decay ('corruption') of both the material and the moral characteristics of beings.

His cosmology/cosmography resembled that of the Greeks (Aristotle plus Ptolemy, i.e. a cosmos of spherical shells centreed on the earth). In this regard, he was not influenced by the Indian conceptions, even though he spent long periods of time in India and became an expert on its history and culture. Biruni was an independent thinker who adhered as closely as possible to the scientific approach and to the results of his investigations – as well as to the Qur'an.

Sufism: Ibn Arabi (1165–1240)

During the centuries following the golden age (tenth to twelfth centuries) of the Islamic civilization, interest in the natural sciences gradually diminished, and the intellectual and spiritual life of Islam got dominated by the Sufi doctrines of Ibn Arabi and the Gnostic *ishraqi* (illumination) doctrines of Suhrawardi (c. 1155–1191)[ii].

The Sufi doctrine can be stated in the following central principle: 'There is no reality outside of the Absolute Reality; that is there is nothing other than Allah', was not explicitly enunciated until the twelfth and thirteenth centuries, when various forms were formulated by Ibn Arabi, al-Qunawi, al-Jili and other masters of the school[53].

Ibn Arabi, who was born in Andalusia and died in Damascus in 1240 after much travelling through the Islamic world, can be considered as diametrically opposed to the Muslim peripatetics (e.g. Ibn Rushd). His doctrine rests on the fundamental principle that knowledge acquired through spiritual experience is the highest form of learning. In fact, objective results obtained exclusively through reasoning can constitute a 'veil' (a term and concept which carries rich symbolism in Sufism), which actually prevents one from seeing the true nature of things.

The central principle in Ibn Arabi's doctrine, which it must be stressed is not widely upheld among Muslims, is that the existence is one and it is identical to the Reality (or essence). Created entities do not exist in themselves; they acquire an existence by way of their relation to the Being. Whether an entity exists or not makes no difference in the 'mind' of God; in other words, entities are merely images, or (in a famous analogy often

---

[ii] Suhrawardi's 'Ishraqi' school became a dominant aspect of the intellectual life of the eastern part of the Islamic world (particularly Persia) after the twelfth century. His cosmology drew heavily on the writings of Ibn Sina, particular his later 'esoteric' works, but also on the Ikhwan's and some of Jabir Ibn Hayyan's (Nasr: 'Cosmological Doctrines', p. 278). Suhrawardi constructed an immensely complex system that attempted to unify all the previous religious insights of the world in one spiritual vision that put God as its goal and guiding light ('*ishraq*' in Arabic word, which literally means 'the first light of dawn'). His cosmology was therefore a combination of pre-Islamic Iranian philosophy with the neo-Platonic scheme of emanation and the Ptolemaic planetary system (Armstrong, p. 230). Suhrawardi was not primarily interested in a description of the universe or even its origin; his natural science considerations were only a gateway to metaphysics. He believed that rationalism and philosophy alone could not lead to any satisfactory answers, they had to be combined with mysticism. Influenced by the Shiite theory of imams, he concluded that the world needed a 'pole' (*qutb*), some spiritual entity perhaps embodied by the Imam, without which the world could not exist.

used in this context) partial mirror reflections (virtual objects) of a more fundamental existence.

The Universe is thus described as a 'theophany', in which God 'sees' Himself in His creation. If existence is totally inaccessible to human understanding, the same must hold with regard to the 99 divine Attributes (omniscience, omnipotence, mercy etc.), because these simply describe the ways in which God reveals Himself to Himself. William Chittick explains this as follows: 'The cosmos manifests the Divine Names in a differential code (*tafsil*); as a result, each and every Name displays its own properties and traces both separately and together with every other name and combination of Names. Hence in its spatial and temporal totality, the cosmos represents an infinitely vast panorama of existential possibilities'[54]. Knowledge of Reality (the 'true' science) therefore consists in understanding these modes of self-revelation (of God), which are infinite in number. Ibn Arabi explains in his visionary treatise *al-Futuhat al-Makkiyah* (*The* Meccan *Revelations*) that knowledge of the cosmos can only be reached by way of a mystical journey through it.

In this doctrine, the question of the *ex-nihilo* creation and the beginning of the universe are mooted. However, this does not necessarily imply that the universe is eternal. For everything depends on God's desire, and how and when entities (or even the whole cosmos) come into existence or disappear will remain unknown. One can also imagine a sort of continuous creation and evolution of the cosmic 'landscape', for God never reveals Himself to Himself in the same manner. When an entity/state is created, it is no longer the object of the divine 'desire', and it is thus annihilated instantaneously; a new state is then desired, and so it appears.

islamic cosmology today

The decline of the Islamic civilization led to the twilight of most intellectual activities, including science, except for some bright spots here and there, which appeared and disappeared after a short while[55]. Except for very few punctual exceptions (e.g. Mulla Sadra, 1571–1640), the twelfth-century era of Ibn Rushd, Ibn Arabi and Suhrawardi was really the sunset of Islamic cosmology.

Today one may at most find a few commentaries, based on the scientific developments of our times[56]; sometimes we find translations of western cosmological works, even though they are rarely accurate enough to allow for a proper digestion of the latest knowledge, let alone produce further thought and innovation. The reason for this is that Muslims have become convinced that the proper remedy for their civilizational decline, and the way to a renewed progress, is the reformation of their religious life and the development of a truly progressive (democratic) political system within the perimeters of the Islamic culture. Science and philosophy have been assigned a very low priority in that analysis and are not seen as determinant in the overall scheme of development.

Still, if one looks more closely at the ideas that have been put forward and popularised in the past half century, one notices the emergence of two distinct, perhaps diametrically opposed, schools of thought: the popular I'jaz school, and the neo-Sufi school.

Chapter 5 was fully devoted to the first 'school', and I shall come back to it briefly below just to note the kind of cosmology its proponents tend to present. The second school (neo-Sufi), subscribed to by an elitist segment of the Islamic society, sees all new cosmological developments and discoveries as pointing to the old global, unified cosmos as proposed and developed by Ibn Arabi and his followers. The main proponents of this neo-Sufi school are S. H. Nasr, Osman Bakar, Muzaffar Iqbal and Abdelhaq Bruno Guiderdoni, to different degrees.

Let me give just one example of the cosmology presented by the I'jaz school. In a book titled *Man in the Universe, in the Qur'an, and in Science*, Abdulrahman Khudr writes[57] the following:

Is there a unity more global and more perfect than the similarity between the atom, the cell, the solar system, and the galaxy? Each one is comprised of a nucleus at the center, holding the secrets of life and controlling various functions, and particles or bodies or fluids revolving around it in a constant unstopping motion?

In a footnote, he adds: 'The electrons around the nucleus are found in shells called levels, and the astonishing thing is that scientists call them "heavens", as their number can reach seven in heavy elements . . . '

He further writes[58] as follows:

The beginning was indeed as told in the Qur'an. There was 'interstellar dust grain'[59] . . . The gathering of these particles together was apparently only to 'separate' the chemical elements in the matter of the universe . . . And so the scientists discovered that the atoms of this gas/'smoke', which the Qur'an told us about some 1,400 years ago, is the only way for the formation of the largest solid bodies in the universe, including stars. Thus Glory be to Allah, the Truest of Speakers . . .

Khudr then concludes: 'I believe that the Qur'an did not leave out any cosmic/natural phenomenon without referring to it!'

As to the neo-Sufi school, we may cite Abdelhaq Bruno Guiderdoni[iii], who has written a number of articles and taken part in several conferences on the topic of Islamic cosmology. His upholding of the neo-Sufi doctrine is apparent in his writings:

- God's self-disclosure is endless . . . The appearance of 'emerging properties' at all levels of complexity, and particularly the appearance of life and intelligence, is another aspect of this continuous self-disclosure.
- Islamic tradition has always taught that God is nearby and continuously acts in Creation. 'Each day some task engages Him' (55:29). God is hidden, but He is also apparent, according to His beautiful Names *azh-Zhahir wa-l-Batin* . . . [60]
- To avoid those dead-ends [struggle against science on the one hand, and I'jaz on the other], we must be aware that Man is called to a knowledge which goes much beyond rational knowledge[61].

scientific cosmology today

How far has cosmology come today from all those medieval, philosophical and metaphysical considerations? Knowledge of the universe has witnessed extraordinary developments during the past century. Seminal theoretical works and crucial observations have led to more and more sophisticated theories (Relativity, Big Bang, inflation, imaginary time, dark energy etc.) and more and more accurate results and constraints.

The very definition of cosmology has been sharpened to reflect that transformation into a rigorous scientific field. A typically simple definition is as

---

[iii] Bruno Guiderdoni is a French astrophysicist who is now the director of the Lyons Observatory. He converted to Islam in 1987 after a trip to Morocco, where he discovered the Sufi approach to life and cosmos. He has since adopted the name Abdelhaq (derived from one of God's Names/Attributes). From 1993 to 1999, he supervised and directed the TV program 'Knowing Islam', which is broadcast every Sunday by the state (public) channel France 2.

follows: Cosmology is the scientific study of the universe as a whole and of its large-scale properties. Moreover, its scientific rigor is today strongly emphasised[62]: Cosmology 'endeavors to use the scientific method to understand the origin, evolution and ultimate fate of the entire Universe. Like any field of science, cosmology involves the formation of theories or hypotheses about the universe which make specific predictions for phenomena that can be tested with observations'.

In order to illustrate the exponential increase in cosmological knowledge that humans have achieved in the last century or even in the past few decades, we may compare the values of a few basic quantities, such as the age of the universe or its size, since antiquity:

Era	Size of the Universe	Age of the Universe	Nature of the Universe
Antiquity	10^8 km	10,000 years	Static
1900	10^{17} km	Infinite	Static
1980	10^{23} km	10–20 billion years	Inflated, expanding, but slowing down
Today	10^{23} km, if not infinite	13.5–14 billion years	Inflated, expanding and accelerating

The same contrast between what we knew a century ago and what we have learned today can be expressed as the cosmologist Edward L. Wright neatly put it[63]: Only one (cosmological) fact was known in 1917, namely that the sky is dark at night – and Einstein ignored it.

The most striking aspect of the recent knowledge that humans have constructed about the cosmos is the staggering scale(s) of the universe's parts: from the earth, to the solar system, the Milky Way Galaxy, the local group of galaxies, the clusters and superclusters of galaxies and the universe as a whole. Indeed, if the earth were a little ball of 1 cm, the moon would be orbiting about 1 m away, the sun would have a diameter of a few metres and would be shining from a distance of 100 m, our Galaxy would have a diameter of a thousand billion kilometres, with the universe about a hundred thousand times larger ... The other stunning realisation that humans have come out with is the time scale (or the timeline) of our existence compared to that of the solar system, our Galaxy and the universe: If the emergence of humans as an intelligent species had occurred 24 hours ago (representing a few hundred thousand years), then the earth and the solar system would have formed a few dozen years before, the Galaxy would have existed for about a hundred years, and the universe would have been created some 150 years ago. We are very tiny, and appeared on the scene very recently . . .

Cosmology today attempts to answer the following questions as precisely as possible:

- How big is the universe and how far apart are objects in it? (the 'large-scale structure' of the universe.)
- What is the (topological, i.e. geometrical) shape of the universe?
- What is its content?
- Is the universe changing and how?
- How old is the universe?
- What will its ultimate fate be?
- Are there other universes?

In addressing these questions, cosmology builds its investigations on two large pillars:

1. Theory: Einstein's General Relativity (a geometrical theory of gravity, which is the only one correct at large scales) with some basic assumptions such as the Cosmological Principle (that the universe is globally uniform and isotropic).
2. Observations, which have become richer and more diverse over the past century or so, that is, since the discovery of the expansion of the universe, the discovery of the cosmic microwave background radiation (a relic of the initial creation energy/radiation outpour), and the measurement of the abundances of the main elements in the universe (hydrogen, deuterium, helium and lithium).

Now, although it would take too much space to give even a summary of the methods and knowledge referred to in the previous paragraph, I should still give a brief rundown of these extremely important discoveries, which have come to drastically change our view and understanding of the universe.

For example, take the expansion of the universe. Most people have heard of it, but they tend to imagine it as objects flying *in* space, whereas the correct way to envision it is as a stretching of the universe's space itself. The best analogy to give in this regard is that of a balloon in which someone is blowing air (see next diagram).

In this case, our universe (which is three-dimensional) is represented by the two-dimensional *surface* of the balloon. And as I am fond of asking my students, the key question for the understanding of this expansion is: 'what is inside (and outside) the balloon?' The answer, of course, is: time! Inside is the past, outside is the future. The universe is not expanding in some already-existing space; it is expanding its own space in time. And one can check that indeed the speed at which objects on the surface (in the universe) 'fly away' from each other is proportional to their distance at any given moment, which is simply Hubble's Law[iv]. From the balloon analogy one also understands that there is no centre of the universe because all the points on the surface are equivalent to one another, none is more special than the other, and each sees the others as expanding away from it.

The rate at which the universe is expanding has long been recognised as one key parameter of any cosmological model. The American cosmologist Alan Sandage often described cosmology as 'a search for two parameters': the expansion rate and the acceleration/deceleration rate. And for several decades the value of this crucial quantity was highly disputed, somewhere between 50 and 100 km/s/Mpc[v]. It has recently settled at 71 km/s/Mpc, with an uncertainty of about 5 per cent – another sign of the extraordinary progress that observational cosmology has made in recent years.

[iv] Edwin Hubble is the great American astronomer who in 1929, with the help of others, including Vesto Slipher and Henrietta Levitt, realized that galaxies were flying away from one another according to a law of proportionality between the recession speed and the distance (henceforth known as Hubble's Law); he thus concluded that the universe was expanding.

[v] The parsec (symbol pc) is a unit of astronomical distance; it amounts to about 3.26 light years, and Mpc stands for Megaparsec, so that km/s/Mpc is a unit of speed per distance, which is equivalent to inverse-time.

Likewise for the cosmic microwave background radiation, which was discovered in 1964 by Arno Penzias and Robert Wilson, essentially by pure chance. Although it had been theoretically predicted with more or less the right features some 15 years earlier, all that was known about it for the next 30 years was that it had a characteristic temperature of about 2.7 Kelvin. By the early nineties (through the COBE satellite), and certainly by 2003, when the WMAP satellite produced a new cosmic radiation map and detailed information that were hailed (by the authoritative *Science* journal) as the 'Breakthrough of the Year', differences of one part in 100,000 in temperature between different parts in the sky could be determined. This was extremely significant, for it allowed scientists to describe 'fluctuations' in the early space-time fabric of the universe, a crucial aspect of the emergence of structures (galaxies and clusters) in time. Furthermore, detailed analysis of the data permitted a quantum leap in the determination of the age of the universe; where one could only give values of around 15 billion years (± 1 or 2 years), WMAP brought greater accuracy: 13.7 billion years (± 0.2 billion years).

In parallel to this extraordinary progress, other approaches in observational cosmology were yielding equally – if not more – stunning discoveries. In 1998, another result was declared as 'the scientific discovery of the year', if not of the decade: the inference that the universe was accelerating in its expansion – whereas it had long been believed to be slowing down, under its own gravity, towards some kind of standstill. That breakthrough was made possible by the ingenious usage of Type Ia supernovae as 'standard candles' (known luminosities) and thus as distance indicators of galaxies in the farthest reaches of the universe (if one knows the standard power of a given light source, then measuring its apparent brightness immediately gives the distance at which it is shining that brightly or dimly).

And when this was confirmed and accepted, it was quickly concluded that there must be some 'force' that has been stretching the fabric of the universe faster and faster; but as no one had any idea about its origin, it was called 'dark energy'. This should not be confused with 'dark matter', whose existence astronomers had also gotten convinced of without any knowledge of its nature. Dark matter is an exotic form of matter; it has some mass, it acts gravitationally, but it has no electromagnetic interaction and thus does not emit or absorb light of any kind – it is totally and forever dark. Scientists have now estimated its amount at about 25 per cent of the total mass-energy content of the universe, compared to about 4 per cent of the 'real' (normal) matter (from hydrogen to uranium, including carbon, oxygen, iron and everything we can find in nature and the universe). The rest of the universe's mass-energy, i.e. about 70 per cent, is this dark energy, which we have only come to know about in the past few years! Do we have any clue or candidate for these 'dark matter' and 'dark energy' things? Yes, quite a few actually: MACHOS and WIMPS (and other propositions, which we will not explain here) for dark matter; 'quintessence', 'decaying dark matter', and the 'cosmological constant' (which Einstein had introduced but soon decided that it was his 'biggest blunder') for dark energy. Here we shall not delve into these questions for fear of straying too far from our overview and its objective, namely to give the reader both an indication of how much progress cosmology has made in the past one or two decades and to present a general overview of what we have learned so far.

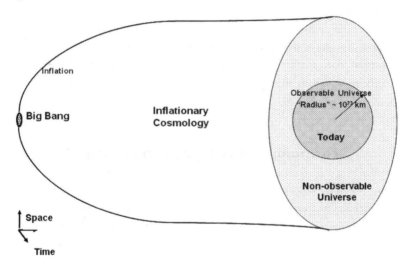

So let me summarise the main findings and conclusions of cosmology from recent years:

1. The universe has been expanding and seems (for almost certain now) to be accelerating due to some 'dark energy', which stretches space-time as if some internal repulsion was pulling the fabric apart. Contrary to the Big Crunch or Standstill that were previously theorised to occur in the far future (assuming that the internal gravity of the universe was large enough to stop and reverse, or at least balance out, the expansion), the acceleration now seems to have convinced everyone that our universe is destined to an 'open' infinite expansion that will lead to a dark, cold Big Chill.

2. Far from being static or eternal, the universe started and evolved from the moment of its creation (or emergence) from a 'singularity', a point of infinite density, energy and temperature (this is now the well-established Big Bang theory). As it expanded, the universe produced particles, then nuclei, atoms and structures (stars, galaxies, solar systems etc.); details of the expansion (including the inflation epoch and its characteristics) have also been worked out by ingenious theorists.

3. The age of the universe (since its creation/emergence) can now be determined quite accurately; this automatically gives us the size of the observable universe (speed of light times age), which is not to be confused with the size of the whole universe, some of which we will be able to see in the future as the universe expands, and the rest we will not (see the diagram).

4. Only 4 per cent of the universe is made of the 'ordinary' matter that we know; an additional 25 per cent (of the matter-energy content) is in the form of 'dark matter'; and 70 per cent is the 'dark energy' that we have come to accept as an important feature of our strange universe.

cosmology, philosophy and theology

In the previous section, I tried to stress that cosmology is a theoretical, observational and empirical science. The universe can and must be known by modern science, I seemed to say. This is true, and it cannot be emphasised enough. We can no longer allow any freewheeling discourse on the cosmos, on creation, on the structure of the universe and our physical position in it, and certainly not on the physical processes that have governed the cosmos and its history (including us) since its birth.

But we must be careful not to allow cosmology to be dressed in a materialistic garb. More than any other branch of science, cosmology relates to us in the personal and collective world views we construct. Indeed, philosophy may have some important contribution to make to cosmology as a human endeavor, in a somewhat wider perspective than modern science tends to allow. And theology can interact with and benefit from cosmology by constructing a reasonable and scientifically consistent belief system. The cosmologist Kim Coble likes to present the following diagram to show the connections that one must see between 'scientific cosmology' and one's personal world view:

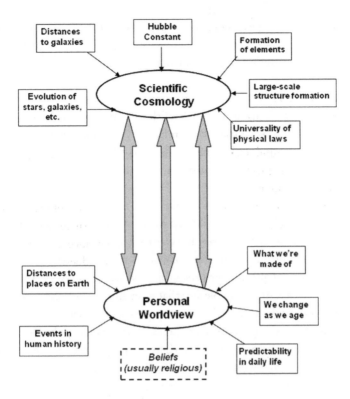

The only box that is missing from her diagram, the one I have added in dashed lines, is the beliefs (usually religious) that are quite often part and parcel of the personal world view area.

The cosmologist Joel Primack and his wife Nancy Ellen Abrams (an artist and philosopher) have recently written a book (*The View from the Center of the Universe*, 2006) precisely for the purpose of constructing a modern cosmology, which, while solidly built on the latest scientific knowledge, takes social and cultural world views (what the authors lovingly call 'myths') into account. They begin by decrying the 'widespread cultural indifference to the universe'[64] in our times, contrary to all previous eras; they ascribe this phenomenon to the culture of separation (the 'social schizophrenia'[65]) that set in – for the first time in history – between the physical world and the world of values and meaning. Primack and Abrams argue that, on the contrary, traditional, cultural and even religious images can be quite useful in presenting today's cosmology, without this implying any compromise in the scientific conclusions or the methodology by which one arrives at that knowledge: 'Many religions have concepts that resonate harmoniously with aspects of the new scientific picture'[66]; 'We don't have to pretend to live in some traditional picture of the universe just to reap the benefits of the mythic language popularly associated with the traditional picture'[67].

William Chittick (the Muslim perennial philosopher) could not agree more; he insists that one of the goals of the Muslim and other intellectual traditions is 'to recognize the proper role of myth in human understanding and, if necessary, to revitalize mythic discourse'[68]. He cites the (greatest) Sufi master Ibn 'Arabi, who described myth and reason as 'the two eyes' of human perception of Reality.

Primack and Abrams's thesis is that while in old cosmologies humans (and earth) were physically considered to be at the centre of the universe, today in a metaphorical way we can construct a modern cosmological 'big picture' where humanity finds itself of central importance in the universe. Decrying the fact that '[m]odern scientific cosmology doesn't even discuss us', they argue that 'it is a simple fact that if science has nothing to say *about* human beings, it will have little to say *to* most human beings'[69] (emphases in the text). They then set out to find a place for humans in the story of the universe. To summarise the chapters where they argue each point, we are at the 'center' in the following essential ways[vi]:

- We are made of the rarest elements that the cosmos has made: stardust.
- We live at the midpoint of our planet and sun's lifetime.
- We have sizes that are (in powers of ten) in the middle of the entire scale of objects.

Primack and Abrams insist that they did not come up with this centrality because they wanted to make sure that humans are forcefully put at the centre; it was rather the structure of the universe that placed us at those centre points. It is now the responsibility of the cosmologists to make sense of this, whether they like it or not.

These authors then find more links between the scientific cosmology and the personal world view (the links that were postulated by Coble in the above diagram). Using scientific and social concepts as metaphors, they find resonances between, for example, cosmic inflation and the exponential

---

[vi] Primack and Abrams expound on seven such 'center points' (which they summarize on pages 270–2), but the three I am citing here are the more convincing ones.

growth of human population, gravity in space and the increasing concentration of wealth in some countries compared to others, the clash of civilizations and what they call 'scale chauvinism' (expecting some objects to behave in certain ways that are actually 'scale-confused'). Primack and Abrams sometimes use cosmology to enlighten certain old traditional concepts (pre-initial vacuum field to explain the 'primeval waters', which one finds in Islam, in Christianity and in Babylonian creation stories) or the Kabbalistic beliefs to illustrate the ideas of 'eternal cosmic inflation'.

In parallel to this kind of discussion, which was quite unprecedented until recent years, a strong interest is gradually emerging around the question of the 'philosophy of cosmology', as can be seen in the September 2009 conference on the Philosophy of Cosmology at Oxford. Indeed, many thinkers, including some leading cosmologists (e.g. George F. R. Ellis) and some philosophers of physics (e.g. Simon Saunders), have started to insist that as cosmology has now developed into a solid branch of science, one must make sure that its foundations and methodology are as strong as they can be and that any weaknesses therein must clearly be expressed, lest anyone draw unwarranted conclusions. Among the main issues being raised which necessitate a serious philosophical exploration and wide interdisciplinary treatment, one must stress the following: the assumptions themselves (e.g. the cosmological principle), the tools (e.g. linear, real and imaginary time), the models (chaotic and/or eternal inflation), the concepts (beginning vs. creation, probabilities and extrapolations to infinite sets, chance and necessity etc.) and a host of others.

Some thinkers have called for a discussion of issues pertaining to the nature of cosmology and the interpretation of its findings, and their meanings. Michael Heller[vii], for instance, reminds[70] us of the necessity to address the question of 'first cause', which has been troubling scientists and philosophers since Aristotle. Heller also remarks that taking the physical laws for a given cause is by no means a trivial premise. Likewise, the renowned cosmologist George Ellis questions the nature and scope of cosmological theories: (1) Are these supposed to deal only with physical issues or should they also query why things are the way they are? (2) Are cosmological theories, like other scientific theories, supposed to be gauged by the usual criterion of falsifiability and their explanatory power? Ellis remarks that quite often cosmological

---

[vii] Michael Heller is a Polish philosopher and theologian who lectures on the philosophy of science and logic; he received the Templeton Prize (in science and religion) in 2008.

theories cannot be supported by direct (or sometimes even indirect) observational and experimental evidences. Furthermore, falsifiability and explanatory power often go in opposite directions: when one increases, the other decreases; which one should we then emphasise and require more? Similarly, other typical scientific approaches, such as experimentation, induction and generalisation from limited cases are often unavailable to cosmologists (there is only one universe to deal with). These thinkers then conclude that first we must be very cognisant of cosmology's constraints and special nature, and then accept that there is a philosophical, perhaps, even metaphysical (in the sense that we defined in Chapter 3) aspect to cosmology and honestly deal with it in that regard.

The philosopher Ernan McMullin has addressed precisely this epistemological question. In *Is Philosophy Relevant to Cosmology?*[71] he starts by pointing out the special nature of cosmology, in particular its strong(er) dependence on hidden assumptions and the frequent existence of many 'degrees of freedom' (he points to the debate between the Big Bang and the Steady State theories as a classic case). He also points to cosmology's particular dependence on concepts such as time, beginning, creation etc. So McMullin asks himself whether philosophy can be brought in for help. He is fully conscious that philosophers may be out of their element here, because cosmology is a science that requires advanced training and mastery of specific (theoretical) tools; indeed, he quotes Paul Davies, who remarked that 'ten years of radio astronomy have taught humanity more about the creation and organization of the universe than thousands of years of religion and philosophy', but hastens to insist that he is talking about philosophy in a general sense, not about philosophers; indeed, he acknowledges that 'much of the liveliest philosophic writing on scientific topics in [the twentieth] century [was] done by scientists' and that '[b]asic science, at its most innovative, merges into philosophy'. Then he gives at least one example (Mach's Principle) where 'philosophic clarification' proved to be necessary for a scientific theory (General Relativity). Finally, he points to serious issues in cosmology (either within it or stemming from it) that call for the intervention of philosophy: the assumed principles; the methods used; the acceptance of dominant paradigms (what he calls 'cosmythology') etc.

One important example McMullin brings up is the question of beginning and creation. He first remarks that the word 'creation' is an explanatory one; it points to a cause (a Creator) for the appearance of the universe. Obviously, this is not a scientific explanation, for science cannot establish 'a sufficiently strong principle of causality'. McMullin then asks: Can philosophy do so?

He goes on to explain that *ex nihilo* creation is not the only way to establish a divine origin of the universe; this is only the most intuitive view; one can, however, almost equally argue for a divine origin of an eternal universe. This is only to say that the issues of beginning and creation are much more philosophical than scientific. (I would like to point out here in passing that this is almost exactly the argument that was made by Averroes in his defense of the eternity of the world against al-Ghazzali's proclamation that this concept had been imported from Greek philosophy in opposition to Islamic creed; Ibn Rushd had indeed explained that the question was a philosophic one and if there was a strong argument, i.e. Aristotle's, in favour of eternity, and if one, that is he, could show that it was consistent with Islamic principles, then one was obligated to accept it, for it constituted the highest knowledge at one's disposal.)

This brings us to the relation between cosmology and theology. Here, despite the tons of literature and pronouncements linking concepts and theories, such as the Big Bang and the expansion of the universe, to religious and theological views, I must note that the relation is much more shaky and risky. It is well known, of course, that the Big Bang theory, in particular, generated heated debate and passionate declarations in relation to God, the Bible, the Qur'an etc. For example, Pope Pius XII in 1951 proclaimed the Church's support for the Big Bang theory, which had first been introduced as Lemaitre's 'primordial atom' idea; he noted the similarity with the *Fiat lux* of Genesis and declared: 'Hence creation took place in time, therefore there is a creator, therefore God exists!' Similarly, Muslims have in recent times insisted that the Qur'an is fully supportive of the Big Bang theory, and the latter 'confirms' the holy verses. Furthermore, the so-called 'cosmological argument' for the existence of God (the *kalam* argument) originated with Muslim philosophers (al-Kindi and later others). On the other side of the fence, some scientists and thinkers have expressed strong views against such inferences; David Bohm, for example, accused 'scientists who effectively turn traitor to science' of 'discard[ing] scientific facts [in order] to reach conclusions that are convenient to the Catholic Church'. Similarly, Stephen Hawking, both in his famous book (*A Brief History of Time*[72]) and paper with Hartle ('Wave function of the Universe'[73]) tried hard to show that the Big Bang's beginning was not necessarily a boundary or edge; so he famously asked: 'what place then for a creator?'

Primack and Abrams attempt to explore[74] the relevance of modern cosmology to the concept of God. Insisting that one can only believe in God as one *understands* 'God', they consider various aspects of the concept and emphasise the importance of the scale at which one regards God; they conclude that God must be relevant to and encompass all scales. Thus, they insist, God must be viewed not just from our human perspective and be declared as the 'all-loving', 'all-knowing', 'all-everything-else-we-humans-do-only-partially-well'[75] but also from the galactic and atomic perspectives. They conclude that a God 'disconnected from this amazing universe that science is revealing would be a God entirely of the imagination' and that 'a God that arises from our scientific understanding is not entirely created by us. Such a God runs deeper than humankind's imagination and is speaking in some way for the universe itself'[76].

Another issue relating cosmology to theology, the 'question of cosmogenesis', was recently addressed by the Muslim philosopher Seyyed Hossein Nasr. We must remind ourselves that Nasr is a perennial philosopher, who believes in the unity of everything: knowledge, tradition, the cosmos (with its physical and metaphysical parts) and the Divine. He has been calling, for almost 50 years now, for that unification, and the field where he sees his principle as most obviously applicable is, of course, cosmology.

In his 2006 article titled 'The question of cosmogenesis'[77], Nasr starts by stating that the cosmos is a legitimate field of study in Islam; indeed, he points to Qur'anic verses that emphasise both the divine creation aspect of the cosmos and its intelligibility; he remarks, for example, the etymological relation between the Arabic term for knowledge (*'ilm*) and the one for the world/universe (*'alam*). He goes on to insist, however, that cosmogenesis is a religious and metaphysical question, 'the answer to which comes from the truth of revelation and not simply from an extension and extrapolation of the sciences of the natural and physical order'. In his view, '[t]he Islamic attitude to this question stands therefore at the antipode of the modern Western scientific view . . . ' He goes even further and finds fault in cosmology's 'extrapolation' of time 'across vast periods of the past and future' and in the changing nature of cosmology and its results: 'Many scientists now speak of the big bang theory while yesterday they spoke of something else, and tomorrow they will point to other theories'.

I do not subscribe to this analysis; indeed, I will argue that the Islamic attitude (as Sardar said in the anecdote I related in the beginning of this chapter) is to start and fully adopt the hard facts of science (all established knowledge), not to pick and choose what one likes and, in the worst cases, reject the scientific knowledge altogether and declare the topic to be the exclusive province of religion, metaphysics and revelation.

Much more modern and scientific is the approach of Arthur Peacocke, who attempts to fuse modern cosmological knowledge with the imagery of the Bible's opening chapter, Genesis. In the prologue, interestingly titled 'Genesis for the third millennium', of his book *Pathways from Science towards God*, Peacocke writes[78] the following:

> There was God. And God was. All That-Was. God's Love overflowed and God said, 'Let Other be. And let it have the capacity to become what it might be, making it make itself – and let it explore its potentialities'.
>
> And there was Other in God, a field of energy, vibrating energy – but no matter, space, time or form. Obeying its given laws and with one intensely hot surge of energy – a hot big bang – this Other exploded as the Universe from a point twelve or so billion years ago in our time, thereby making space.

Peacocke goes further: 'Vibrating fundamental particles appeared . . . the universe went on expanding and condensing into swirling whirlpools of matter and light. . . . Five billion years ago, one star in one galaxy – our Sun – became surrounded by matter as planets. One of them was our Earth . . . [S]ome molecules became large and complex enough to make copies of themselves and became the first specks of life . . . '[79]

Clearly, Peacocke has integrated all the main elements of modern science, from cosmology to biological evolution, into his story of creation. His approach is very similar to, but less 'mythical' than, Primack and Abrams's, who have summarized their viewpoint as follows: 'Scientific cosmology is not the last word: it's the first word. Scientific accuracy has to be our minimum standard'[80].

summary and conclusions

Cosmological models and theories about the universe's history and content have always led to great philosophical debates, where references to religious beliefs and teachings were often inevitable. For example, the question of whether the Big Bang theory gives support to the idea of a Creator or not has been widely discussed by many scientists and thinkers. At the same time, the public at large, of all religious and cultural backgrounds, was always captivated by those metaphysical discussions, especially when they emerged from scientific developments. Philosophers, however, have only recently been called upon to help address some of the conceptual issues inherent in cosmological studies.

During the golden era of the Islamic civilization, the Qur'an had been a primordial source and reference for cosmic information and doctrines. But Muslim thinkers quickly realised that the Qur'an was too ambiguous and ambivalent to construct any cosmology from. The other major influence was then the Hellenistic philosophy that the Muslim thinkers encountered and absorbed quite early on (ninth century). Far from the rehashing of ancient Greek theories, the body of 'Islamic cosmology' is a rich fresco of worldviews, fusing other cultures' views with Islamic principles.

The twentieth century, however, saw the growth of a renewal attachment to all religions (in various ways), particularly in the Islamic world. In this culture, where the Qur'an is the paramount, über alles reference, new scientific results were quickly compared to the sacred texts, so that a rather amateuristic cosmology that liberally mixed science and religious texts, started to take shape and root in the cultural landscape.

But at the start of the third millennium, and after the modern scientific revolution, can one still discuss subjects such as cosmology, which evidently owe much more (some would say entirely) to physical science, with a religious perspective and present an 'Islamic cosmology'? Can there be an Islamic cosmology or a Jewish cosmology or a Hindu cosmology? Hasn't modern science closed the door on such muddling?

As Primack and Abrams insist: 'Traditional cultures' cosmologies were not factually correct, but they offered guidance about how to live with a sense of belonging in the world'[81]. Similarly, the discussion of Islamic cosmologies may be useful and even needed today in light of the identity crisis that the oriental world seems to be undergoing, particularly after the domination of the western methodologies of inquiry and the globalisation of culture. Many opponents to these trends have thus been advocating the formulation by various peoples of their own cultures and even their own sciences.

But this is really not the reason for my attempt to defend a cosmological viewpoint that is both fully consistent with the modern theories and discoveries and in harmony with the fundamental religious principles. This can apply to any and all religious traditions; here I only take the Islamic tradition because it is my own and I feel the need to draw attention to its rich areas as well as to some of the misguided ventures performed under its title.

For me, the main reason why scientists and philosophers should collaborate on such an ambitious, albeit controversial project is the inherent inability of modern science to find meaning in many of its discoveries. Can one discuss the 'big bang singularity' (a technical phrase meant to replace the religiously loaded term 'creation') without discussing its philosophical meaning and its theological implications? Can one discuss the nature of the physical laws in the universe and not relate them to the existence of life, intelligence and humanity? Can one discuss the existence of other causally disconnected universes and not explore the meaning of it? Clearly there are many issues on which science admits not to be in a position to say a word and where paradoxically human interest and thirst is heightened.

It is becoming increasingly clear to many thinkers and scientists that a purely scientific approach to the cosmos is not satisfactory; in modern cosmology, the universe (which is totally material by definition) is supposed to be self-consistent and not requiring a Creator. Modern science makes no room for transcendence or meaning, unless theistic science – as I have advocated in Chapter 3 – is allowed to develop and prosper.

Many scientists still believe science alone holds the answers to questions pertaining to the universe, life and everything else. But a 1997 survey showed that about 40 per cent of American scientists believe in God, and 45 per cent do not, while 15 per cent are agnostics[82]. The 40 per cent value has remained surprisingly constant over the twentieth century, because similar polls started being conducted (in 1916)[83]. And many scientists are now openly insisting that the hard line approach to nature and life be softened to include other non-formal approaches, including religion, metaphysics and philosophy, which might enrich the human perception and understanding on various issues.

The challenge, however, is how to construct a theology that marries the religious conceptions of God (as a personal god) with a 'natural theology', which identifies God with the origin of the underlying orderliness of the cosmos, the basis upon which the universe was built. To be sure, we cannot accept theologies that clearly clash with or contradict rational methods and scientific results; we cannot compromise our intellects.

Although I am far from able to formulate a full and self-consistent theistic cosmology, I believe some synthesis, perhaps similar to the one that some of the Muslim medieval philosophers (Averroes in particular) produced, is still possible. In order to achieve that synthesis – or at least the harmony – of the theistic principles with the modern scientific methods and results, I believe that a double programme must be pursued: (1) Some new theology must be proposed that would be consistent with modern science even if it does not adhere to the sacred beliefs and writings in a literal way; (2) a less materialistic cosmology must be produced, one that would allow for some meaning and spirit to be found in the universe and in the existence.

In my view, the greatness and power of the creation lies in its absolute elegance and perfection. God is the perfect abstraction of all being and Reality; He is the underlying principle upon which everything is built, and rests. This principle then 'sustains' the universe like a spirit pervading all of the existence, like a necessary but undetectable field.

The alternative to this viewpoint is the neo-Sufi concept whereby God and the universe are one (or at least intertwined) and the Creation is continuous, because God never ceases to disclose Himself (to Himself) without being directly apparent.

In conclusion, let me reiterate what by now should be a rather obvious idea: An 'Islamic cosmology' cannot limit itself to a pseudo-scientific exegesis of the sacred texts; it should rather provide ample room for creativity and freedom of the mind and for the wide horizons of the Islamic culture. The latter was able to – a thousand years ago – and should still be able to absorb humanity's knowledge, science and progress and produce an interesting and valuable synthesis.

A modern Islamic/theistic cosmology, which is fully compatible with science, is – I think – possible to construct, provided that it is intellectually open and creative, and far from rigid in either the religious or the scientific directions.

7

Islam and design

'O people! A parable is set forth, therefore listen to it: surely those whom you call upon besides Allah cannot create (even) a fly, though they should all gather for it! And if the fly should snatch away anything from them, they could not take it back from it; weak are the invoker and the invoked. They have not regarded Allah with what is due to Him; most surely Allah is Strong and Mighty.'

Qur'an 22:73–74

'The way the artifact proves the existence of the artisan is in the existence in the artifact of an order in its parts, that is the way that some of its parts have been built in relation with the others and the way that the whole has been made adequate for the intended usage of this artifact; this proves that the artifact is not a product of nature but rather the creation by an artisan who ordered each thing in its place . . . '

Ibn Rushd[1]

'If we survey a ship, what an exalted idea must we form of the ingenuity of the carpenter, who framed so complicated, useful, and beautiful a machine? And what surprise must we feel, when we find him a stupid mechanic, who imitated others, and copied

an art, which through a long succession of ages, after multiplied
trials, mistakes, corrections, deliberations, and controversies,
had been gradually improving?'

David Hume[2]

the return of the design argument

Well before the twentieth century, science and scientists had managed to
impose a naturalistic approach to and conception of our world to erase the
old ideas of design and purpose from our vision of the universe. Naturalism
was usually considered and presented as a methodological standpoint, which
simply insists that all explanations of world phenomena must be strictly
'natural' (or naturalistic) and be limited to the laws of matter. Sometimes
naturalism was equated with materialism, when thinkers went one step
further and concluded that nature and the world were nothing but matter,
and there was never any need for any metaphysical principles, indeed that
there was no such thing as metaphysics.

In parallel to this trend, the Copernican revolution was bringing about
its full effect. Indeed, it had first displaced the centre of the world from the
earth to the sun, then it declared that no point was a centre in the whole
universe, and soon it was erected into a 'principle of mediocrity', insisting
that the earth – and thus the man – were of no importance whatsoever in the
scheme of things. This seemed more than logical, as the size of the universe
grew and was soon found to be staggeringly larger than the earth, not to
mention man. Why would such a dust speck have any value, much less be the
purpose of any grand design? And, once the beautiful diversity of creatures,
big and small, began to be understood on the basis of evolutionary schemes,
the whole idea of design was expiring at last.

That is how many generations of scientists, philosophers and educators
were brought up, in the certainty that science had established the 'facts'
that (1) the universe is only matter; (2) the world is without purpose; (3)
the scientific method excludes any idea or principle of design; and (4) man
is of no importance at all in the general picture.

More than a century passed in this general atmosphere and in this conceptual framework. But then the arguments of design started to reappear, in
a very remarkable, albeit not necessarily welcome, series of developments.
Indeed, we first started to hear of a 'fine tuning' of the universe, that the
building bricks of the universe (parameters and laws) had to be precisely

what they are if life, consciousness and intelligence were to appear at some point and place. Then an 'anthropic principle' (putting man in the centre of what were purely cosmological or at least physical considerations) started to be discussed. Further to this, the old argument from design began to make a comeback; you will recall that this argument, which I introduced in Chapter 1, was one of the main classical 'proofs' of the existence of a creator, but modern science and philosophy thought they had laid it to rest with Darwin and Hume. Finally and most recently, the design pendulum was pulled to the extreme with the Intelligent Design (ID) theory (barely a hypothesis, and not very scientific at that, as we shall see), which claimed that it could display cases of direct divine conception and creation. I shall come back later to critically describe and assess this 'theory'.

Now, I do not subscribe to the new design trend wholesale, and certainly not to the ID viewpoint. Still I consider the phenomenon of the return of the design argument to be quite remarkable, and think it is interesting and instructive for Muslims to review this concept and to see to what extent the Muslim discourse and world view can accommodate itself with such approaches. Needless to say, we must try to strictly uphold the scientific spirit when looking at the world.

I shall defer discussions of the much more important anthropic principle and the theory of evolution to the next two chapters. In this one we shall limit ourselves to the question of design. ID will be addressed towards the end of this chapter, not because it is such an important theory but simply because it has started to take hold in some Muslim creationist circles, and it raises a few interesting questions about the nature of science and its method.

a brief history of the design argument

The idea that the world (nature and humans) exhibits 'obvious' signs of design and therefore points to a creator has been expressed and developed by many cultures for a long time. This observation, and the inference from it of a creator, is usually referred to as the design argument or the argument (for the existence of a creator) from design. A closely linked, but somewhat different idea is the teleological argument, which stipulates that one can observe and deduct (by various means of deduction) that the world and its creatures carry features that appear to point towards a goal or a purpose (telos), as though a plan has been drawn and is being executed through this glorious construction.

Both of these arguments, in one form or another, appear in the Islamic traditions, from the core Text itself (the Qur'an) to the latest writings and productions of Muslims in the twenty-first century, including the important philosophical and theological debates that raged during the golden era of Islam, with the active participation of such towering intellectual figures as al-Kindi, Ibn Sina, al-Ghazzali and Ibn Rushd. Unfortunately, this rich tradition seems to be almost completely unknown today, to both the western and most of the Muslim intellectuals. To give just a couple of striking examples, I can mention that two recent publications that purported to present a (brief) history of the argument from design (Anna Case-Winters' 'The Argument from design: what is at stake theologically?'[3] and Michael Ruse's 'The argument from design: a brief history'[4]) ignored any contribution of Muslim thinkers to the issue and jumped from the Greeks (represented by Cicero and Lucretius in the first article and Socrates, Plato and Aristotle in the second) to Thomas Aquinas. In fact, Case-Winters wrote: 'It was not until the thirteenth century that long-lost classical philosophy and science were rediscovered. With this turn the argument from design reemerged [. . .] and received its classic formulation'[5]. This statement could not be farther from the truth.

Indeed, one can easily find multitudes of examples of the ubiquitousness of this argument in the Islamic traditions. I must note, however, that even during the classical era (ninth–thirteenth centuries) of Islam civilization, the design and teleological arguments were often considered as rather simplistic and valid more for the general public than for the philosophers. (We shall see, however, that Averroes, on the contrary, liked the argument precisely because it could appeal – at different levels – to both intellectuals and laymen.) In recent times, Muslims have made strong usage of this argument and don't seem to have been much affected by the Darwinian revolution's impact on this idea.

Now, as I explained briefly in the first chapter, within the classical philosophical traditions of humanity (including and particularly those of Christianity and Islam), arguments for the existence of God have usually fallen into three to five categories: (1) the cosmological argument; (2) the ontological argument; (3) the design or teleological argument; (4) the moral law (within) argument; and (5) the spiritual experience argument. (See the presentation of each argument and the general discussion in Chapter 1.)

The design argument was greatly developed and given widespread credence before the advent of Darwin's theory, which largely shushed the fervor of the proponents for design. This argument went through several versions or formulations, which I can briefly mention in historical order: (1) the Greeks' early formulations; (2) Aquinas's 'fifth way', or the arrow and archer argument; (3) the 'simple-analogy' argument; (4) Paley's watchmaker metaphor; (5) the guided-evolution argument; (6) fine-tuning; and (7) ID.

Let us review briefly these formulations, starting with the early formulations of the design argument by the Greeks. Case-Winters cites Cicero's book *The Nature of the Gods*, where he presents the Stoics' teleological argument: 'When we see a mechanism such as a planetary model or a clock, do we doubt that it is the work of a conscious intelligence? So how can we doubt that the world is the work of the divine intelligence?' The counter-argument, made by the Epicurean camp as well as the atomist Lucretius, is also cited: 'The world is made by a natural process, without any need of a creator. [. . .] Atoms come together and are held by mutual attraction'. In fact, Lucretius considered the world to be 'badly' made. Plato was impressed with the end goals that objects and phenomena seemed to display, for instance, the growth of human beings: why should bodies grow? In his view the Ordering Mind made and put everything in its best place, and if we want to understand something, all we need is to look for its best place in nature. Aristotle, while still believing in a god or gods, introduced and emphasised final causes (purposes) as an internal mechanism in nature. Because of this distinction between the views of the two great philosophers, Plato's teleology is often referred to as 'external' (i.e. with the emphasis on the designer), while Aristotle's is 'internal' (i.e. with an emphasis on the principle and process that govern phenomena in nature).

Let us jump for now over the Islamic contributions to the formulation of the argument, because I will be devoting large sections to them in the second half of this chapter.

In medieval times, Thomas Aquinas, imbued with the Christian scriptures[i] but also fully familiar with the foregoing philosophical discourse (Greek, Muslim and Christian), produced a solid formulation of the teleological proof; he argued that since we see many natural bodies acting towards a purpose, these bodies lack knowledge and intelligence, they must be

---

[i] For example: 'The heavens declare the glory of God, and the firmament showeth his handiwork' Psalms 19:1 (Old Testament). 'For what can be known about God is plain to them, because God has shown it to them. Ever since the creation of the world his eternal power and divine nature, invisible though they are, have been understood and seen through the things he has made. So they are without excuse' Romans 1: 19–21 (New Testament).

directed towards an end by an intelligent being 'like an arrow is directed by an archer'; this being he identified with God.

The argument from simple analogy was developed in the seventeenth and eighteenth centuries by John Ray[ii], Richard Bentley[iii] and William Derham[iv]. Mainly basing their arguments on observations of nature, but also on the scientific developments of their times (Newton's gravitation etc.), they stated that features of the world and its bodies, living and nonliving, including, for instance, the eye socket, the ear drum, the digestive system, the orbits of planets etc., displayed such perfection that clearly pointed to an intelligent designer.

Paley's watchmaker metaphor[v] is a similar but more sophisticated argument than the above simple analogy. Indeed, he first attempted to identify what should be considered as reliable indicators of ID in a watch: (1) the fact that it performs some function, which gives it value; and (2) the fact that this function could not be performed if some of its parts were made or

[ii] John Ray (1628–1705): One of the most eminent naturalists of his time, he was also an influential philosopher and theologian and is often referred to as the father of natural history in Britain (http://www.ucmp.berkeley.edu/history/ray.html)

[iii] Richard Bentley (1662–1742): English theologian, scholar, and critic.

[iv] William Derham (1657–1735): Anglican clergyman, he was interested in a variety of sciences including natural philosophy and history; he published a number of books and numerous papers in the Philosophical Transactions.

[v] Here's Paley's watchmaker metaphor as described in his own book (*Natural Theology*; or, *Evidences of the Existence and Attributes of the Deity*, 1809, p. 1): 'In crossing a heath, suppose I pitched my foot against a *stone*, and were asked how the stone came to be there; I might possibly answer, that, for any thing I knew to the contrary, it had lain there for ever: nor would it perhaps be very easy to show the absurdity of this answer. But suppose I had found a *watch* upon the ground, and it should be inquired how the watch happened to be in that place; I should hardly think of the answer which I had before given, that, for any thing I knew, the watch might have always been there. Yet why should not this answer serve for the watch as well as for the stone? Why is it not as admissible in the second case, as in the first? For this reason, and for no other, viz. that, when we come to inspect the watch, we perceive (what we could not discover in the stone) that its several parts are framed and put together for a purpose, *e.g.* that they are so formed and adjusted as to produce motion, and that motion so regulated as to point out the hour of the day; that, if the different parts had been differently shaped from what they are, of a different size from what they are, or placed after any other manner, or in any other order, than that in which they are placed, either no motion at all would have been carried on in the machine, or none which would have answered the use that is now served by it'.

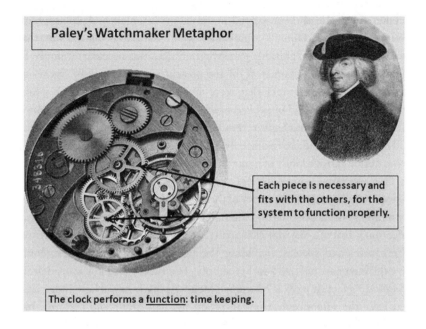

Paley's Watchmaker Metaphor

Each piece is necessary and fits with the others, for the system to function properly.

The clock performs a <u>function</u>: time keeping.

put differently (in size or in arrangement). Paley then made the following reasoning:

- Nature displays the same kind of functional complexity as the watch.
- There must clearly be a reason or explanation for this fact.
- All materialistic explanations completely fail in this regard.
- The only remaining, logical, and acceptable explanation is therefore that of an intelligent designer, a deity.

This kind of reasoning has come to be known (and often resorted to in theological discussions) as 'inference to the best explanation' (IBE). Michael Ruse explains IBE simply by referring to Sherlock Holmes (the detective), who famously told Dr. Watson: 'How often have I told you that when you have eliminated the impossible, whatever remains, however improbable, must be the truth'.

However, it should be noted that Paley's argument was immediately recognised by its critics as an analogy, and not as a proof, so its validity and potency remain subjective and dependent on the attractiveness and applicability of the analogy (to a machine? to an animal?). More importantly, it should be stressed that although this argument received its most crippling blow by the Darwinian theory of evolution, especially in its 'blind', chancy version, it had already been sapped on similar evolutionary and philosophical grounds by David Hume's arguments, who essentially pointed out that even a 'beautiful' or 'glorious' construction does not necessarily imply an omniscient and perfect creator, for the present situation may be the result of many previous attempts that were much less glorious and from which the Artisan learned and improved upon his work (as explained in the quote given at the top of this chapter). Hume also brought up the argument of pain and evil in the world, and pointedly asked what benevolence could possibly come out of various kinds of pains and killings.

Then, of course, came Charles Darwin, who showed that there was a natural explanation to the complexity of the world and of its biological organisms. By offering both a general programme (evolution) and a mechanism (natural selection), Darwin showed that many, if not all, of the observed features of creatures and species, including their geographical distribution, could be explained naturally and rather simply. His theory was not immediately accepted; in fact, it took at least until Mendel's discovery of hereditary laws and later the role of mutations in producing new traits and features, but the theory's impact on the design argument was readily apparent. Darwin himself was troubled by the implications of his discovery: On the one hand he could see the complexity and beauty of 'this wonderful universe and especially the nature of man' (his words) and found that difficult to accept as a chancy development, on the other he could not accept the idea of design in everything around us, and certainly could not reconcile the widespread misery in the world with any preconceived and intelligent design. He compromised and concluded that the world's features can be explained as the result of 'designed laws, with the details, whether good or bad, left to the working out of what we may call chance'[6]. But even this did not satisfy him fully, noting that the whole issue is probably 'too profound for the human intellect'[7].

The 'guided evolution' (or what is more commonly known as 'theistic evolution') argument is an attempt by the theistic camp to digest the Darwinian revolution and integrate it in a design world view. It stipulates that evolution is God's ingenious way of bringing about the glorious diversity and dynamism of nature, and that the means of the evolution process (natural selection etc.) do not preempt the existence of a design plan or even of a covert guiding mechanism by which God ensures that each creature fulfils its purpose. The first proponent of this viewpoint was Frederick Robert Tennant[vi],

---

[vi] Frederick Robert Tennant (1866–1957): English clergyman and philosophical theologian who espoused an empirical approach to theology and insisted on the harmony between science and religion.

who argued for design to be seen in the 'quality of the evolution process' and not on the functional complexity of the organisms. I shall have much more to say about evolution, in both its classical random (naturalistic) and theistic versions in Chapter 9.

The fine-tuning argument was exploded on the scientific, philosophical and religious scene in the last quarter of the twentieth century; it argued that the very fabric, structure and mechanisms (building blocks, laws etc.) of the universe are found to be fine-tuned to the appearance and existence of intelligent life (humans and possibly other advanced species). The argument then goes on to imply that such a (teleological) preparation and steady progress of the universe towards the appearance of such species can justifiably be interpreted as a predetermined design plan, a 'cosmic blueprint' (in the words of Paul Davies).

Finally, the most recent and controversial version of the design/teleological argument is the ID hypothesis. I devote a section of this chapter to this controversial topic, with both the main ideas put forward by its proponents and the recent reactions by Muslims.

the design argument in Islam

The Qur'an relates Abraham's story as the first instance of philosophical and rational contemplation of nature and the extent to which it could lead to God[8]. In the story, although God encourages Abraham to observe the world and make use of his faculties, the latter does not succeed in his rational discovery of God; indeed he exclaims: 'Unless my Lord guides me, I surely shall become one of the folk who are astray' (Q 6:77). One should also note that the Qur'anic verse seems to point to the design or order of the world ('When the night covered him over, He saw a star: He said: 'This is my Lord' [. . .] When he saw the moon rising in splendour, he said: "This is my Lord" [. . .] When he saw the sun rising in splendour, he said: "This is my Lord, this is the greatest of all"'. Q 6:76–78) as a confirmatory argument and not as an inductive one.

Probably the earliest version of the design argument goes back to al-Kindi who, referring to the *dalil al-'inayah* (argument of 'providence' or 'benevolence'), proclaimed that 'the orderly and wonderful phenomena of nature could not be purposeless and accidental'[9]. The prominent classical Asharite theologian and jurist al-Baqillani (950–1013) expressed the argument as a simple analogy: the world must have a Maker and a Fashioner (*muhdith wa musawwir*) 'just as writing requires a writer, a picture must have a painter, and a building a builder'[10].

al-Ghazzali too expounded on the argument in his largely unknown book, titled *al-hikmah fi makhluqat-i-Llah*[11] (*The Wisdom in God's Creation/Creatures*), a book edited and published for the first time in 1978, which is fully devoted to exploring the world and finding insights into its workings and its utility and adequacy for human life. In the introduction he writes

> Do know, [dear reader], May God Bless you, that if you contemplate this world, you will find it like a house where everything needed has been placed, for the heavens are raised like the ceiling, the earth is extended like a carpet, the stars are hung like lamps [. . .], and man is like the master of the house, who has prerogatives to rule over it, where vegetation is made for his needs, and various animal species are subjected to his exigencies . . .

He adds: 'These are clear arguments that point toward their maker, their masterly and eternal design point to the extent of knowledge of their creator, and their arrangement aspects point to the intentions of their producer'[12].

Abu Hamid al-Ghazzali (1058–1111): If today one takes a survey to determine the most influential Muslim scholar of all time, al-Ghazzali

would most likely rank first. Indeed, I recently heard Sheikh Ali Gomaa, the current grand Mufti of Egypt, answer the question, 'Which book (other than the Qur'an) would you recommend for all Muslims to know as well as possible?' He quickly replied: 'al-Ghazzali's *Ihya 'Ulum al-Deen* (*The Revival of the Religious Sciences*). It has been said that all Islamic knowledge can be reconstructed from this book, should it all be destroyed at any moment . . . '

Abu Hamed al-Ghazzali was born in 1058 in Tus, in the Persian province of Khorasan, which is a part of today's Iran. He was educated in the classical religious sciences, in which he excelled. He impressed the prime minister of the time and was soon appointed to the Nizamiyya school (university) of Baghdad, where he became very popular among students and professors for his willingness and ability to engage in debates. He later came to be known as the *Hujjatu l-Islam* (The 'Proof' of Islam), the *Sharafu l-'A'imma* (The Honor of the Scholars), and *Zain u-Deen* (The Glory of Religion).

At the age of 37, however, he underwent an intellectual crisis when he started to doubt the bases of his knowledge. His subsequent epistemological explorations and attempts to rebuild knowledge on what was certain (divine enlightenment, according to him) are often compared to Descartes' *Discourse on Method*. During his crisis, Abu Hamed stopped his academic career and undertook a journey (literally and spiritually), which landed him in Sufism. His intellectual autobiography *al-Munqidh min al-Dalal* (*The Deliverance from Error*) is a small masterpiece, though it won't convince any rationalist or scientist today. Indeed, al-Ghazzali starts with his famous strong critique of Islamic/Hellenistic philosophy, decrying both its methods and results. In one of his most renowned and unfortunately influential books, *Tahafut al-Falasifah* (*The Incoherence of Philosophers*), he judges al-Kindi, al-Farabi and Avicenna (especially the last two) and finds them guilty of heresy on at least three counts: the eternity of the world, God's knowledge of particulars, and the non-physical nature of resurrection. Many have blamed al-Ghazzali for the subsequent decline of philosophy, science and intellectual life in Islam. Today we know that many other and more important factors are responsible for that decline.

Indeed, in the western parts of Islamdom, in Andalusia in particular, philosophy thrived for at least another century, with Ibn Baja, Ibn Tufail and Ibn Rushd, the latter writing a full rebuttal to al-Ghazzali's *Tahafut at-Tahafut* (*The Incoherence of the Incoherence*), where he showed that Abu Hamed had misunderstood Hellenistic philosophy (partly because Avicenna had misunderstood Aristotle to some extent), and that – as I have explained – philosophy and religion (both truths in different ways) cannot be in disagreement, much less in opposition.

al-Ghazzali gave both the orthodox Muslim scholarship and Sufism their firm groundings and their glorious days. The Ash'ari version of Islamic theology thus became the dominant system of belief for Sunni Muslims in most of the world. Never mind that Abu Hamed introduced and insisted on the 'occasionalist' view of God's action, namely the events that occur because God wills them at every instant, and *not* because any natural cause happened to occur at that moment. His famous example, cotton burns not because it came in contact with fire, but because God decided so at that instant, to him and to orthodox Ash'ari Muslims after him, preserved God's omnipotence and explained miracles.

al-Ghazzali wrote over 70 books and essays on theology, Sufism, philosophy and jurisprudence. He remains an icon of Islamic classical scholarship, although for philosophy and science his legacy and influence were minimal, if not negative.

In the various chapters of this book, each devoted to one creature/body in nature, al-Ghazzali cites Qur'anic verses to support his teleological anthropic principle; for instance, in the chapter on the wisdom behind the creation of the earth he cites: 'And He has set up on the earth mountains standing firm, lest it should shake with you; and rivers and roads, that ye may guide yourselves' (Q 16:15); in the chapter on the seas he cites: 'In the sailing of the ships through the ocean for the profit of mankind [. . .] indeed are signs for people who are wise' (Q 2:164); on the human body he writes: 'Then see if you find in the creation of the human body something that has no meaning, wasn't vision created except to perceive objects and colors, then if colors existed and vision did not, would there be any value in colors [. . .], and wasn't hearing created except to perceive sounds, so if sounds existed and there was no hearing to perceive them, would there be any benefit in them, and so it is for the rest of the senses'[13]. al-Ghazzali thus finds it logical to draw a general design conclusion: 'All of these [examples] are different testimonies and mutually consolidating proofs, eloquent signs with regard to their creator, speaking clearly for the perfection of His powers and the wonder in His wisdom [. . .] It is the work of the Omnipotent, the Guardian and Mighty'[14].

It is always interesting and instructive to find similarities and reverberations in opinions and writings of thinkers from different cultures and across the ages. Indeed, echoing al-Ghazzali, we find the clergyman – naturalist John Ray[15] writing (six centuries after al-Ghazzali)the following:

> There is no greater, at least no more palpable and convincing argument of the Existence of a Deity, than the admirable Art and Wisdom that discovers itself in the Make and Constitution, the Order and Disposition, the End and uses of all the parts and members of this stately fabric of Heaven and Earth[16].

In our times, Michael Denton[vii] echoes further: 'It is the sheer universality of perfection, the fact that everywhere we look, to whatever depth we look, we find an elegance and ingenuity of an absolutely transcending quality, which so mitigates against the idea of chance'[17].

[vii] Michael Denton is a British-Australian geneticist (currently at New Zealand's University of Otago) who accepts evolution in general but finds important flaws in the theory as it is presently formulated; his main contribution to the field is his book: *Evolution, a Theory in Crisis* (1985).

al-Ghazzali finds numerous Qur'anic verses that point to design in nature, for instance, 'Do they not look at the sky above them, how We have made it and adorned it, and there are no flaws in it?' (Q 50:6), and 'It is He who sends down rain from the sky: from it ye drink, and out of it (grows) the vegetation on which ye feed your cattle. With it He produces for you corn, olives, date-palms, grapes and every kind of fruit: verily in this is a sign for those who reflect' (Q 16:10–11).

(Averroes explains that the reason the design argument is used so widely and strongly in the Qur'an is because the Revelation wants to reach and convince people of all mental capacities, and design lends itself so well to this exercise.)

But al-Ghazzali does not content himself with the examples given in the Qur'an, he explores nature for more phenomena and signs of design. He mentions, among many examples, the following:

- Stars help people in determining directions and time. The moon was placed so as to help people in the activities they may need to undertake at night.
- Wood is lighter than water in order for boats to float and carry people on the seas.
- The multitude, diversity, and harmony of animal species is an indicator of both the Creator's omnipotence and the role each species is to play in nature (including aesthetics), even if the roles are often difficult to identify[18]. al-Ghazzali concludes: 'Look how the wisdom of the Creator preceded the art of the creation.'[19]

Finally, al-Ghazzali gives his general conclusions: 'What do you (reader) then think of a creation that has reached such levels? Raise then your sight and infer the might of this Creator, His majesty, power, and knowledge, the necessity of occurrence of His will, and the perfection of His wisdom with regard to His creatures'. He further adds,

Whoever looks at all this with his mind and heart benefits knowledge of his God and due regard to Him and His matters; there is no escaping this implications for those who reflect, for every time the mind reconsiders the wonders in the design and the splendor in the creation, it (the mind) deepens its knowledge and conviction [of the Creator] and achieves submission to and exaltation of the Master . . . [20]

Ibn Rushd's design considerations

Ibn Rushd too has referred to the teleological argument in several ways in his major works. In his treatise *Fasl al-maqal* (see Prologue), he begins his demonstration of the lawfulness (and for a class of people, the obligation) of executing philosophy in Islam by defining it as the 'teleological study of the world'; indeed, the second paragraph starts as follows:

> We say: the activity of 'philosophy' is nothing more than study of existing beings, and reflection on them as indications of the Artisan, i.e. inasmuch as they are products of art – for beings only indicate the Artisan through our knowledge of the art in them, and the more perfect this knowledge is, the more perfect the knowledge of the Artisan becomes[21].

It is immediately clear that Ibn Rushd subscribes not only to the design argument but also finds in it a raison d'être for a kind of 'natural theology' research program.

The great philosopher makes the design argument a bona fide proof for the existence of the Maker more explicitly in *al-Kashf*, as can clearly be read in the quote I put in the beginning of this chapter[22]. His adoption of the design and teleological arguments could not be more explicit. In fact, this is almost exactly Paley's version of the argument, expressed here more than six centuries earlier. Elsewhere Ibn Rushd refers to these as 'the argument of artisanship' (design), or alternatively as 'the argument of providence' (*dalil al-'inayah*), i.e. towards humans.

Our great philosopher insists that the recognition of design as defined earlier, that is, 'the existence in the artifact of an order in its parts', is a greater channel of knowledge of God than the mere belief that objects have been made; he makes his reference more explicit by stating that the philosopher (or for us, the scientist) not only knows God better than the general public but also better than the theologian who ignores this argument.

Ibn Rushd argues further that the existence of design or a cosmic blueprint is more in line with the wisdom of God than the Ash'arite theological conception of God as a sustainer (in a puppeteer way) of all the acts in the world, the latter – in his view – practically amounts to denying the wisdom of God. Our philosopher sees in the search and identification of design and teleology of all creation in the world a major goal of the philosopher (scientist), who is called upon to receive some of God's wisdom (in other words, what Stephen Hawking called 'knowing "the mind of God"'[23]). Marc Geoffroy in his (French) translation of and commentary on Ibn Rushd's work, emphasises this relation between the Wisdom of God and the wisdom that the philosopher receives in undertaking such reflections and searches; he writes: '"The Wise" (*al-Hakim*) is one of the ninety-nine names of God in the Qur'an. A God who has thought out the whole cosmic order on the basis of absolute unity is the Wise par excellence, to Whom the wise philosopher gets close in a certain way when he attempts to absorb the intelligible structure of the universe insofar as the human intellectual capacities are able to'[24].

Averroes insists that the above two arguments (providence and creation) deviate from the Ash'arite theology of divine creation and sustenance of the world but also have the merit of being so clear and explicit as to be useful to both the general public and the elite; the difference between the apprehensions of these two arguments by the two camps is in the details that each will have at its disposal and in the depth to which each can go in considering the observations. He concludes that this method (based on those arguments) is indeed in harmony with both the religious approach (*shar'iyyah*) and the natural (scientific) one; indeed, he insists, this is the method that has been taught by the prophets and revealed in the sacred books.

design/telos in modern Islamic traditions

For the sake of simple, less than formal, classification, modern-day Muslim thinkers can be divided into three classes: (a) religious scholars of all specialties; (b) philosophers, with whom I will lump social scholars from all fields; and (c) commentators of any other discipline or orientation, including scientists. The first group will thus comprise scholars like Mohammad Abduh, Rachid Rida, Mohamed Tahar Ben Achour and others; the second group will consist of thinkers such as Muhammad Iqbal, Mohamed Talbi, Seyyed Hossein Nasr and others; and the third group will include a variety of people, like Maurice Bucaille, Zaghloul al-Naggar, Harun Yahya and others. Covering the views and beliefs held by all these thinkers, even if one focuses only on a single concept, namely the design/teleological argument, will take time and effort. Thus, here I will only brush briefly over the views of such a spectrum of thinkers.

Twentieth-century Islamic scholars, especially such modernists and rationalists as Abduh or Ben Achour, have strongly upheld the design argument because it can be supported by the Qur'anic world view and it lends a hand to science and to the modern, rational and natural approach to theism.

Among the philosophers, an important voice in modern times is Muhammad Iqbal (see the sidebar). Iqbal gave little regard to any of the aforementioned philosophical arguments for the existence of God; indeed he disposed of all three in less than two pages in his philosophical testament *The Reconstruction of Religious Thought in Islam*, the edited series of lectures on the status of philosophical and theological thought in the contemporary Islamic nation.

Muhammad Iqbal (1877–1938): is one of the greatest poets and philosophers of modern Islam, of the Indo-Pakistani subcontinent, and of the

religious tradition of humanity. Having received a rich education in the traditional religious and literary heritage of Islam in his home province of Punjab (in today's Pakistan), he discovered western scholarship in Lahore under Thomas Arnold, who was then teaching at the Government College. Iqbal soon followed his teacher back to England and enrolled in Cambridge, which was reputed for the study of both European philosophy and the Arab and Persian traditions. From his early years, Iqbal's talent in poetry was noted, and indeed his first collection, *Asrar-e-Khudi* (*Secrets of the Self*), written and published in Persian in 1915, was soon translated into English and hailed; it would later earn him a knighthood.

He obtained BA and barrister's qualification from Cambridge and then travelled to Munich; there he earned a doctoral degree with his thesis *The Development of Metaphysics in Persia*, which was published in 1908. His interest in philosophy and metaphysics thus solidified, he turned back to his first love, poetry, and started to address many of the great themes of human anguish and artistic expressions: head and heart (intellect and emotion), motion and rest (change and constancy), choice and force (freedom and determinism), climbing and falling (progress and decadence), earth and stars (body and spirit), spring and winter (life and death) etc. No surprise then that many of his poems would be written as dialogues and struggles, where the outcome is never clear.

Iqbal would also display a refined spiritual sense; indeed, he later related the great spiritual experience he went through in London in 1907, though he only realised the meaning of his vision many years later. His greatest icon, one who would often appear as a guide in his poems, was the supreme Sufi master Jalal ud-Deen Rumi. Iqbal also showed some influence from Ibn Arabi (the Sheikh ul-Akbar of Sufism). One should not, however, try to describe Iqbal in any one dimension, for indeed the influence of western philosophy on him was also quite evident. Indeed, he acknowledged his debts to Kant, Nietzsche, Bergson and Goethe, in particular, and one can find in some of his poems parallels to Dante's *Divine Comedy* as well. His most important work of philosophy is the series of lectures he delivered in Madras, Hyderabad and Aligarh in 1930, which were published in 1934 as *The Reconstruction of Religious Thought in Islam*, a few years before his death, which I shall be referring to several times in this book.

Finally, and to find some link with our Averroesian thread, let us mention that shortly before his death, Iqbal, who came to be known as '*Allama* (Superior Scholar), visited Spain, where the past glories of Cordova and Seville inspired him to write *Bal-e-Jibril* (*Wings of Gabriel*), widely considered to be his finest Urdu poetry collection.

First, Iqbal strongly rejects the cosmological argument, claiming that 'a finite effect can give only a finite cause, or at most an infinite series of such causes. To finish the series at a certain point, and to elevate one member of the series to the dignity of an uncaused first cause, is to set at naught the very law of causation on which the whole argument proceeds'[25]. He concludes, 'Logically speaking [. . .] the movement from the finite to the infinite as embodied in the cosmological argument is quite illegitimate; and the argument fails in total'[26]. (Kant had denied the validity of the argument, considering it a mere variant of the ontological argument.)

As to the ontological argument, Iqbal dismisses it even more quickly, deeming it 'somewhat of the nature of the cosmological argument which I have already criticized. [. . .] All that the argument proves is that the idea of a perfect being includes the *idea* of his existence. Between the idea of a perfect being in my mind and the objective Reality of that being there is a gulf which cannot be bridged over by a transcendental act of thought'[27].

And for Iqbal,

the teleological argument is no better. [. . .] At best, it gives us a skilful external contriver working on a pre-existing dead and intractable material the elements of which are, by their own nature, incapable of orderly structures and combinations. The argument gives us a contriver only and not a creator; and even if we suppose him to be also the creator of his material, it does no credit to his wisdom to create his own difficulties by first creating intractable material, and then overcoming its resistance by the application of methods alien to its original nature. [. . .] The truth is that the analogy on which the argument proceeds is of no value at all. There is really no analogy between the work of the human artificer and the phenomena of Nature'[28].

And to make his views very explicit, Iqbal then concludes: 'I hope I have made it clear to you that the ontological and the teleological arguments, as ordinarily stated, carry us nowhere'[29].

Despite the virulence, hastiness and absoluteness of the criticism, Iqbal's viewpoints are not that surprising. Indeed, one must recall the central principles of his philosophy: (a) Iqbal was very skeptical and suspicious of the rationalist approach and philosophical enterprise, identifying it with the materialistic atheistic philosophy, which he rejects at all levels; (b) he was trained in and influenced by the European traditions that broke away from the 'pure reason' approach; (c) he was to a large extent seduced by the mystic religious experience, which he saw not only as a valid thought process and experience but further considered as a liberator from the strait-jackets of rational thought; and (d) his programme for the reconstruction of religious thought in Islam consisted of digesting as much as possible the intellectual and scientific developments in the West and integrating those in a newly merged holistic mental approach. It seems, however, that Iqbal did not have the proper training to digest at least the scientific aspects of the western contributions[viii] and that he got carried away in both his suspicions of rational approaches and his attempt at adopting and validating mystical experiences.

And among the prime representatives of the third group of modern thinkers as presented above, that is among scientists and other commentators, one cannot but mention Harun Yahya and Zaghloul al-Naggar; the first is the Turkish writer and leader who over the past decade or so has

---

[viii] For example, even though Iqbal's book was first published in 1930, we still find him speaking of 'ether waves' (when the ether had been dropped from scientific vocabulary for some 30 years) and discussing the concept of time in Einstein's Relativity theory in a very confused way (pp. 36–43).

been waging an aggressive campaign against 'atheistic' theories like evolution, Marxism and materialism, and organizations such as free-masonry and others; the second is the retired geology professor who now devotes his time entirely to the theory of *I'jaz* (see Chapter 5). Let me note that Yahya has made the design argument a centrepiece of his philosophy and campaign. Indeed, several books of his contain the words 'design' or 'creation' in their titles; among them are *The Design in Nature*; *The Perfect Design in the Universe Is Not By Chance*; *Signs of God Design in Nature*; *Tell Me About the Creation*; *The Miracle of Creation in DNA*; *The Miracle of Human Creation*; *Wonders of Allah's Creation* etc.

Going back to the concept of design in Harun Yahya's vision of nature, I may cite a few passages from his book *Allah is Known through Reason*[30]:

- Take a look around you from where you sit. You will notice that everything in the room is 'made' [. . .] A person who is about to read a book knows that it has been written by an author for a specific reason. [. . .] And not just works of art: even a few bricks resting on top of one another make one think that they must have been brought to rest just so by someone within a certain plan.
- When we examine the structure of atoms, we see that all of them have an outstanding design and order.
- In fact, a flawless order has prevailed at every point since the beginning of existence. [. . .] First, electrons find themselves a nucleus and start to turn around it. Later, atoms come together to form matter and all these bring about meaningful, purposeful and reasonable objects.

It quickly becomes clear that Yahya's design arguments are of the most simplistic kind.

Zaghloul al-Naggar is more interested in proving the scientific miraculousness of the Qur'an than in upholding any rational or naturalistic argument for the existence of God[31]. Within that framework and research program, al-Naggar seems to endorse the design argument, although he never refers to the argument per se or even to the concept of design. At most, we find in his writings general pronouncements such as, 'No rational man can ever imagine this masterly creation of the universe is anything but a command from The Creator (Exalted be He) Who Says [in Q 36:82] "Verily, when He intends anything, His Command is only to say to it: 'Be' and it is!"'[32].

intelligent design and Islam

Let me now briefly review the 'theory' (or rather hypothesis) of Intelligent Design (ID) and investigate the extent to which it has penetrated into Muslim minds.

The leaders of the ID movement, which is a cultural and political one draped in 'scientific' garb, purport to show that there exists in nature at least a number of cases that exhibit aspects of divine design. They try to avoid clashing with the American separation of religion and state and thus always use 'intelligent' for 'divine', as their aim is to have this 'theory' taught in schools. This movement's goal is first to oppose Darwin's theory and ultimately to destroy the foundations of modern science's materialism, which rests on naturalism, as we have seen.

The first thing that one must explain about ID is that its leaders mislead the public by trying to elevate their idea to the rank of a theory. Indeed, ID has had no aspect of it confirmed by science; in fact, it makes no predictions whatsoever and does not even explain any phenomenon scientifically; hence, many have declared it non-scientific, for it violates the essential falsifiability criterion of Popper. For the purposes of our discussion here, let us consider it a hypothesis, although even as that it is somewhat confusing.

Until recently, ID was built around two main arguments: (a) irreducible biochemical complexity (IBC, developed by Michael Behe); and (b) specified biological information (SBI, presented by William Dembski)[33]. Recently Behe introduced a slightly different argument, i.e. the limited capacity (or 'edge') of evolution in its action, or more precisely of random mutations.

The IBC argument (first presented by Behe in 1996 in his book, *Darwin's Black Box*) states that particular biochemical parts of certain organisms are 'irreducibly complex', meaning that they can only function (play a role) when they are put together with other irreducibly complex parts; that they by themselves do absolutely nothing; and, most importantly, one cannot show that they would have evolved according to Darwinian processes. Behe and his followers thus conclude that an intelligent designer must have built these parts (directly) and assembled them. I think it will be clear to all that such reasoning is very much like the old 'god of gaps' approach, which instead of looking for an explanation to a mystery, simply plugs in God (or a designer). Even theists have long dropped such a useless approach. The clever cartoonist Sydney Harris nailed this point perfectly with his cartoon in which a scientist responds to what could be an IDer showing a series of statements with 'Then a miracle occurs . . . ' in the middle, and the skeptical colleague undercuts him with 'I think you should be more explicit here'.

In parallel to that, Dembski developed a probabilistic approach to try to prove that the information content of DNA nucleotides can best be explained by an ID hypothesis and not from blind approaches. In my opinion, this reasoning is much more scientific; it is negative in the sense that it finds flaws in a given theory and does not present a solution to the problem it identifies, but it is at least scientific. Should Dembski be able to prove his thesis in a number of cases (a very doubtful prospect), he would then have given a hard blow, though not necessarily fatal, to some essential parts of the Darwinian theory.

I think it is this route that Behe decided to follow when he published his recent (2007) book *The Edge of Evolution*, where he tries to show that the probability for random mutations to lead to the development of complex organisms and their current features is extremely small. Leading biologists[34] have shown that his calculations are seriously flawed and his thesis/argument does not make the slightest dent to the standard theory of evolution or any part of it. Let me mention in passing that Behe in this book makes some surprising statements (for an IDer); he accepts 'common descent' (that living creatures, such as humans and chimpanzees, have descend from common ancestors) as 'trivial'; he has no problem with the 'modest concept' of natural selection; he concedes that evolution works very well at 'limited' levels, such as drug development; and, of course, he accepts the long geological ages of the earth (4.5 billion years) and rejects the ridiculous young-earth creationist views.

Now, what do Muslims think of all this? First, the ID controversy has not mobilised the media and commentators in the Muslim world as it did in the West. In Turkey it has generated a strong supportive and active movement, led (initially) by Mustafa Akyol (who has since dropped his support for ID) and (independently) supported by Harun Yahya, but elsewhere, very few people even know the basics of ID. Most of those who have heard about it have the impression that it is a strong scientific rebuke to Darwinism, and as such they welcome it warmly. We will thus not be surprised to find a book by Harun Yahya titled *The Collapse of Darwinism and the Triumph of Intelligent Design*[35]. Akyol, a young Turkish intellectual, was initially seduced by ID and thus wrote at length and spoke (very knowledgeably) about it[36], organised conferences on the subject in Turkey[37], and testified as an expert on the ID movement in the USA. One of the first pro-ID articles to target the Muslim audience directly is Akyol's 'Why Muslims should support Intelligent Design'[38]. A few other media articles have appeared on the subject, most of the time showing minimal and superficial knowledge and understanding of the topic[39].

To my knowledge the only Muslim to have written a long and serious article[40] against ID is Ahmed K. Sultan. He, who at the time (2004) was a doctoral student in space sciences, radio science and telecommunications at Stanford University, wrote 'The non-science of intelligent design', in which he shows a clear and solid understanding of the IDea, dissects it to expose its methodological flaws, and warns against the mirage of any impending 'victory over Darwinism'. Another Muslim, Rizwana Rahim, a medical doctor in the USA, has written an interesting article on the subject. She wrote 'Darwin vs. Intelligent Design', of which only the first part can be found on the web[41]. In it, Rahim shows a keen understanding of the biological details of both the theory of evolution and the ID critiques; she also seems to adhere to Darwin's theory, but the article ends abruptly; perhaps, the editors did not like the logical pro-evolution, anti-ID conclusions the writer must have drawn in the second part of her article.

Clearly a strong effort to inform and educate the Muslim public is urgently needed. Troublesome developments are taking place in this regard; for instance, in October 2006, the Turkish Minister of Education expressed support for ID, calling for the inclusion of the ID theory into Turkish high school biology textbooks[42].

conclusions

At the end of this review, and having noted some serious disagreements as to the significance, the relevance and the importance of the design argument, we must ask ourselves: What value is there in this argument – in general and within the Islamic tradition, especially today?

In my view, the design argument must be considered as one of the main pillar(s) of the natural approach to belief, i.e. the approach that stresses reason, as opposed to the revealed theological approach that stresses faith, acceptance and religious authority. But then how do we counter the modern consensus that design, or at least the simplistic considerations of al-Ghazzali and Harun Yahya, has been rendered obsolete and meaningless in the theological debates?

In my opinion, design has no reason to be opposed to evolution, especially if the latter is not totally 'blind'. Indeed, and as Case-Winters notes, design does not have to be understood in such a constraining mode; she writes: 'What if it is part of the design that some things happen by necessity, others by chance, and others in open interplay of relative freedom? A design might include a whole range on the spectrum: contingency as well as regularity, chaos as well as order, novelty as well as continuity'[43].

Indeed, several modern and progressive theologians have reformulated the concept of design to imply a general directionality or 'vector' of continuous, evolving creation and not as a ready-made and detailed blueprint[ix]. In the most liberal interpretation, design can even be regarded as simply the underlying conditions that make life, consciousness and intelligence possible, something akin to the fine-tuning of the universe's parameters that lead to the anthropic principle. Obviously, such a concept of design is much closer to standard evolutionary theory than to divine providence or divine intervention.

The idea/argument of design seems to be taken for granted in the Islamic traditions, with its potency and relevance accorded varying degrees of importance. In fact, we have found a rich tradition of design argument discussions in the classical Islamic philosophies, including the original and potent idea of Ibn Rushd that this argument (a) is the clearest one in favour of the existence and adoration of God; (b) is clearly present in the Qur'anic text (e.g. Q 88:17; Q 22:73); and (c) allows for layers of appreciation to be found in understanding and appreciating it between the laymen and the elite.

I have also exposed an important connection in contemporary Muslim writings between a strong emphasis on design and a near-total rejection of Darwinian evolution, although the evolutionary scenarios of matter in the universe, that is, the Big Bang cosmology and the very long ages of the universe, the solar system and the earth, are commonly accepted.

The fact that the Darwinian revolution has been and continues to be disregarded by the Islamic theologians and thinkers has prevented the latter

---

[ix] Among western contemporary scientists who have upheld this idea, one may cite Paul Davies, who has proposed a principle or law of increasing complexity – see his article 'Emergent complexity, teleology, and the arrow of time' in W. A. Dembski & M. Ruse (ed), *Debating Design* (2004), p. 191. Among Muslim thinkers, we may cite Mohamed Talbi who promotes the idea of an 'oriented vector'; he writes: 'Man must be positioned in that creative momentum that goes from the first amino-acid to our days, a creative spirit that constitutes, in the formula of Teilhard de Chardin, the ascending arrow of creation'. (Mohamed Talbi, *Plaidoyer pour un Islam Modern*, de l'Aube edition, 2004, p. 67, my own translation).

from exploring the design argument and trying to find those aspects of it that can still withstand both the Islamic theological framework and the evolutionary biological paradigm. This has also kept Muslims unaware and outside the strong debates (intellectual and socio-political) raging around in the West regarding such topics as ID, from a scientific, theological or sociological point of view.

8

Islam and the anthropic principle: was the universe created for man?

'A life-giving factor lies at the centre of the whole machinery and design of the world.'

John Wheeler[1]

'God created all things in the correct proportion and harmony, and the world is dominated by this remarkable harmony, which is the imprint of unity upon the domain of multiplicity.'

Seyyed Hossein Nasr[2]

a very special universe

Imagine that you have been sentenced to death and are about to be executed. A dozen shooters are facing you from a short distance, and your eyes have been covered. The order to shoot is given; you hear the fire shots. But a few seconds pass, and you realise that you are perfectly fine; the shooters have all missed you – or maybe they had blanks . . .

What would be your reaction? Do you tell yourself: there is nothing to explain; I am fine; case closed . . . ? Or do you begin to think: there is something behind this event; somehow my life has been saved for a reason . . . and perhaps for a purpose?

This nice allegory was first used by the philosopher John Leslie[3] to describe the situation that scientists have found the universe to be in after discovering that many features of the cosmos are astoundingly fine-tuned to our existence, or to the emergence and evolution of life, more generally. Indeed, if the physical parameters which make up the universe had been drawn at random, the probability that they would have values allowing for life and intelligence to appear (at some point in time and space) would be ridiculously small, one in billions of billions of billions . . .

Many thinkers have recognised the huge significance of this discovery, perhaps the most important one of all time unless it is simply the result of a bias factor, as we shall discuss. In fact, evidence for the 'specialness' of our universe has been accumulating for at least the past century, with the pace accelerating in recent times. As a result, an anthropic principle was formally proposed about thirty years ago, declaring the universe to be particularly fit for life or even for humans ('anthropo' meaning 'human' in Greek). Indeed, over the past few years many significant publications (magazines or books) in the West have devoted quite a number of pages to this hot topic. To name just a few, the serious French magazines *Ciel et Espace* and *Philosophie* each devoted a whole special issue and a dossier to the subject in Fall 2006 and May 2007. And Paul Davies, perhaps the foremost scientist – philosopher[i] of this generation, recently published *The Goldilocks Enigma: Why Is the Universe Just Right for Life?*[4]

Davies refers to the anthropic principle as 'nothing less than a revolution in scientific thinking'. Nicola Dallaporta[ii] says that it should be considered as 'a decisive moment in the development of science, having opened new avenues toward unknown aspects of the universe'[5]; and George V. Coyne[iii] emphasizes the 'exciting meeting point that has appeared between theology and the sciences, particularly in that the human factor has been reintegrated after having been excluded from the physical sciences for centuries . . .'[6]

I should quickly note that not everyone is thrilled with this development. Malcolm S. Longair (Professor of Natural Philosophy at Cambridge

---

[i] Paul Davies has published, among many others, *The Cosmic Blueprint*, *Are We Alone?* and *The Mind of God*; in 1995 he was awarded the Templeton Prize for progress toward research or discoveries about spiritual realities.

[ii] Nicola Dallaporta is an astrophysicist who was professor at the University of Padua and member of the board of directors of the Pontifical Academy of Sciences in the Vatican City.

[iii] George V. Coyne is a Director Emeritus of the Vatican Observatory and head of the observatory's research group based at the University of Arizona.

University and author of many books), for instance, has been quoted as saying[7]: 'I hate the anthropic principle theory; I consider it as an absolutely last recourse, in case all physical arguments failed. The whole essence of the anthropic argumentation seems to go against the flow of what we aspire to achieve as scientists'. More recently, Michel Paty, the emeritus director of research at the French National Center of Scientific Research (CNRS), declared, 'The strong anthropic principle is a useless metaphysical concept'[8].

The anthropic principle is indeed a very hot and controversial topic today for several reasons. First, it obviously tends to bring up the ideas of design that we discussed in the previous chapter, and thus many scientists and philosophers tend to be incensed with the subtle (or not so subtle) implications of a creator, a designer, a cosmic blueprint etc. (The whole *Ciel et Espace* special issue was titled 'Does the Universe need God?' and one of its main articles was titled 'Is "the Anthropic Principle" the new name of God?') Secondly, the very idea of anthropic principle can be regarded as a rolling back of the Copernican Revolution, which had displaced earth and man from the central place they had always occupied in the universe and in the old philosophical and religious considerations. Indeed, the more we discovered about the universe, with its immense scales of space and time and its innumerable and humongous objects of all kinds, the more we realised that we are nothing but a speck of dust in this near-infinite cosmic landscape. And the more we explored life and the human and animal species, the more we convinced ourselves that we are nothing special. As Wheeler sarcastically commented in the foreword to the most important (and imposing) work on anthropic principle (Barrow and Tipler's)[9]: 'Man? Pure biochemistry! Mind? Memory modelable by electronic circuitry! Meaning? Why ask after that puzzling and intangible commodity? [. . .] What is man that the universe should be mindful of him?'[10] But barely a paragraph later, Wheeler explains, 'No! The philosopher of old was right! Meaning is important, is even central. It is not only that man is adapted to the universe. The universe is adapted to man'.

fine-tuning in the universe

As we saw in the previous chapter, the idea that the world is well designed and made adequate for humans is both old and ubiquitous among many cultures, including the Greek, the Christian and the Islamic. The design argument did suffer serious blows in the wake of the Copernican, Darwinian and cosmological revolutions, but one may say that contrary to the elites, laymen overwhelmingly continued to believe that humans are special and that the earth and perhaps the universe were designed (directly or indirectly) as cradles for humanity. So what was new with the discovery of the fine-tuning of the universe?

First, although the 'principle of mediocrity' had been widely accepted, it came as a shock to the (western) elites to realise that the universe instead of being completely oblivious to humans, was in fact particularly suited for life, consciousness and intelligence. Secondly, it was no longer a matter of seeing beauty and harmony in nature, nor even a set of smart observations such as the temperature, pressure, gravity and environment of the earth being 'just right'[iv] for our existence and activity; it was now a question of the very foundations of the universe, the parameters and physical laws upon which everything was built, all of which were found, time after time, case after case, to be finely tuned to the existence of life in general, con-sciousness in particular and higher intelligence above all. In short, if the characteristics of the physical universe (and in physical, one includes the chemical, the biological, the geological etc.) had been just a little different from their actual values, obviously we would not be here to wonder about them, but most importantly the universe itself would most likely not have developed its various forms, i.e. galaxies, stars and planets would not have formed.

Let me give a few simple examples of this fine-tuning.

If gravity had been slightly weaker in the universe, stars would have never formed and hence carbon, oxygen and other crucial ingredients of life would never have been made (except for hydrogen, all these elements are made inside stars); if gravity had been slightly stronger than what it is in the cosmos, the universe would have collapsed upon itself sometime after the Big Bang but before the formation of galaxies, stars and planets.

[iv] Barrow and Tipler (*The Anthropic Cosmological Principle*, p. 143) make a special reference to the two seminal books by Lawrence J. Henderson (Harvard professor of biological chemistry), *The Fitness of the Environment* (1913) and *The Order of Nature* (1917); indeed, Henderson had noted the very special regulation of acidity and alkalinity in living organisms; CO_2 dissolved in water is the regulator of neutrality, and water is absolutely unique as a regulator and conductor of heat (having a very large specific heat capacity and conductivity), in its surface tension, in dissolving other substances, and in many other properties.

Let us now consider electricity; if its basic brick, the electron, had a slightly smaller charge, chemical reactions would have been too slow, and it would have taken forever to produce any complex molecules (like DNA, which carries the code of life); if the electron's charge had been somewhat larger, chemical reactions would have been more difficult to take place, as large amount of energy is often not available in the cosmos.

Likewise, the relative strengths of the nuclear and electromagnetic forces and also the ratio of the photons to protons had to be finely constrained for carbon atoms upon which all life is built, and to exist in nature. Indeed, if the value of the ratio G/e (the constants of universal gravity and the charge of the electron) had been different from its present value by more than 10^{40} (the tiny value of 0, decimal point, 39 zeros and then 1), life would not have appeared!

I should not give the impression that these arguments are very general and vague; there are in fact very detailed and specific calculations and constraints. For example, astrophysicists[11] have shown that if the nuclear interaction had been just 4 to 6 percent stronger, then the nucleus ^2He (proton + proton) would have been stable, while ^2H, ^3H and ^3He would not, and this would have consumed all the hydrogen in the Big Bang nucleosynthesis. If the nuclear interaction were slightly weakened, then deuterium (^2H) would not be stable, and this would break the chain of energy production inside stars such as the sun.

Calculations have shown that if the ratio of the amount of matter and energy to the volume of space, the quantity known as Ω in cosmology, had deviated by $10^{-60} - 10^{-56}$ from the 'critical value' that cosmologists can easily determine, the Universe would have either collapsed back upon itself or suffered amplified relativistic effects. In that case it would never have produced the cosmic objects or the amazing creatures that can be found at least on the earth; certainly no life could have appeared in those cases[12].

The emergence and existence of human life on earth depends strongly on the stability of the Sun over several billion years (the time interval required for biological evolution to lead to the appearance of intelligent beings), so that the planet's star must remain stable over such long periods, and this sets strict constraints on the fundamental forces (gravitation, nuclear and weak), which determine the life and death of any star. Lee Smolin, a professor of Physics at Pennsylvania State Universe, has calculated the probability of formation of stars that will be stable over long time periods by assuming random values of the fundamental parameters of the Universe (masses of basic building blocks, such as electrons and protons etc.); he obtained the astounding result of 1 chance out of 10^{229}! To put this value in perspective, he reminds us that the Universe contains (only) 100 billion galaxies, each one comprised of roughly 100 billion stars, so there are in total no more than 10^{22} stars . . .

My favourite example is the realisation that our universe could only have three dimensions of space and one dimension of time. Indeed, nothing else would have worked. Concerning the dimensionality of time, Demaret and Lambert[13] refer to the work by Dorling showing that in a universe of more than one time dimension, no particle of mass greater than zero would have been stable! Regarding space, Barrow and Tipler recall that in history the three dimensions (n = 3) of our world had elicited commentaries and explanations by several thinkers, going at least back to Ptolemy, and Kant had realised that there was a direct relation between n = 3 and the inverse square law of gravitation, although he considered the three-dimensionality of space to be a consequence of the form of Newton's law and not the other way around. William Paley (famous with his watchmaker argument of design) had also done some analysis of the possible mathematical forms of the law of gravity and concluded that only in n = 2 or n = 3 could planetary orbits be stable; he thus concluded that the inverse square law in nature was another example of divine 'programming'. But it was Ehrenfest who in 1917 showed that the stability of orbits of planets, atoms and molecules, as well as the coherent propagation of waves, were possible only in a three-dimensional space.

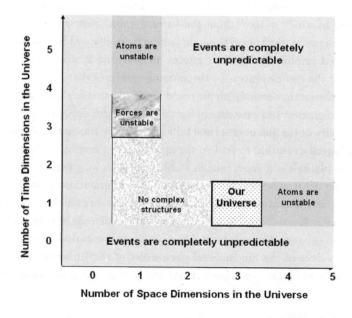

And yet, while he points out 'other fine-tuning marvels'[14] (including the values of the speed of light and the expansion rate of the universe), Paul Davies considers another example as 'the biggest fix in the Universe'[15]: the amount of dark energy (see our brief discussion in Chapter 6). Davies explains why our physics requires such a mysterious, black thing to exist out there, and cites advanced calculations that show the natural value of that quantity (considering all we know about the physics of the universe) to be greater by 120 orders of magnitude than its actual value; in other words, there is just the tiniest amount of it (no analogy from our everyday example can be given to represent how tiny that amount is), something that is still necessary for our existence and yet unimaginably small. Davies puts it nicely: 'The cliché that "life is balanced on a knife-edge" is a staggering understatement in this case: no knife in the universe could have an edge *that* fine'[16].

Now, how many such parameters had to be fine-tuned? This is an important question because obviously if this set of physical factors is limited to just a few, then one could either strike them off as a set of coincidences or, better yet, claim that these values will probably be explained in the future through some unified theory (such as the much sought-after Grant Unified Theory (GUT), or the Theory of Everything (TOE)). If there are too many of them, then one must first ask whether they can be related to one another, thereby reducing the set of constraints to just few and hoping that such relations will be explained later. But if the parameters are too many, apparently highly unrelated, and finely tuned (as the cosmological constant is), then the problem becomes more worrisome.

Davies has counted about 30 such fine-tuned parameters; Demaret and Lambert, referring to Rozental, suggest that the fundamental structure of the universe, both microscopically and macroscopically, can be characterised with just seven dimensionless numbers, which include the coupling constants of the four fundamental forces, the ratios m_e/m_p and m_p/m_n (the masses of the electron, the proton and the neutron), and the number of space dimensions $n = 3$.

So which characteristics of the universe are fundamental, necessary and finely tuned, and how many such features are there? As Paul Davies jokingly remarks in *The Goldilocks Enigma*, one might wish to get rid of giant black holes, irregular galaxies, super giant stars or ugly asteroids for whatever personal reasons, and that may not (at first sight) wreck the universe or prevent our existence from being fulfilled, but playing with more fundamental aspects of physics and chemistry may not be a good idea at all. At most, Davies points out, one could remove from the periodic table a few chemical elements, but the first few rows, at least from hydrogen to iron, are a must. And the more

Effects of weaker/stronger fundamental forces

Force	Weaker	Stronger
Gravity	Stars can't form, so carbon, oxygen and other ingredients of life can't be made.	The Universe would have collapsed upon itself sometime after the Big Bang, and before the formation of galaxies, stars and planets.
		The sun burns too fast, giving life too little time to evolve on the earth.
Electromagnetism	Chemical reactions are too slow to produce complex molecules like DNA, the code of life.	Repulsion between protons in nuclei is too large for stability.
		Chemical reactions necessitate too much energy to proceed.
Strong nuclear force	^{2}He is stable; hence all hydrogen is consumed in the Big Bang.	Neutrons decay faster during the Big Bang, hence too little He is made (which is needed to make nuclei of C, N and O in stars later).
Weak nuclear force	^{2}H is unstable, hence there is no production of energy and elements inside stars.	Neutrinos interact inside the dying stars and do not reach outer layers to make stars explode (supernovae) and allow chemical elements (C, O, Fe) to pour out.
	Neutrinos are unable to explode stars in supernovae, thereby allowing chemical elements (C, O, Fe) to pour out.	

we explore the universe at deeper levels, the more we find that we better not change anything.

This is why Paul Davies concludes: 'This is for me powerful evidence that there is something going on behind it all . . . It seems as though somebody has fine-tuned nature's numbers to make the Universe . . . The impression of design is overwhelming'[17]. And Roger Penrose was even more striking: 'The accuracy of the Creator's aim . . . would have to have been one part in [a truly unimaginable number] in order for our universe to exist'[18].

other (astronomical) special circumstances

More recently, serious claims have started to be made to the effect that this fine-tuning is found not only in the fundamental features of the universe but also in the cosmic, astronomical, geological and other aspects of our circumstantial presence on the earth. Indeed, Gonzalez and Richards in 2004 published such a bold thesis in their book *The Privileged Planet: How Our Place in the Cosmos Is Designed for Discovery*[19]. These authors extend the observations of fine-tuning from the cosmos in general to the galaxy (we are appropriately located), the sun (a very stable and finely moderate star) and the earth (perfect size, perfect moon, perfect distance to the sun, perfectly inclined and stable axis etc.). Here are a few examples of their thesis:

1. The earth is special ('just right' for life to flourish there):
   - Earth had to be of just the right size to retain a substantial atmosphere, to maintain plate tectonics, and to keep some land above the oceans.
   - A less massive planet would have quickly lost its atmosphere, its interior would have cooled too much, and it would not have generated a strong magnetic field. Smaller planets also tend to have more erratic orbits.
   - Bigger planets would lead to too much atmosphere, greater surface gravity and less surface landscape, i.e. smaller mountains and shallower seas; larger planets would also make for bigger targets and constitute greater impact threats.
   - The orbit of the earth around the sun had to be within the 'circumstellar habitable zone', where the temperature is moderate enough to allow water to exist in liquid form at least part of the time (and hence complex molecules to assemble and life to evolve in oceans or ponds).
   - The earth had to have an inclined axis for seasons to occur and thus allow life to prosper through cycles.

2. Our moon is vital:
   - Our moon keeps earth's axis tilt, or obliquity, from varying over a large range. Small moons like Mars's would make the tilt vary by more than 30° (instead of 3°) over long periods, hence prevent evolutionary conditions from existing.
   - The moon also helps life by raising earth's ocean tides, which mix nutrients from the land with the oceans.
3. Our sun is just right:
   - A larger star (more than 1.5 solar masses) would imply rapid, unstable burning.
   - A smaller star would require the planet to be close, hence overly strong tides, flares and other effects would make life very difficult.
   - The sun is a highly stable star; its light output varies by only 0.1 percent over a full cycle, hence preventing wild climate changes on the earth.
4. Our galaxy (and our place in it) is (are) just right:
   - A very large majority (about 98 percent) of galaxies are less luminous than the Milky Way and thus metal-poor, so many of them could be devoid of the earth-like planets.
   - Similar to the 'circumstellar habitable zone', one can define a 'galactic habitable zone' under the following constraints: (1) terrestrial planets cannot form in regions of poor metallicity, they thus must be close enough to the central galactic bulge; (2) life cannot exist too close to the bulge, where X-rays and gamma radiations are intense.

I should also add that another important thesis of Gonzalez and Richards is (as noted in the subtitle of their book and thoroughly argued therein) that the special circumstances briefly mentioned above (which they describe in fuller detail in their book) are not just for our existence but also for our ability to explore and discover the universe. Without all those characteristics, they argue, there are many aspects of the world that we would not have been able to find out.

This bold book, highly interesting in many regards, however remains controversial. First, some scientists have pointed out that Gonzalez and Richards have been too optimistic and lax in their analyses and conclusions. For instance, Nikos Prantzos has shown[20] that if one takes all factors into account, the galactic habitable zone at least, at present, turns out to be essentially unconstrained and covers the whole galaxy; we cannot claim to be in a special location in the Milky Way. To be fair, however, one should state that the authors themselves admit the speculative[21] nature of some of their considerations and point specifically to the circumstellar habitable zone and the galactic habitable zone.

Secondly, many have dismissed the work (quite unfairly in my judgment) on the basis of Richards's affiliation with the Discovery Institute, which is the intellectual base of the Intelligent Design (ID) movement; Gonzalez is not formally affiliated with that institute, but he has expressed sympathies for ID theses.

Lastly, I must stress one important remark: Except for some of the new ideas that have been suggested by Gonzalez and Richards, and which they themselves have labelled as speculative or uncertain, the largest part of the fine-tuning observations I have reviewed are firm established facts. In the next section I will go over the more philosophical – and sometimes metaphysical – interpretations that people have proposed on the basis of the previous observations, but it is extremely important to distinguish between the cosmic facts, the semi-scientific speculations, and the free philosophical interpretations. The fine-tuning of the universe is a set of facts; the fine-tuning of the earth, the solar system and the Milky Way is a varied collection ranging from accepted

Do not confuse the following:

- *Argument from design*: One can infer the existence of God from the adequacy and perfection of all creation (nature). The various aspects of this argument were discussed in the previous chapter.

- *Fine-tuning*: In its fundamental properties, the universe has a set of physical parameters in just the narrow range allowing the emergence of biological complexity and life. It is a mater of dispute whether this effect is 'real' or is due to a selection effect (our universe, or the part that is ours, displays such characteristics, hence our fortuitous existence).

- *Anthropic principle*: A metaphysical principle proclaimed with the intention of explaining the fine-tuning; various versions, ranging from 'weak' to 'super strong', can be formulated, and each then is more or less scientific, teleological, philosophical or theological.

- *Intelligent design*: A hypothesis subscribed to by creationists mostly, according to which living creatures are too well 'designed', i.e. display some complex features that cannot be explained by the standard theory of evolution, so that the only recourse is a direct intervention by the Designer (i.e. God).

facts to disputed assertions and speculations. The anthropic principles, which I am about to review, constitute a set of philosophical or

metaphysical propositions (in the sense of 'metaphysical' that I have discussed in Chapter 3).

anthropic principles

It is commonly stated that the first formal proposition of the anthropic
principle was made by the British physicist Brandon Carter in 1973 among
the celebrations of the 500th anniversary of the birth of Copernicus, who had
removed the earth – and thus man – from the centre of the world. And that
is true if we underline the word 'formal'. Before Carter, however, the idea
that the very existence of humans automatically led to certain constraints on
physical theories had already been raised by several scientists. For example,
in 1957, Robert H. Dicke had written: 'The age of the Universe "now" is not
random but conditioned by biological factors . . . [changes in the values of
the fundamental constants of physics] would preclude the existence of man
to consider the problem'[22]. Earlier still, in 1903, Alfred Russel Wallace had
written (in his book *Man's Place in the Universe*): 'Such a vast and complex
universe as that which we know exists around us, may have been absolutely
required [. . .] in order to produce a world that should be precisely adapted
in every detail for the orderly development of life culminating in man'[23].

Carter, however, had the merit of formalising these ideas into clear principles. Moreover, he realised that this important idea could be formulated
in two substantially different statements, which he labelled as 'weak' and
'strong'. As we shall see, the weak version is commonly accepted by the
largest majority of scientists and philosophers as being rather neutral and to
some extent scientifically useful; the strong version is very controversial, it
is often attacked as metaphysical, teleological, theological, useless and even
dangerous. To make things worse, Carter had the misfortune of describing
this principle as anthropic, by which he only meant that one had to take into
account the existence of life creatures that were conscious and intelligent,
and the only such creature known to exist in the whole universe is man,
hence the adjective anthropos. But many cried wolf, seeing in this a return of
pre-Copernican and even religious anthropocentrism and thus smelling some
disguised religious push to rehabilitate man's central place in the universe.
And that is why the weak and the strong anthropic principles have since
been followed by other formulations, which differ from the original ones by
varying degrees, from the principle of complexity to the misanthropic and
the thanatotropic principles.

Carter's weak anthropic principle is generally stated as follows: 'The observed values of all physical and cosmological quantities are not equally probable, they take on values restricted by the requirement that there exist sites where carbon-based life can evolve and by the requirement that the Universe be old enough for it to have already done so'[24].

This version is often thought of as a tautology, meaning a trivial statement that simply asserts an obviously true thing. Of course the universe must have been such as to produce carbon-based life that could evolve; we are here, aren't we? But the tautology is removed by the first part of the statement, where the emphasis is placed upon the *observed values of all physical and cosmological quantities*, which – because of the anthropic requirement – could not be equally probable. In other words, when one constructs a physical theory of any kind, its parameters are not a priori free to take all kinds of values; they are indeed restricted by the above requirement. One simple and nice example of this is that the universe cannot be young or small. It must be at least a few billion years old (for life to have had the time to evolve enough for anyone to inquire about it), and it cannot be small, for its expansion during that period automatically makes it huge.

Similarly, Demaret and Lambert[25] speculate that as the emergence of life seems to be an extremely improbable event on any planet (and for support of this idea they refer to the great evolutionary biologists T. Dobzhansky, G. Simpson and E. Mayr), perhaps the fact that there are 10^{22} stars in the universe was a statistical necessity for life to have a reasonable chance to appear somewhere.

So the AP, even in its weak version, allows us to understand why the universe is so old (compared to our timescales) and so incredibly huge (compared to our sizes) and also why there are so many stars and planets in the galaxy and in the universe . . .

Another example of how scientifically useful WAP can be is Fred Hoyle's realisation in 1953 that the nucleus of carbon-12 must have a special energy excitation level at about 7.7 MeV: without such a level the production of carbon from helium (the triple-alpha reaction) would be too slow (non-resonant), and thus far too little carbon would be produced in stars; and yet we are here, said Hoyle; hence, the triple-alpha reaction must be resonant, so carbon must have a special energy level at about 7.7 MeV, and indeed later experiments confirmed the existence of such a level at 7.644 MeV. Similarly, the production of oxygen by helium and carbon must not be resonant, or else too little carbon would be left in the cosmos, and hence no complexity or life. And indeed, oxygen was found to have an energy level at 7.162 MeV, below the value that would make the reaction resonant. Clearly, AP reasoning can be scientifically useful.

Carter's strong anthropic principle is generally stated as follows[26]: 'The Universe must have those properties which allow life to develop within it at some stage in its history'. Barrow and Tipler, unfold this controversial version by presenting three ideas[27] that can be inferred from this strong principle, depending on one's philosophical or metaphysical background or agenda:

SAP-1: 'There exists one possible Universe "designed" with the goal of generating and sustaining "observers"'. This is the 'theological' interpretation, the one most strongly fought against by materialistic scientists and thinkers.

SAP-2: 'Observers are necessary to bring the Universe into being'. This is the quantum-mechanical interpretation ('the observer plays a central role in bringing reality into existence').

SAP-3: 'An ensemble of other different universes is necessary for the existence of our Universe'. This is the multiverse interpretation, which I shall expound on later.

There are many other formulations of the anthropic principle. When Barrow and Tipler published their outstanding book on the subject, they presented an interesting new version of anthropic principle, one they called the 'final' anthropic principle: 'Intelligent information-processing must come into existence in the Universe, and once it comes into existence, it will never die out'[28]. This version takes the strong anthropic principle to its (teleo)logical conclusion: If Man (or some such advanced species somewhere in the universe) was the 'goal' of the universe, then it would not make sense for that goal to later disappear; on the contrary it must continue to grow, thrive and become ever more powerful until it has taken control of everything – at some Omega point, which one can assimilate with Teilhard de Chardin's idea[v].

There are additional interesting versions of the anthropic principle. In his recent book, Jean Staune identifies nine propositions, including – to cite only the most striking ones[29]:

---

[v] Teilhard de Chardin and his philosophy will be presented in some detail in the next chapter.

- 'The really anthropic principle' (with which Staune characterises the work of Gonzalez and Richards): Earth is the best place for the emergence of a mind capable of understanding the universe.
- 'The super-strong anthropic principle' (which Staune attributes to himself): The universe is pre-adapted to the existence of not just intelligent beings like us but to creatures far superior to us.
- 'The thanatotropic anthropic principle' (which Staune attributes to Jean-Pierre Petit): Complexity is not a viable solution; self-destruction of civilizations is built into the universe.

One version that Staune does not cite is Polkinghorne's 'Moderate Anthropic Principle', is the one that 'notes the contingent fruitfulness of the universe as being a fact of interest calling for an explanation'[30].

Finally, I should add that the anthropic principle has impressed many non-science thinkers, writers and philosophers. For example, Vaclav Havel had this to say about it:

> The universe is a unique event and a unique story, and so far we are the unique point of that story. But unique events and stories are the domain of poetry, not science. With the formulation of the Anthropic Cosmological Principle, science has found itself on the border between formula and story, between science and myth. In that, however, science has paradoxically returned, in roundabout way, to man, and offers him – in new clothing – his lost integrity. It does so by anchoring him once more in the cosmos[31].

the multiverse 'solution'

Needless to say, many scientists and philosophers, denounced what they considered in the anthropic principle, especially the strong version, to be an attempt to bring back the Designer, insisting that science had gotten rid of such teleological thinking; in their view, giving man any special importance in the universe is a biased move, a scientifically heretical one. After all, they said, man may not be the epitome of evolution, and our universe may not even be the only one 'there' . . . And so, the holders of the materialistic (or naturalistic) philosophy came up with the idea of the multiverse: There must be zillions of universes out there, separated by voids and thus impossible to detect, much less contact; they are all different in their physical parameters and laws, with our universe being the only good one. And so, the argument goes on, and there is no wonder that we find ourselves in a universe so hospitable to life: One of the 'tickets' had to be the winning ticket – nothing amazing about it.

This idea of a Multiverse is not new either. It has got the scientific support, however, which is recent. It should be emphasised, though, that the scientists who defend this hypothesis, and who at present represent the majority in the community, do it as much on the basis of a philosophical rejection of the anthropic principle as on the basis of some cosmological principles which lend it support. George Ellis, the prominent South African cosmologist, puts it simply: 'In my view, belief in multiple universes is just as much a matter of faith as any other religious belief'[32].

The cosmological principles that argue in favour of a multiverse are in order of decreasing importance: (a) the 'eternal inflation' theory, which predicts the formation of 'bubble universes' that are causally separated from each other and thus are physically independent; and (b) the possibility of getting a 'baby universe' each time a black hole produces a singularity, which amounts to a hole in the fabric of space time, thus creating a wormhole (a kind of space-time tunnel) towards another universe. In either case (bubble universes or baby universes), there would have been, by now, billions of universes, where the physical parameters and laws would in principle be different in each one, such that they would all be sterile and lifeless, perhaps even totally empty of anything, except for one or two, and our universe would thus be the winning ticket. If there are zillions of universes, then one of them is bound to be amazing.

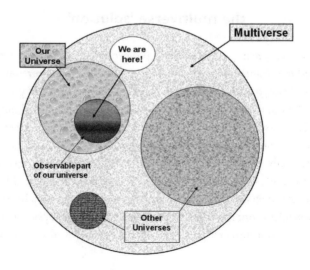

But how many universes do proponents of the multiverse expect it to contain? This is part of the crisis that physics finds itself in these days (an example of which we have encountered in the prediction of dark energy exceeding by a factor of 10^{120} what observation reveals). Indeed, the latest fad, known as 'the landscape' version of the string theory, predicts 10^{500} solutions (universes), each one of which represents a plausible universe with its own physical parameters. How far apart would such universes be in a mega-multiverse? 10 to the power 10^{29} metres[vi], in the estimate made by the physicist Max Tegmark! In comparison, the size of the observable part of our universe is (only) 10^{26} metres. So in order to explain one (amazing) universe, one ends up postulating 10^{500} (invisible) universes!

Tegmark has also proposed that all logically possible universes exist – another metaphysical principle. But this proposition is not philosophically satisfying; indeed, Lee Smolin has noted that such reasoning is defeatist because it gives up the search for a real explanation; if this attitude had been adopted in biology, he stresses, the mechanism of natural selection would have been never identified.

Smolin is fond of natural selection; in his book *The Life of the Cosmos*[33], he proposed to explain the special disposition of our universe towards life by the same combined 'mutation and natural selection' process with which Darwinian evolution explains the appearance of new, more complex species. In Smolin's (speculative) hypothesis, black holes produce baby universes, with a mutation, a slight reshuffling of the physical parameters of the universe, and a selection process favouring those which produce more black holes, i.e. those that have many stars and thus produce the chemical elements (C, N, O, Fe etc.) upon which life can be built. So in his scenario, our universe emerged from a black hole in a previous universe, which was less conducive to life, and so on in a long series of improved universes. Needless to say, there is no evidence for any such process or principle in nature.

Interestingly, the most important argument against the whole multiverse idea was first made by Paul Davies in a short article in the *New York Times*[34], which was based on no physics or cosmology at all. (Davies has since reproduced his idea in his latest book[35], though in my view he and others have not given it the emphasis that it deserves.) First Davies raises the same objection to the multiverse hypothesis as raised by Ellis ('just as ad hoc'), then goes further. He waves the spectre of the loss of our reality and the faith in our very own existence in any such multiple universe system. How?

[vi] Yes, 10 to the power 10^{29} (in metres), not 10 to the power 29 (metres)!

If we have zillions of such universes, it is then more than probable that in one of them some civilization must have reached such a degree of scientific and technical advancement that it can simulate, or create (virtually or even really, by way of black holes) other universes where the actors look and behave in ways that are practically indistinguishable from the real ones. (This is essentially the basic idea of *The Matrix* movies.) The Oxford philosopher Nick Bostrom has specialized in this topic; he puts it simply: 'There is a significant probability that you are living in [a] computer simulation. I mean this literally . . . '[36] In fact, in those created worlds, the creatures could start their own simulation games and create other universes, which would be as real looking as the parent or grandparent ones[vii]. And so on and so forth in an ad absurdum succession of sub-worlds, all as real/virtual as the others. We can then be the result of such a simulated/created world, where the gods (the creators) are themselves creatures of other gods. Davies points out that such fake universes are much cheaper to produce than the real ones, because processing information requires far fewer resources than putting together physical matter and making it function; moreover, a fake experience can be produced with a limited local setting (or façade), without the need for a whole, full-blown universe.

Ironically, the multiverse hypothesis brings back not only the idea of a creator but also produces a whole host of gods. Most importantly, it makes us lose any semblance of Reality. It is a mult-existential crisis of the greatest cosmological order.

I should note that there are many opponents to this multiverse idea; some of them hate it with passion. There are those who find it totally unscientific, for it does not have any supportive evidence and is moreover most likely unprovable. Others point out some fallacious reasoning in the usage of probability theory[37]. Interestingly, some critics have attacked the multiverse because it appears to be giving up on the dream of finding an ultimate, unique and elegant explanation, one which does not need to postulate 10^{500} universes in order to explain one universe's (apparent) fine-tuning; other critics have described it as a materialistic desperate recourse, invented simply to avoid having to face the extraordinary evidence that seems to point, if not to a Designer, at least to a special place of life, consciousness and perhaps humanity. As Davies asks in his recent book: 'Could it just be a fluke? Might

---

[vii] Davies, in *The Goldilocks Enigma*, p. 204, gives as an analogy the fact that our dreams often seem very real while in progress; he adds that sometimes we (at least he) have (has) dreams within dreams . . .

the fact that the deepest level of Reality has connected to a quirky natural phenomenon we call "the human mind" represent nothing but a bizarre and temporary aberration in an absurd and pointless universe? Or is there an even deeper sub-plot at work?'

And this is how scientists, philosophers and theologians have found themselves implicated in fascinating discussion, with scientific papers, semi-philosophical books and theological arguments being contributed almost every month.

philosophical issues

As I have somewhat briefly explained, the perception and interpretation of the anthropic principle ranges from 'tautology' to 'stealth religious principle'. Debates have continued unabated over whether the principle has real scientific value or is at most an interesting metaphysical, philosophical or even theological viewpoint. In a previous section, I gave at least two examples of how the weak anthropic principle has allowed scientists to make either specific (correct) predictions (energy level in the nuclei of carbon and oxygen) or produce simple explanations (the age and size of the universe). Still, most thinkers tend to present anthropic principle as a metaphysical principle, one which can guide scientific thinking and research into various directions and modes but is not itself subject to falsification. Freeman Dyson has written[38] that he finds the anthropic principle 'illuminating' and recommends that both anthropic principle and the argument from Design be 'tolerated in meta-science'.

But whatever its status may be, what does the anthropic principle actually tell us? Are humans somehow back at the centre of the universe? Is life a fundamental property of the universe or are we biased in our analysis? Is anthropic principle pointing towards a Designer or rather towards a greater principle of existence, something that we have yet to discover and understand, a more general and unifying explanation of why things are the way they are?

The ghost of Copernicus still hovers over us. How would it not when the more we learn about the universe, the more we find it unimaginably larger than us? Many scientists, however, have remarked[39] that the human size is not only in the middle ('central') position between the largest scales and the smallest ones (quarks) but also this position is a 'privileged'[40] one as it allows us to explore and discover everything, large and small. Indeed, our size is also important in what technology has permitted us to produce, and without telescopes, microscopes and accelerators – which ants will never be able to construct – we would never be able to discover much of the nature and the cosmos. Michael Denton has summarised it thus: 'It would appear that man, defined by Aristotle in the first line of his Metaphysics as a creature which "desires understanding", can only accomplish an understanding and exploration of the world, which Aristotle saw as his destiny, in a body of approximately the dimensions of a modern human'[41]. Still, other scientists insist on the cosmological principle, which is essentially a generalised version of the Copernican principle, as it declares any point in the cosmos, including our own position, to be no more nor less significant than the any other; these scientists turn this physical assumption (of uniformity and homogeneity) into a philosophical principle by calling it a 'principle of mediocrity'[42] or 'principle of indifference'[viii].

So what implications can be inferred from the fine-tuning? The most crucial – and controversial – one is whether the fine-tuning of the universe implies the existence of a Designer. This is clearly a non-scientific issue, for it moves the explanation from the natural to the supernatural. And that is partly why many scientists reject it outrightly. (Many also reject it because of its theological overtones, even though many insist that Designer is not necessarily synonymous with Personal God.) Still, Davies stresses that 'it is still a rational explanation'[43]. Gonzalez and Richards insist that the design that one infers 'is embedded or encoded in the laws and initial conditions themselves'[44]. Davies adds: 'Intelligent design *of the laws* does not conflict with science'[45] (emphasis in the text). He points out that such a God, because He 'is *responsible* for the universe and might be thought of as upholding its existence at every moment, but does not tinker with its day to day operation . . . comes close . . . to what many scholarly theologians – and for that matter quite a few scientists – profess to believe in'[46]. Davies also cites[47] Andrei Linde and Heinz Pagels (non-believing scientists I should note), who have speculated that a super-civilization or a demiurge could have configured the birth of such a 'life-encouraging' universe in order to send a message to its future inhabitants.

---

viii The 'principle of indifference' was initially a principle of mathematics, more specifically of probabilities and statistics, going back to Bernoulli and Laplace, then known as 'the principle of insufficient reason', later renamed as 'indifference' by the economist John Maynard Keynes.

Related to the question of design is the concept of purpose. Gonzalez and Richards remark that 'detecting design is often easier than discerning purpose or meaning . . . Nevertheless, detecting a purpose usually enhances our confidence that something is designed'. But then how do we convince ourselves that there is a purpose in this universe? The question is far from simple, and when it was recently[48] put to 12 distinguished scientists/thinkers, three answered 'yes' (David Gelertner, Owen Gingerich and John Haught), one said 'certainly' (Jane Goodall), one replied 'indeed' (Nancy Murphy), one said 'perhaps' (Paul Davies), one answered 'very likely' (Bruno Guiderdoni), one said 'not sure' (Neil deGrasse Tyson), one replied 'unlikely' (Lawrence M. Krauss), one said 'no' (Peter Atkins), and one expressed his view as 'I hope so' (Elie Wiesel).

Let us go back to the general philosophical implications of the observations of fine-tuning and the anthropic principle formulations. In his recent book, Davies nicely summarised[49] all the possible conclusions that can be drawn from the 'Goldilocks (just-right or fine-tuning) Enigma'; let me here rephrase (and rename) them briefly:

- The absurd universe: One might be tempted to skip this option, but Davies's first sentence reads: 'This is probably the majority position among scientists'. He explains this position as follows: 'It could have been otherwise . . . Had it been different, we would not be here to argue about it . . . The fact that it exists, seemingly against vast odds, is attributed to an extraordinary accident'.

- The God-designed universe: This is the religious (monotheistic) view, which declares fine-tuning to be consistent with the fact that God created the universe and thus chose its parameters for good reasons, possibly with man as a goal.

- The unique (or explainable) universe: This is the explanation that optimistic physicists hope will emerge from their quest for a final theory or theory of everything, which will presumably tell us why the universe (or multiverse) is the way it is. They imagine that all parameters will be determined by some underlying principle. The problem, as Davies and others have pointed out, is that even in that (highly optimistic) case, one would still have to explain why *that* theory was the final one and not another.

- The life-laden universe: An overarching principle is postulated, whereby the laws and parameters of the universe are underlied by something that requires them to combine to lead to the emergence of life.

- The self-explaining universe: Some 'closed explanatory principle' would make the universe both self-creating and self-explaining. It is unclear how this would be achieved, much less why the universe would be so. Further explanation would then be required.
- The simulated universe: This is the Matrix-like scenario that I have mentioned – and decried – in the multiverse section.
- None of the above: By this Davies wisely points out that despite the apparent extensiveness of the above options, we could still be missing the main ideas and reasons for why the universe is the way it is.

anthropic principle(s) of Islam

So where does the Islamic tradition stand with regard to these observations (fine-tuning), principles (anthropic ones) and interpretations (multiverse, unique universe, centrality of man etc.)? The Islamic literature (sacred and profane) has always strongly adopted the Argument from Design. So for Muslims, the fine-tuning of the universe did not seem like anything new. Indeed, the anthropic principle is essentially the modern, sophisticated version of the 'argument from design', hence the reason why Muslims were very slow to react to the (important) emergence of anthropic principle.

There are, however, other reasons to this slow reaction. First, some of the Islamic philosophical schools are not interested in any theology that is grounded in nature; indeed, some defend a sacred science, which both clearly separates itself from modern (western) science and fuses physics with metaphysics, gravity with angels and cosmology with spirits. Secondly, most of today's Muslims, including the educated elite, frown at any scientific paradigm that is based on an evolutionary scheme – and the anthropic principle certainly espouses evolution at every level and every stage, from the Big Bang to planets (physical evolution), and from mud to man (biological evolution).

This is not to say that no Muslim intellectual has presented a positive Islamic position with regard to the anthropic principle. In fact, several writers have focused on the Qur'anic word/concept '*taskheer*' (subservience/disposal of creation to men), e.g. 'Do you not see that Allah has made what is in the heavens and what is in the earth subservient to you, and made complete to you His favors outwardly and inwardly?' (Q 31:20). From this many authors, from classical (e.g. Ibn Rushd) to contemporary ones (e.g. Jaafar Sheikh Idrees) have constructed an 'argument of providence' (*dalil al-'inayah*), emphasising how God has made nature so adequate for humans that this is both a sign of His intention and love towards us and an indication of our obligation of thanks and submission to Him.

Among the many verses in which the Qur'an upholds the Argument from Design, a few can be interpreted as alluding to the fine-tuning of the universe or to the importance of humanity as a goal/purpose of creation. Let me cite those that I find relevant:

- 'And the heaven, He raised it high, and He set the balance' (Q 55:7).
- 'Verily, all things have We created in proportion and measure' (Q 54:49).
- 'And everything with Him is with (due) measure' (Q 13:8).
- 'Do you not see that Allah has made subservient to you all that is in the earth and the ships running in the sea by His command? And He withholds the heaven from falling on the earth except with His permission; verily Allah is Compassionate, Merciful to men' (Q 22:65).
- 'And He has made subservient to you whatever is in the heavens and whatever is in the earth, all from Himself; surely in this are signs for people who reflect' (Q 45:13).

The above verses are of two types: the first three speak of the 'measure' (*miqdar*) and balance (*mizan*) in all that was created, in fact. even in all that is 'with Him' (in His plan); the last two speak of creatures in the heavens and in the earth and seas as being all subservient (*musakhkharah*) to humans. It is these verses that have led some recent commentators (Adi Setia[50] and M. Basil Altaie[51]) to quickly identify this concept of *taskheer* with the anthropic principle, even though the latter – with what I have explained above – is significantly different from that.

Commenting on Ibn Rushd's understanding of the above verses on 'measure', Marc Geoffroy writes[52]: 'The idea of "measure" [*qadar, miqdar*] must be understood both as "measured quantity" and as "decision", as when we say for example "taking measures", the measure given to something implying a destination or purpose for that thing'.

What is important to me in the above set of verses is – without sliding into some self-serving *I'jaz*-type identification of modern scientific concepts with Qur'anic statements – the resonance with the idea of balance in the universe and the measured (or fine) tuning of the whole creation. Although this idea is rarely, if ever, raised in the modern Islamic literature, in searching the corpus one does come across a few references to this idea. Fakhr al-Din al-Razi (the great twelfth-century scholar), in his masterful *Mafatih al-Ghayb*

Qur'an commentary, emphasised the idea that each object in nature has its proper mode of physical existence among many possible ones, a mode that is precisely predetermined (*muqaddar bi maqadir makhsusah*) by the Creator; for example the celestial objects have precise orbits and space-time coordinates that show a complete ordinance (*tadbir kamil*) and a profound wisdom (*hikmah balighah*)[53]. al-Razi insisted that all objects, in the heavens and in the earth, living or inert (*jamadat*), have been decreed by Allah in a manner that serves the interests (*masalih*) of humans. Finally, he drew our attention towards the complex interconnections between the benefits lying in the cosmic horizons (*al-ni'am al-afaqiyyah*) and those that can be found within ourselves (*al-ni'am al-anfusiyyah*). He concluded that such benefits are for drawing humans to attain deeper and everlasting spiritual benefits.

As to the Qur'anic idea of the subservience of all creation to humans, Ibn Rushd notes that (1) all that exists is in harmony with humans; and (2) this can only be the result of an Agent who wants it so. The night and the day, the sun and the moon and all other bodies, have been placed to serve us, and it is on the basis of the order and the design which the Creator put in their motions that our existence and that of things down here is maintained. And if one of them were to be removed or given another position, dimension or speed than those assigned by God, beings on the surface of earth would be prevented from existing[54].

In the same line of thought, the contemporary Muslim philosopher Jaafar Sheikh Idrees, who has shown some interest in topics at the intersection of science and Islamic theology[55], sees in the harmony existing between the many creatures of the world an argument for the existence of God. He adopts al-Kindi's and Ibn Rushd's argument of providence or benevolence (*dalil al-'inayah*) in the following words:

> In the argument of benevolence, the aspect which shows us the existence of a Creator is the relation between the creatures, and also the relation between the various parts of one body. Anyone who contemplates creatures sees that they are not a bunch of random objects, they are rather well ordered and purposefully planned, thus showing that they come from a knowledgeable and wise constructor[56].

Further, he supports his view with the following Qur'anic verses:

> Have We not made the earth as a wide expanse? And (set) the mountains as pegs? And (have We not) created you in pairs? And made your sleep for rest?

And made the night as a cloak? And made the day as a means of livelihood?
And (have We not) built over you the seven strong (heavens)? And placed
(therein) a shining lamp? And do We not send down from the clouds water in
abundance, That We may bring forth therewith grain and plant, And gardens
of luxurious growth? (78:6–16)

Idrees notes and stresses the finalist intentions in these verses, seen most
clearly in the pronoun 'That' ('That We may . . . ').

Hence, Idrees sees an anthropic aspect in these verses in the following
words:

> Even those faraway stars have some bearing upon us (humans), for as the
> earth is for us a bed, the sky is a built-up roof, the sun is a lamp which
> provides us with light and heat, without which there could not be any life,
> whether human, animal, or vegetal[57].

Continuing with modern Muslim thinkers, I should briefly mention the
viewpoints of two distinguished philosophers, Badi'uzzaman Said al-Nursi
and Muhammad Iqbal, both of whom we have met earlier (the first one
briefly in Chapter 1, and the second one in Chapter 7). al-Nursi was very
impressed with the 'ubiquitous balance' and the 'cooperation' that one can
clearly see throughout the cosmos, and this for him constitutes 'material
evidence' of the divine unity[ix]. On the other hand, Iqbal having (as we have
seen) found no value in any of the rational arguments for the existence of a
Creator, particularly the teleological arguments, rejected any predetermined
goal for the creation, for that in his view would deprive the cosmos from
any creative quality or originality. It is indeed a rather surprising position
considering all that we have found in the Qur'an and in the classical Islamic
literature, and indeed Mehdi Golshani (whom we have met and listened to
repeatedly in previous chapters) adopts a diametrically opposing position.
Golshani considers the principle of telos to be explicit in the Qur'an, so much
so that Muslim thinkers have never given up such final-cause reasoning even
after western science and philosophy had removed it completely from their
general methodology.

[ix] Adi Setia cites Nursi in *The Supreme Sign: The Observations of a Traveller Question-
ing the Universe Concerning His Maker* (trans. Hamid Algar, Istanbul, Turkey:
Sozler Nesriyat, 1993), with statements like: 'universal co-operation visible throughout the cosmos' and 'the comprehensive equilibrium and all-
embracing preservation prevailing with the utmost regularity in all things'.

One last viewpoint that I would like to mention, not because it comes from a particularly illustrious thinker or represents a widely held opinion but because I wish to show the very large diversity of interpretations that can be found within the Islamic tradition, old and new. Faheem Ashraf[58], a professor at the Government Degree College (in India), finds some similarity between the multiverse theory and the Qur'anic verse 'Allah is He Who created seven heavens and of the earth a similar number . . . ' (65:12). Ashraf first identifies the term 'heavens' (*samawat*) with 'universes', then he interprets the term 'seven' as simply an indication of multiplicity, and finally he considers the utilization of the present tense as an indication of the current existence of the many universes, each (according to him) with 'their own values of physical constants and nature of the physical laws'[59].

conclusions

Let me first summarise the main ideas that I have developed in this chapter. First, one must distinguish between the fine-tuning part of the topic, something that is accepted by all scientists as incontrovertible, and the anthropic principle, which has different versions and constitutes a metaphysical or a philosophical principle, with or without theological overtones. In all cases, these need to be distinguished from the Design Argument, and certainly from the Intelligent Design theory (or, more precisely, hypothesis).

Carter had originally presented two versions of the anthropic principle. The weak anthropic principle simply states that all parameters in nature are adequate for the emergence of humans; this may appear as a tautology, but, in fact, it is a very useful *scientific* principle, for it allows us to weed out incorrect theories that do not mesh well with the bio-friendly features of the universe. The strong anthropic principle is the more controversial version, one which *requires* the universe to be set up anthropically. Since then, several more versions have been proposed. I have also reviewed the various metaphysical inferences that one may draw from such considerations.

In reviewing the Islamic viewpoints with respect to fine-tuning and the anthropic principle, I first noted the slow response of Muslim thinkers to take full measure of this important development. I explained that this is due to several reasons: (1) anthropic principle was probably seen as only a slight variation on the Design theme, which the Islamic tradition already fully espoused; (2) some important contemporary Islamic philosophical schools are not at all interested in seeing the development of any strong science-based natural theology; and (3) most Muslims, including many in the elite, refuse to accept any principle, such as anthropic principle, which takes full account of the evolution of the universe, physically, chemically and biologically.

Still, the huge popularity of the topic in the West has forced some Muslim commentators to deal with it, but unfortunately it has often taken the shape of one-to-one correspondences between such modern concepts (as anthropic principle) and specific Qur'anic terms like *taskheer*. I am very surprised not to find any reference to the (more fundamental) issue of fine-tuning of the universe in Islamic literature.

I have tried to show that at least in some readings of a few Qur'anic verses, one can find resonances between the idea of fine-tuning of the cosmos and the concepts of *taqdeer* (ordering in proportion), e.g. 'It is He who created all things, and ordered them in due proportions' (Q 25:2), and balance: 'And the heaven, He raised it high, and He has set the Balance' (Q 55:7).

As to the concept of *taskheer*, I am inclined to refer to the Islamic viewpoint as an 'ultra-anthropic principle', for it puts humans squarely at the centre of its world view: not only is man the purpose of the whole creation, everything has been created and made 'subservient' (*musakhkhar*) to him!

This ultra-anthropic aspect of the traditional Islamic viewpoint is important to discuss. In reviewing Ibn Rushd's position with regard to the teleological argument, Taneli Kukkonen notes that the argument from providence in Averroes's *al-Kashf* treatise 'appears almost embarrassingly anthropocentric in character'[60]. Kukkonen then argues that if one reviews all of Ibn Rushd's works, including and especially his commentaries on Aristotle, one realises that this argument occupies only a mid-level position, the highest one being reserved to the truly demonstrative ones, i.e. the logically deductive philosophical proofs. Indeed, Kukkonen reminds us that Averroes always stressed that mankind is inferior to the cosmos, and the superior cannot exist 'in first intention' (Ibn Rushd's words) for the sake of the inferior. Still, Averroes insists that the very fact that one can find in the heavens final causes that are of benefit to us down here is an indication that there is a higher intelligence at work.

Ibn Rushd attenuates the traditional Islamic anthropocentrism even more by noting that not everything in nature is obviously of benefit to man. Clearly there are many animals that do not benefit us, directly or indirectly. For this reason specifically, Barrow and Tipler also cite Averroes as a prime example of a non-anthropic stand[61]. Kukkonen explains that Ibn Rushd considers creatures to fulfil the principle of 'great good . . . tainted by little evil', and this secondarily turns out to benefit mankind: 'man's well being itself comes as a corollary to the primary good of universal design'[62].

Now Muslims may insist that humans were created to worship God, and nature is here to facilitate this (physically, emotionally and spiritually) by helping one to at least reflect upon it and perhaps come to know God through it. Muslims may then read these FT and anthropic principle developments and be tempted to see them as confirmation that we are 'evidently' at the centre of the universe, and perhaps the universe was indeed created for us. But one must always remember that the purpose of creation is a divine reason, which will largely remain outside our understanding.

9

Islam and evolution
(human and biological)

'Although the Creator could, no doubt, have brought into being a ready-made world, God has in fact done something cleverer than that, bringing into being a creation that could "make itself".'

John Polkinghorne[1]

'Darwin did not want to deny the argument for a designer, but the designer had to work at a distance, through and only through the law of natural selection.'

Michael Ruse[2]

'If a plant reaches this [high] level of development, only one step separates it from the animal range, and that is to take off from the ground and move in search for food . . . if the plant then moves and takes off and no longer becomes tied to its place waiting for its food, new organs form in it to allow it to perform its new tasks, it then has become an animal.'

Ikhwan as-Safa (tenth century)[3]

> 'Even research into the beginning of creation [is allowed in Islam], as long as one keeps in mind that we are looking into creation, meaning that there is a Creator; even if we assume that species evolved from species, this is only by the will of the Creator, according to the laws of the Creator . . . If Darwin's theory is proven, we can find Qur'anic verses that will fit with it . . . .'
>
> Sheikh Yusuf al-Qaradawi[4]

Pokemon and evolution fatwas

Around the year 2000, the Pokemon craze hit the world. It started out as a fantasy-genre children's cartoon from Japan, but soon turned into a multifaceted commercial venture, including toys, dolls and a card game. In the game, children bought cards representing one character/creature, with Pokemon being the cute lead character in the series, where each creature had specific points of strength and weakness that allowed it to 'compete' in the world with other characters. The crucial part in the process was that these creatures could evolve from one form to another within their species, thereby acquiring new characteristics that allowed them to struggle better.

In the spring of 2001, one American colleague approached me and asked why the Muslim 'authorities' had seen this aspect of evolution of Pokemon as a danger to the point of banning the series altogether. As I had not heard of any *fatwa* (religious ruling) on the game, I commented that it must have been banned for other reasons. Indeed, shortly before that, I had heard a sermon by a local Imam, who had found it important to discourse on the subject to the faithful, insisting that 'Pokemon is a Jewish-influenced Japanese cultural virus', and remarking that 'Po-ke-mon is a Japanese linguistic construction which means "I am God" . . . '

But with all the discussions going on about Pokemon, I found it necessary to go for personal search for what had been written about it and its cultural influence. It was then that I found a news piece about a fatwa banning Pokemon, with the justification that, first and foremost, Pokemon instills the idea of evolution, struggle and survival of the fittest in the minds of the Muslim children, which – it was claimed – constitutes a threat to their belief and creed. Other reasons were added: (a) the cartoon and game involve flights of fancy, which are too remote from God's laws (ways) in nature;

(b) there is too much emphasis on the idea of struggle, which induces a wrong education in the minds of children; (c) there are hidden symbols of Zionism, Masonry and Shintoism in the game; (d) too much money is spent on the cards, and the trade often turns into 'gambling' (the sheikh's word) when 'strong' cards are bought at huge prices.

There is no doubt that the idea of biological evolution constitutes a major cultural blockage in the Muslim world today. Even cartoons and card games are judged with that mindset. And even highly educated Muslims expound negative views towards evolution, as we shall see further. One serious problem is the wide gap between the correct definition of evolution and the general misconceptions prevailing in people's minds. For this reason and others, I shall devote full sections of this chapter to explaining the theory(ies) of evolution and to the evidence for it.

But let me first provide some more information on the Muslims' attitudes towards evolution, both from some field survey data and from literature reviews.

During the fall of 2007, I conducted a survey on Science–Religion issues among professors and students at my university, i.e. the American University of Sharjah. About 100 students and 100 faculty members (about one-third of the faculty body) responded to the survey, so the results may not be statistically very significant, but they give a good indication on the views of educated Muslims about these issues today. A small fraction of the respondents were non-Muslims, and they provided for an interesting comparison with their Muslim peers. The results of the whole survey and their analysis are presented in Appendix C; here I only wish to briefly consider the questions pertaining to the evolution:

- Regarding Darwin's theory of evolution: 62 per cent chose 'it is only an unproven theory, and I don't believe in it'; 25 per cent chose 'it is correct, except for humans'; and only 13 per cent chose 'it is strongly confirmed by evidence'.
- Concerning human evolution: 56 per cent chose 'humans did not evolve from any earlier species'; 32 per cent chose 'humans evolved only as humans, not from animals'; and only 12 per cent chose 'it is now a fact that humans evolved from earlier species'.
- On the teaching of the theory of evolution: 24 per cent said 'evolution is against religion; it should not be taught'; 58 per cent said 'it should be taught but as "just a theory"'; and only 18 per cent chose 'it should be taught as a strong theory'.

A western reader may not be too surprised at the existence of such strong negative views and attitudes towards evolution; after all, similar surveys (at least among the general public) in the USA reveal similar levels of distrust or outright rejection of the theory. Indeed, many polls investigating public attitudes towards the evolution in the USA have been conducted[5] for the past 15 years; they have produced the following consistent results: 45–50 percent of people believe that humans and all species were created *in their present forms*; about 30 per cent believe that evolution did occur, but that *God set the process* (theistic evolution); about 15 per cent believe evolution occurred *without any divine involvement*. Another set of questions in such surveys has often asked the public whether evolution or creationism (or both) should be taught in public schools[6]. Responses have again been consistently showing a wide desire to have both of them taught. Last but not the least, a 1991 Gallup poll[7] presented results according to the education level (and the income) of respondents; it showed that among college graduates, 25 per cent held creationist views, 54 per cent subscribed to theistic evolution, and 16.5 per cent insisted on a materialistic version of evolution. A similar Gallup poll conducted in 1997 showed[8] that among scientists, 5 per cent held the creationist view, 40 per cent subscribed to the theistic evolution, and 55 per cent insisted on the naturalistic version. Among the general public, the results were 55 per cent for the creationist view, 27 percent for the theistic evolution, and 13 per cent for the naturalistic evolution. Finally, a 2007 survey[9] showed that 74 per cent of people with postgraduate degrees believe in evolution, and another survey conducted among medical doctors in the USA in 2005 found that 34 per cent of respondents agreed with Intelligent Design (ID) more than with standard evolution, and 73 per cent of Muslim doctors favoured ID[10].

The major difference between survey results in the USA and those in the Muslim world is that in the USA the educated elite is by and large pro-evolution. In the survey I conducted, the responses of Muslim professors were as negative as those of the students towards evolution, which are indeed similar to those of the general public (as seen from very recent surveys)[11].

Evolution is such a sensitive issue in the Muslim culture that fatwas are issued on it. One can certainly imagine and welcome philosophical and theological critiques of it (or at least of its interpretations). But unfortunately, one finds very little discussion of the scientific strengths or weaknesses of the theory; the discussion seems to be confined to a (passionate) religious attack.

Indeed, a quick web search for 'Islam and Evolution' brings out a good number of pages with fatwa in the title, pages which can be found in generally serious websites, such as IslamOnline. For example, in the section on 'Ask About Islam' of that important portal, one finds the question, 'Can I Believe in Evolution?'[12] and in the Fatwa Bank section, the following questions are posed: 'Where does Islam stand on the theory of evolution?'[13] and 'I believe in Adam and Eve, from whom human life started. But what about the theory that we are descendants of the apes, according to scientists. Didn't they prove this theory? . . . '[14] In the answers to these questions, usually given by religious authorities (e.g. the former president of the Islamic Society of North America), we have the following: 'We do not accept the theory (it is important to keep in mind that this is only a theory and not a fact) that says that all living organisms came from matter and man evolved from lower living organisms'[15]; 'Darwinism [. . .] has turned out to be pure dogma, far from being a possibility, much less a scientific fact'[16]; 'what the Darwin theory wants to prove runs in sharp contrast to the divine teachings of Islam'[17]; 'the claim that man has evolved from a non-human species is unbelief, even if we ascribe the process to Allah or to "nature", because it negates the truth of Adam's special creation that Allah has revealed in the Qur'an'[18].

The following three important ideas transpire from the above-selected quotes: (1) Evolution is at best a 'theory' and at worst a dogma; it should not be regarded as an established fact of nature; (2) man is special, and in no way should Muslims accept the idea that man evolved from a non-human ancestor; (3) any phenomena in nature (e.g. transformations) must be looked at as actions caused and affected by God, not as naturalistic or random occurrences. The first idea is totally erroneous, as I will show in this chapter; it stems from the following double ignorance: (a) of the known facts about species in the past and in the present; and (b) of the meaning of the term 'theory', which contrary to popular understanding signifies a largely confirmed general description of some part of nature (like planetary orbits or geological plate tectonics etc.). The second idea also runs contrary to today's knowledge, and I shall show that we must accept that not only man did evolve from lower species, and all species go back to a common starting point but also that a non-literal understanding of the Islamic Revelation can accommodate the scientific facts. The third idea that the evolutions and transformations of species must not be regarded as random and purposeless is a much more respectable one; it falls within what is today referred to as "theistic evolution", and Muslims should insist on it and join the many non-Muslim scientists and thinkers who subscribe to it. Indeed, while this approach rejects none of the scientific results of evolution (including the fossil records, the intermediate forms, the mutations that lead to the appearance of new characteristics and thus to new species etc.), it still upholds the idea of divine creation, although in a subtler way.

The above issues with evolution can also be found in more formal treatments of the subject by Muslim intellectuals. In a lecture on evolution given a decade ago, Zaghloul al-Naggar struggled to adopt a moderate position with regard to the theory; in fact, he sees the existence of that idea in one form or another in the ancient Egyptian, the Greek and the Hindu civilizations as a sign that the devil was 'speak[ing] in the ears of people as if to make them disbelieve in the presence of a creator . . . '[20] al-Naggar insists that while 'we must consider the theory critically [. . .] any aspect of the theory that can be proved by observation must be accepted', so he accepts 'the gradualism of creation', stating that 'things were not created suddenly, [even though] Allah has the power to say "Be" and it "is"'. He cites evidence for this in the geological record. He stresses the importance of seeing in this the action of the creator: 'To assume that simple forms evolved to higher forms by themselves without any divine guidance is completely incorrect'. He adds: 'It is unscientific', though this comment is surely without any foundation.

But one soon understands that al-Naggar does not really accept evolution, not even the theistic version of it; indeed, he only accepts microevolution, that is, the evolution of each species separately; for example, dogs or horses may have changed over time, but no transformation of one species into a new one (by successive mutations). He says: 'Does this imply that these life forms developed from one another, from the simpler to the more complex? No'. Why? Because, according to him, the fossil record does not show this – which is false, as I will show later.

Finally, al-Naggar strongly rejects the idea that humans evolved from more primitive forms or species. He cites Björn Kurtén's book *Not From the Apes* (the lecture transcript spells his name as Curtin and al-Naggar refers to him as 'a very famous anthropologist and geologist actually studying the organic remains of Man'), who 'said in the book that throughout his research he has observed that the oldest records of man on Earth are much older than the oldest records of monkeys on Earth, and if there is a relationship between the two, then apes were men and not the other way around'. al-Naggar explains that according to the Swedish paleontologist 'man appeared suddenly on this planet within a very short span of time'. So, our lecturer concludes: 'Man is not part of this general schema and was created separately and is not linked to the process [of evolution]'. I have had no chance to check the book in question, but simply consulting Kurtén's biography, which states that he was 'the founding father of an important scientific movement that united Darwinian theory with empirical studies of fossil vertebrates'[19], and browsing through the list of his books (for which he received the UNESCO Kalinga Award) are enough to convince me that Kurtén was anything but a creationist. As to the title of Kurtén's aforementioned book, one review explains it simply as 'a catchy title'.

The American Muslim author Nuh H. M. Keller has also tried to be critical, yet moderate, in his treatment of evolution[21]. This is a difficult exercise, especially for people who have limited knowledge of evolution, with its principles, mechanisms and results – as we have just seen. Keller first tries to attack the foundations of evolution by painting it as non-falsifiable, hence non-scientific: 'If evolution is not scientific, then what is it? It seems to me that it is a human interpretation, an endeavor, an industry, a literature'. (Needless to say, this line of attack is completely mistaken.) Then, like al-Naggar, he moves on to accept micro-evolution ('which is fairly well attested to by breeding horses, pigeons, useful plant hybrids, and so on') and examines macro-evolution at least from a theological angle: 'With respect to evolution, the knowledge claim that Allah has brought one sort of being out of another is not intrinsically impossible [. . .] because it is not self-contradictory. And [. . .] it would seem to me that we have two different cases, that of man, and that of the rest of creation'. And so again, we find this strong distinction in the treatment of humankind in opposition to the rest of the species, for which some theistic evolution is accepted: 'Belief in macro-evolutionary transformation and variation of non-human species does not seem to me to entail *kufr* (unbelief) or *shirk* (ascribing co-sharers to Allah) unless one also believes that such transformation came about by random mutation and natural selection, understanding these adjectives as meaning causal independence from the will of Allah. [. . .] From the point of view of *tawheed*, Islamic theism, nothing happens "at random", there is no "autonomous nature"'.

The above examples already show that there is a variety of understandings and attitudes towards evolution in the Muslim cultural and theological landscape. In fact, the Muslim positions are even more diverse.

the evidence for (the fact of) evolution

First and foremost, it is important to insist on the distinction between evolution as a fact that one can infer from many observations in nature and experiments in biology and evolution as a theory, which has parts that are much stronger than others. It is of utmost importance for Muslims and others to understand this fundamental difference. And that is why I will devote this long section to citing much evidence in support of evolution as a fact of nature. In subsequent sections, I will summarise the theory of evolution and first describe the (standard) Darwinian version, which general philosophy – particularly in its naturalistic, purposeless propositions – may be in serious clash with the Islamic world view; afterwards I will present the lesser known (and less widely accepted) non-Darwinian versions, which we shall find much more friendly towards theistic world views like that of Islam.

the fossil record

When animals or plants die, their remains usually disintegrate or get eaten up by scavengers. On rare occasions, however, they get buried in mud or soft rocks, and if mineral salts do permeate into the bones or some other parts of the dead organism, the latter gets hardened and preserved as fossils for long periods of time. Dead bodies can also remain quite intact in ice, in tree resins (amber), tar and other such milieus. Hence, most of the remains that paleontologists uncover in deserts are bones or partial skeletons, but in ice sometimes an animal can be found surprisingly well preserved. Fossilisation takes place even more easily in aquatic environments[22], where the precipitation of sediments and minerals occurs at greater rates, so bodies that had fallen to the bottom of the sea would more easily fossilise. But the chances of non-sea animals falling there are often small. This explains why biologists have the best fossil records for marine animals and the worst ones for flying creatures.

Now, geological layers (or strata) are usually ordered chronologically, from the oldest at the bottom and the most recent ones at the top. Moreover, radioactive dating of various geological or biological parts allows one to determine the time at which particular organism was buried. For this reason, fossils constitute extremely important pieces of evidence for the tracing the history of species. Unfortunately, despite the millions of fossil pieces uncovered throughout the world, representing some 250,000 species, the fossil record is patchy at best, and one finds large gaps in the reconstituted chronology of most species. A particular big gap is the one that separates the early Cambrian period (550 to 500 million years ago) when an explosion of invertebrate life forms occurred and the vertebrates that appeared about

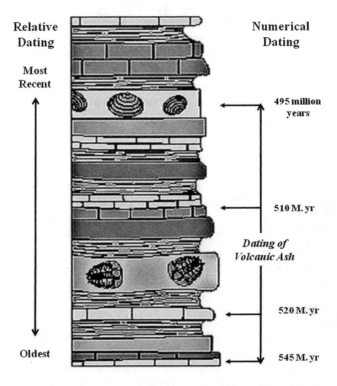

Relative Dating

Most Recent

Oldest

Numerical Dating

495 million years

510 M. yr

Dating of Volcanic Ash

520 M. yr

545 M. yr

100 million years later; few intermediate species have been found, which represents an unfortunate weak point in the theory, or at least in the support that can be provided to it. Those gaps are often referred to as 'missing links', particularly by doubters of the theory of evolution (see the next sidebar).

---

missing links

One of the main arguments against the theory of evolution, from its beginnings, has been that any such transformations should produce transitional forms, but that few, if any, could be found in the fossil record. Darwin realised that the fossil record, due to its accidental history (a fault of geology, not biology), was necessarily spotty, which explains the rarity of unusual species (unusual in the sense that they would have existed only for the periods when the environment was favourable to them), and as Darwin had predicted some would be found sooner or later. Critics of evolution have continued to contend that such links are still missing.

Proponents of evolution, on the other hand, adamantly insist that more and more transitional forms have been found, including some showing the transitions from fish to (quasi-)reptiles, from reptiles to birds and from apes to humans. Francisco Ayala, the distinguished Christian-faithful evolutionist puts it clearly: 'Gaps in the reconstruction of evolutionary history from all living organisms back to their common ancestor no longer exist'[23]; regarding human evolution, he adds, 'The missing links are no longer missing. Thousands of intermediate fossil remains (known as "hominids") have been discovered since Darwin's time, and the rate of discovery is accelerating'[24].

Here are a few examples of 'missing links', or what biologists prefer to call transitional forms:

- *Archaeopteryx* was first discovered in fossil form in 1861 (two years after the publication of Darwin's *On the Origins of Species*), though it produced more controversy than evidence, as not everyone was convinced that it really exhibited a transitional form between a reptile and a bird. Since then, at least seven skeletons have been found, and – more importantly – fossils of birds that were uncovered in Spain and China and found to be 30–40 million years younger than Archaeopteryx were more evolved than their predecessor in the way they would be expected; for instance, they had short bony tails and reduced hand claws. Below is an image of the species in its fossil form and a drawing of how it is imagined to have looked.

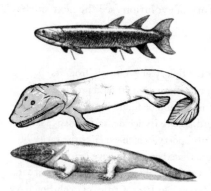

- *Tiktaalik* is the name given to a species that was found in fossil form in the Arctic region north of Canada in 2004. It was hailed as a major discovery, for it showed with stunning clarity features of a fish and an amphibian simultaneously. Below are drawings of Tiktaalik (375 million years old) as well as of its predecessor, Eusthenopteron (the fish), 385 million years old, and its descendent, Ichthyostega (the amphibian), 365 million years old.

- *Aetiocetus* is the transitional form between the early ancestral land whale (before the mammal went into the sea) and the modern aquatic whale. Note, in particular, the gradual shift of the nostrils from the front to the top of the skull.

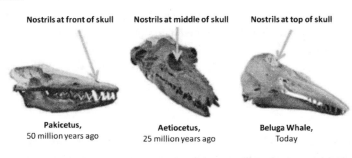

Nostrils at front of skull — Nostrils at middle of skull — Nostrils at top of skull

Pakicetus, 50 million years ago — Aetiocetus, 25 million years ago — Beluga Whale, Today

One should point out that in some cases, however, the fossil record of some species is complete enough to practically trace its entire history. That is the case of the horse, for example, for which a very rich and extensive record of the past 50 million years has been found in North American sites, providing a very important piece of observational evidence in support of evolution.

The fossil record is extensive enough to reveal that life (of very simple form) started in the sea more than 3 billion years ago and that for more than a billion years only single-cell organisms related to bacteria existed[25]. This makes sense from an evolutionary viewpoint as we know from other fields (geology and astronomy) that early in the history of the earth the atmosphere contained too little oxygen, which is why the ultra-violet radiation could not be stopped by ozone (O_3), thereby preventing life from appearing and developing on land. But the early bacteria and algae released substantial amounts of oxygen (through photosynthesis), which prepared the environment outside the seas for the appearance and development of life on lands.

comparative anatomy

Comparisons between the body parts of apparently unrelated animals also point to their emergence from a common ancestor and thus lend strong support to the evolution scenario. Consider, for example, the bone structure in the limbs (feet, hands, arms and wings) of various tetrapod (four-legged) animals like the monkey, the horse, the mole, the dolphin and the bat. Checking their bones will easily reveal that they all have five 'fingers' (or phalanges) despite their very different usages: grasping with the hands by the monkey, running on the feet by the horse, digging with the hands by the mole, swimming with palms by the dolphin, and flying with the wings (which connect the bones of the arms) by the bat. This is then interpreted as evidence that these limbs are the results of an evolution from some past common species.

The same similarity is found in the mouthparts (labia, mandibles, maxillae etc.) of insects, despite the differences in their usage: biting and chewing by the grasshopper, ticking and biting by the honey bee, sucking by the butterfly, piercing and sucking by the mosquito etc.

Likewise, one finds that many animals have developed organs like the eye, despite the great disparities in their life conditions, which explain the differences in the exact shape and structure of the organ (e.g. the eye). This is a case of what some evolutionary biologists have termed 'convergent evolution'[26], and it is a tremendously important issue. Indeed, it is often claimed that the eye in particular, is too complex to be the result of any gradual evolution; 'what good is a quarter of an eye?' some people often ask. I address that question in the next sidebar. In fact eyes of one form or another have been evolved by numerous species in many different situations.

'what good is a quarter of an eye?'

William Paley had famously wondered whether anyone who finds a watch in a field could possibly believe that it assembled and started to give accurate time by itself. Likewise – and even more so – opponents of evolution have long argued that complex organs like the eye showed that no random process could explain the design, installation and switching of this amazing high-resolution, colour camera. How could different parts of the eye (the lens, the retina, the pupil etc.) have evolved (separately) and assembled themselves to complement each other so well in producing a sharp image?

Darwin himself was fully aware of this difficult case to explain but did not consider it a fatal blow to his theory. Indeed, biologists soon came up with a reasonably simple explanation and later found evidence to support it. In short, the explanation states that vision does not have to be complex, that being merely sensitive to light (with no ability to form any sort of image) is in itself an important environmental advantage (in recognising the presence of light sources, i.e. openings or the surface of the sea), one that would quickly be naturally selected. This can be achieved by the pigmentation of the skin (no eye as yet), with nerve fibers connected to the spot carrying the information back to the central nervous system, and later to whatever brain there would be. Changes (mutations) could then curve the spot into a 'cup', gradually turning it into a socket like the eye's, with the light-sensitive area being now enclosed and slowly turning into a retina. The last component to form is the lens, which would come either from the epithelium developing and taking quick advantage of some newly appearing refractive ability or from a double layer of skin containing (refractive) water. (See the diagram taken from the *Encyclopedia Britannica*.)

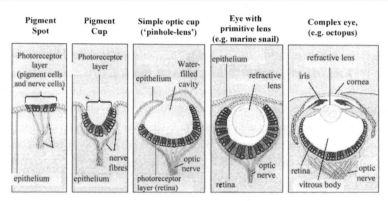

adapted from Encyclopaedia Britannia, 2005

What lends strong support to this scenario is that eyes corresponding to each one of these stages have been found in nature and in fossils going back to over 500 million years[i].

Evolutionists further argue that some flaws in the design of our eye lend further support to its evolutionary origin; for instance, the fact that the retina has a blind spot and blood vessels running across its surface, making it an imperfect organ.

This comparative anatomy argument is such a strong one in favour of evolution that proponents of the theory have often used it to attack the design argument. Indeed, they ask: if the characteristics of various organs in humans and in most of other animals were the result of a direct well-planned design (by a Creator) and not the product of a long and complicated evolution

---

[i] Examples of organisms with various types of 'eyes' include euglena (have only light-sensitive patches); planaria (have cupped light-sensitive spots); nautilus (have a pinhole with no lens); arthropoda (have eyes that vary from simple to compound); and vertebrate animals (have camera-lens eyes).

that was affected by random mutations and various local conditions, then how would one explain the many imperfections that one can easily notice in many cases? Some of the examples they give are quite strong indeed:

- Human head is disproportionately too large compared to the woman's vaginal opening, making birth extremely painful, if not dangerous.
- Our useless wisdom teeth; hollow bones in flightless birds like penguins; sightless eyes in many cave-dwelling animals; such organs are referred to as "vestigial".
- The tail in newborn humans, and extra toes occasionally found on horses, usually duplicating the third toe etc.

Direct-design (or creationism) advocates cannot explain this; they only say: just because we can't explain some of these features, does not mean they are design flaws. Theistic evolutionists, however, adopt an infinitely creative nature viewpoint, stating that life was allowed to develop in all possible directions, with the selection process gradually moving things forward; this then allows one to explain any 'bad' or 'useless' features that one may encounter in some species or in the world more generally.

universal biochemical organization

Another line of evidence for the occurrence of evolution in nature and the ability of the theory to explain the biological diversity (and similarities) that we observe around us is the fact that all organisms have their biochemical organization based on the same genetic encoding in DNA (or RNA in viruses), which is then translated into proteins by ribosomes[27]. In fact, the same piece of DNA will represent a given amino acid in a bacterium, a human cell, a plant cell or an animal cell. Indeed, in 2001 the Nobel Prize in Medicine was given to scientists who showed that a gene involved in the control system of yeast cells is the same one that mutates in some human cancers.

This then allows biochemists to compare and relate various species to each other. If one compares cells from two organisms without knowing whether they are from the same species or not, one will be able to determine the degree of relationship between them: If the genetic code is found to have the same composition, then the two organisms are from the same species, but if they differ, then the amount of difference will tell us how close they are to each other. And that is how we find that human DNA sequences differ by only 1.2 per cent from those of chimpanzees, by 1.6 per cent from those of gorillas and by 6.6 per cent from those of baboons[28]. Moreover, knowing the rate at which specific mutations occur allows us to determine approximately when the divergence between those relatives took place.

genetic evidence

A new and a very strong line of support for evolution is the evidence geneticists have been able to gather recently from the decoding of the genomes of humans and other species, such as chimpanzees and gorillas, something that has only recently been made possible. The prominent biologist Kenneth R. Miller (a believer) gives some striking arguments in his recent general-public book *Only a Theory*."[29] To mention just a few points[30]:

- We humans have 46 chromosomes, while all apes have 48. Examination of the genomes shows that our chromosome 2 turns out to be exactly the juxtaposition of chromosomes 12 and 13 of the chimpanzees, and indeed the latter two have now been renumbered as 2A and 2B. Sometime in our evolution two earlier chromosomes fused into one.
- Similar 'typos' between genomes point to their common antecedent. For example, humans need vitamin C to help build collagen, an important protein which prevents scurvy; other mammals are able to make the vitamin in their own bodies, but we cannot, because we lack a critical enzyme, gulonolactone oxidase (GLO), due to a bad mutation sometime in the past. It was found that primates like gorillas and chimps lack this enzyme too, whereas other more distant primates do not.
- The hemoglobin protein, which makes blood red, is made of two genes, one of which ('a pseudo-gene') is on our chromosome 16, and gorillas and chimpanzees have it in exactly the same configuration within their genomes.

(micro-)evolution occurring in the present

Finally, evolution can be observed in the laboratory, but only at the micro level; for instance, in the frequent mutation of bacteria that often becomes resistant to some antibiotics, or of some plants and insects that becomes resistant to pesticides. That is why doctors always tell us to make sure we take the full course of antibiotic that has been prescribed, because stopping the treatment after only few days (when most but not all of the bacteria have been killed and we start feeling better) will leave some bacteria in the sick organ, and as some mutations will invariably occur among them, others will ('by chance') appear with some resistance to the antibiotic and start multiplying, thus requiring a new treatment with different antibiotic. Those who do not believe in the mutation–evolution scenario would not take the full course of their treatment and would face the consequences.

Two other present-day examples can quickly be mentioned here[31]. One is the case of the HIV virus, which also mutates in AIDS patients and thus defeats the treatment unless a cocktail of drugs is prescribed, because the probability of a mutation occurring and resulting in resistance to all drugs is very small. Another example is the fast increase in the appearance of tuskless elephants because these are not hunted while tusked ones are killed. This constitutes a good example of how special environmental (in this case human-imposed) conditions lead to the special (in this case unnatural) selection of an 'abnormal' species over the previously dominant one. Of course, artificial selection is nothing new, it has been for a long time practised by farmers and breeders to select particular species (e.g. cauliflower, broccoli, Brussels sprouts etc.) and grow them preferentially.

There are additional lines of evidence in support of evolution; for example, the geographical distribution of species, but space does not allow me to explore all those avenues.

Darwinian (orthodox) evolution theory

It is actually *not* quite fitting that Darwinism, natural selection, and survival of the fittest are often used as synonyms for evolution. But such identifications say a lot about the importance of these terms, which I would like to explain in this section.

The first principle that I need to establish and insist upon is the distinction between evolution, which, as I have laboured to show in the preceding section, is a solidly observed fact of nature, and between the theory(ies) that may be constructed to account for it. Darwin's theory, of which natural selection is one essential part, is the theory most widely used and accepted by biologists, but it was not the first one, and today it is not the final word, for new ideas – which have so far remained rather marginal – have appeared in the past decade or two, and they have promised to revise the canon, perhaps in ways that are much more harmonious with the theistic world view. But that is a discussion I shall leave for the next section and the conclusion.

Now before I present the orthodox Darwinian theory of evolution, perhaps we should seek to meet its father, Charles Darwin (1809–1882), who for having come up with that theory is widely considered as one of the most important scientists of all time. He has also become one of the most controversial figures in human history, hailed as a hero by many, despised by others as a devilish corruptor of man's faith in God and view of creation. Even his wife believed he was destined to hellfire. And yet upon his death the Anglican Church bestowed on him the honor of being buried in Westminster Abbey.

There are a number of interesting facts about Darwin, many of them largely unknown and rather surprising for the general public. The first one is that Darwin intended to become a priest in his youth. It is not quite clear whether he truly wanted to become a priest or fell back on it as a reasonable and socially respectable choice when no other prospects opened up to him. There are, however, a number of anecdotes about him that seem to argue for his strong faith, at least early on: One is the fact that as a student he read Paley's famous book on the design argument, *Natural Theology*, and was very impressed with it; another anecdote is that during his famous trip onboard the Beagle ship, which ended up transforming his scientific and theological views, his shipmates used to joke about his very frequent reading and literal interpretation of the Bible.

Another important fact about Darwin is that what shook his faith in God and pushed him from a strong believer to a doubting agnostic is the problem of evil in the world; indeed he could never reconcile the Bible-proclaimed omnipotence of God with the widespread destruction in nature and evil in human history. The coup de grace for him came with the loss of his 10-year old daughter Annie, for whom he had a special love, for she resembled his sister, the one who had replaced their mother when she died early on and took care of him and his father for many years. One should always keep this story in mind when discussing Darwin's religious views towards the end of his life; in a nutshell he came to strongly reject Paley's design argument as applied to animals, organisms and organs, for his theory fully accounted for all that now, and he could at most see the hand of a designer in the general laws and features of nature, but he could only accept a deistic view of God, one who created at the beginning but then withdrew from the world, for he could not see the presence of (an active) God when so much destruction and malevolence exist in the world.

But let us go back a little and retrace his construction of the theory of evolution. The idea that organisms changed over time had been known and accepted to a large extent among many cultures (western and eastern) for centuries. In the West, for example, his own grandfather, Erasmus Darwin, had suggested that species descended from common ancestors, but he did not offer any explanation or evidence for that. The French naturalist Jean-Baptiste Lamarck (1744–1829) had expounded a theory of his own, one that remained popular at least until Darwinism gained full acceptance. What Lamarck had proposed was the 'inheritance of acquired characters', which declares that physical traits that one acquires in life will be passed on to one's children. For example, if your job is a physically demanding one, then you will develop strong muscles, and will pass on that character to your children, who will be born with strong muscles. (Modern biology, in its current paradigm, totally rules this out.)

In December 1831, Darwin, then only 22, sailed on board the *Beagle* ship as a young naturalist to take part in an expedition that was supposed to explore the coast of South America for two years but ended up going around the world in five years. This allowed Darwin to make truly revolutionary discoveries. The Galapagos Islands have become famous since Darwin's stay on them and his observations of variations between some animal species from one island to another. Darwin kept asking himself: Why would God, if he did indeed create each species directly and separately, devote so much attention to these uninhabited islands, and put slightly different but similar species on each one? He came to conclude that those variations were the result of a gradual evolution over time, and that all the similar but different species must have descended and evolved from one ancestor. Generalising this idea to species that were close enough but not that close, say horses, donkeys and zebras, he came to conclude that all animals must have diverged from a common origin.

Still, he needed to explain what made a given animal form split up from its ancestral lineage, or what allowed it to diverge and develop into more and more different species.

The inspiration came to him in 1838 upon reading Robert Malthus's book, *Essay on the Principle of Population*, which was first published in 1798. In this book Malthus explained that because populations increased by multiplying (1, 2, 4, 8, 16 etc.) while food production only increased by fixed increments (1.5, 3, 4.5, 6, 7.5 etc.), there was bound to be some hard competition among different populations (of humans or animals) for food. It was then that Darwin realised that nature was going to favour only a limited number of animals in any given species, and those which nature favoured must somehow (due to special characteristics) have a greater ability to survive – and pass on those characteristics to their offsprings. So the idea of natural selection and survival of the fittest (an expression that Thomas Huxley, not Darwin, came up with later) quickly developed. Darwin himself, and many others after him, identified this idea as the cornerstone of the new theory of evolution. There still remained to explain how a species split up to produce new forms, some which would then be selected, split up again somehow, perhaps be selected again, and so on. But the splitting up would be explained (much) later. That mystery would be resolved with genes and mutations.

But while the discovery of natural selection occurred in 1838, Darwin did not sit to write down his ideas until 1842. He circulated his theory among a select few and opted to shelve the matter until he had decided what to do with such an explosive theory. And it was only in 1858, when he received an essay by an unknown young naturalist named Alfred Russel Wallace, who had described a new hypothesis for explaining evolution, that Darwin realised that this young fellow had somehow hit on exactly the same idea; Darwin thus needed to announce to the world that he had discovered it first. It was then arranged to have the two scientists present their idea(s) to the Linnean Society shortly thereafter, and Darwin finally resolved to write *On the Origin of Species*[ii], which he published the next year.

Darwin's theory of evolution is often described as 'descent with modification', it is usually presented as resting on the following three pillars: (a) replication; (b) variation; and (c) selection. Let me try to explain each of these briefly and simply:

a. Replication refers to the reproduction process by which organisms multiply and produce offspring.
b. Variation indicates that sometimes the offspring are slightly different from their parent(s); as I have mentioned, modern biology, with the merging of Darwinism with genetic theory in the 1930s (a merger known as 'the great synthesis' or Neo-Darwinism) will identify mutations in genes as the source for such variations.
c. Selection is referred to the central mechanism of natural selection by which environment-fit animals or plants find themselves surviving more

[ii] It is important to note that the title of Darwin's book is *On the Origin of Species by Means of Natural Selection, or the Preservation of Favored Races in the Struggle for Life*, which emphasises the central role of natural selection in the creation of new species.

than those that are not endowed with those slightly different but crucial characteristics.

Although some authors (e.g. Francisco Ayala) insist (with Darwin) that natural selection is the central piece in the theory, I (a non-biologist, I should stress) personally see mutations as the more crucial part. Before I explain why, I would like to point out that Darwin considered natural selection as his ingenious discovery when he had no idea how new characteristics occurred or were transmitted. Genes and mutations were not to be discovered for decades. Had he lived to see the full synthesis, it is my conjecture that he might have seen the variation pillar in the theory as the most important one. Let me briefly explain this part.

Mutations are changes that occur in the genetic material (DNA sequence) of a given organism. Despite the error-correction mechanisms that are built into all cells, such typing mistakes do occur. It is like having one letter (and, in rare occasions, a few letters) changed in the organism's features description book, which the DNA represents. Mutations are the result of 'copying errors' that occur during the division process of a cell, of exposure to radiation or to chemical mutagens; they can also sometimes be due to viruses. (I have put quotes around 'errors' because one can easily interpret such changes more positively; one may in fact see them as divine twists that lead to evolution and progress . . .)

Mutations can have some potential evolutionary effect only if they occur in the reproduction (sexual) cells. For only then can such modifications be passed on and possibly lead to changes in the organism. Changes always start in one individual, they then to spread by reproductive multiplication, and if the environment is suitable, they may become a widespread or dominant feature of that species. If the change is significant enough, or if further changes build upon it, the species may get transformed.

Most biologists insist on the random nature of mutations, and this issue will have important consequences, which I shall discuss later. Although very rare, mutations have been carefully observed and recorded in labs for numerous types of organisms, ranging from bacteria to human cells, where thousands of mutations have been catalogued by medical geneticists[32].

In humans, the rate of (random) mutations has been found to be about one per 1 million sex cells[33]. Such frequencies (which can be much higher in other species) are too low to be noticed macroscopically (directly on animals) but high enough to affect bacteria on short time intervals or species on long time periods.

It is important to emphasise that mutations and natural selection work in tandem. First, if a mutation is not selected (due to its greater adequacy to the environment), it will be forgotten quickly and have no lasting effects. Secondly, the environment itself may lead to specific mutations by applying special actions (radiation, catastrophes, climate upheavals etc.). Lastly, and perhaps most importantly, most favoured mutations tend to be of the small kind; indeed, major sudden changes in an organism usually lead to quick death. For all these reasons, evolution proceeds very slowly and gradually, by small adjustments rather than major modifications; therefore, no species with previously unseen features appears all of a sudden[34].

non-Darwinian evolution

Jean Staune has spent the past 20 years (most of his adult life) exploring the implications of the 'new paradigms' of science (incompleteness of mathematics, non-locality and uncertainty principle in quantum physics, non-Darwinian processes in biological evolution etc.) on our world views. Having studied philosophy of science and paleontology (among other things), he has focused special attention on the new non-Darwinian theories of evolution. Staune has just released a thick book[35] in which he reviews all the evidences from the new paradigms and concludes that this points to and calls for a new attitude with regard to the spirit. In particular, he devotes some 150 pages to the topic of non-Darwinian evolution, for he insists that there is mounting evidence that the standard theory of evolution is more and more often hitting cases that it cannot explain, much like Newton's theory of gravity could not explain mercury's perihelion precession and other such strange cases, and it is now time for a new theory of evolution. Needless to say, Staune is careful to state, in the strongest terms, that he fully accepts the fact of evolution; indeed, he accepts that much of the standard theory of evolution (Neo-Darwinism) is correct, but certain elements of it (such as the insistence on the randomness of the process or the central part played by natural selection) are apparently in need of revision. I will be drawing most of the information in this section from Staune's book, which so far has only been published in French. (There was also a marginally useful dossier on Non-Darwinian theories of evolution in the December 2005 issue of the French popular science magazine *Science et Vie* as well as several articles on the 'essential' randomness inherent to the standard theory of evolution in the August 2007 issue of the same magazine.)

First, what is the difference between Darwinian evolution and Non-Darwinian evolution? The first term describes the standard theory of evolution, which is accepted today by the overwhelming majority of biologists (probably because they are not aware of the problems and the ideas developed by the non-Darwinians); it insists that evolution is driven by the combined effect of random mutations and natural selection. It further insists that the development of life was completely blind and unplanned; life went into various directions, produced all kinds of forms which end up dying (more than 90 percent of species that ever appeared no longer exist), and if the history of life were run again, like when scientists run a computer simulation by drawing new random numbers to start with, a completely different world would emerge, one where the existence of conscious and intelligent beings would be very unlikely. Non-Darwinian evolutionists, however, accept mutations (maybe random, to some extent, maybe not at all) and natural selection, though the latter is often not accorded a predominant role, but they also notice new effects in nature (such as self-organization of biochemical structure or the channeling or even the orderly progress of life forms) and introduce them in their theories.

Staune divides Darwinians into two categories: the strong Darwinians and the weak Darwinians. In the first group, he puts people like Richard Dawkins who consider the random, blind, purposeless version of the theory as absolutely correct and needing no discussion and certainly no tinkering with; such scientists tend to be dogmatic and fueled with a materialistic agenda, even if the scientific arguments they usually bring are for the most part correct. Among the weak Darwinians, Staune puts such illustrious scientists as Stephen Jay Gould, the famous (and late) American evolutionary biologist and popular science writer. In scientific circles Gould is mainly known for his (rather controversial) theory of 'punctuated equilibria', which claims that evolution goes through long, quiet, slow periods where practically no development occurs, but these are then followed by (relatively) very short, accelerated periods of fast evolution where new species appear; the famous Cambrian Explosion would be a perfect example of such occurrences.

Then there are several important scientists (biologists, paleontologists and biochemists) who are Darwinists to some extent. Few examples are as follows:

1. Christian de Duve, the Belgian Nobel Prize winner (in 1974) for his work on cell structure and processes; Staune calls him 'the most Darwinian

of the non-Darwinian evolutionists'[36]. De Duve insists that even though evolution proceeds by way of mutations and natural selection, something in the process makes it 'programmed' such as to lead to intelligent beings. In his nice metaphor, he describes the dice of evolution as 'rigged', he has smartly quipped: 'God plays dice because He is sure to win'.

2. Simon Conway-Morris, a Cambridge University paleontologist with a very solid scientific reputation; he has written a few books in recent years with bold theses, the most recent one being his claim[37] – substantiated with solid evidence and much academic argumentation – that evolution follows (preset) channels; this, for example, explains why the invention of the eye was made by countless organisms and species of very dissimilar types and in completely different environments. He then argues that the appearance of humans was 'inevitable' (the word he uses in the book's subtitle), even though intelligent life is extremely improbable in the universe.

3. Stuart Kauffman, an American biologist who is best known for insisting on the primary importance of self-organization in the appearance and development of complex biological systems and organisms; for him self-organization is probably as important as, if not more than, natural selection.

These scientists, and other less illustrious ones, should be considered as Darwinian evolutionists (although De Duve and Kauffman can be pictured as bookends to the group) because they still give the standard theory and its processes (mutations and natural selection) a prime place and role in the overall evolutionary scheme. They are 'weak' ones because they want to supplement the standard processes with new factors and effects that may change the overall picture we draw of life's history, its underlying principles and its ultimate outcomes.

Staune then considers the non-Darwinian evolutionists and similarly divides them into weak non-Darwinians and strong non-Darwinians: the first group accepts mutations and natural selection as principal elements of evolution, but, contrary to strong Darwinians, insists that these elements would – in all cases (whatever the values of the random numbers) – have led to the appearance of a conscious and intelligent species; the second group (strong non-Darwinians) do not accept the principle that evolution was directed – even mainly – by mutations and natural selection.

The most important (and to some extent famous, one way or another) non-Darwinians are as follows:

1. Michael Denton, a British-Australian biochemist and senior researcher at the University of Otago in New Zealand, who insists on the existence of 'archetypal forms' that have guided evolution from the start.
2. Brian Goodwin, a Canadian biologist, and Mae-Wan Ho, a biochemist of Chinese origin, who has had a rich and diverse career in the West; they both insist on self-organization and reject the usual reductionist approach of standard biochemistry; they consider the main Darwinian processes as non-dominant.
3. Remy Chauvin, a French ethologue, who adopts a teleological (finalistic) approach and claims to infer from the evolutionary history of life forms an 'internal logic' and even a 'programming'.
4. Anne Dambricourt-Malassé, a French paleontologist, whose research on human evolution produced an earthquake in France; she claims to be able to show that evolution proceeds by step-jumps and does not see natural selection as a determining factor in the unfolding of evolution.

It is clear that with these scientists, all of them active researchers in the field of biology and evolution, one is getting farther and farther from the standard paradigm that 'random mutations and natural selection of favourable characters produce new species, few of which will end up thriving and developing'. In fact, Staune mentions a number of other researchers whose hypotheses are even more radical, but with the ones I have picked and mentioned briefly here, one already notes that many interesting ideas have been put forward with the aim of reviewing and perhaps revamping the general theory of evolution. Staune talks about the status of the Neo-Darwinian theory of evolution as being 'somewhere between Ptolemy's and Newton's' (which I think is rather exaggerated, but we get the idea) and that we are eagerly awaiting a new 'Einsteinian'[iii] theory to replace the 'Newtonian one' (read the Neo-Darwinian one). Staune believes that the shift from the old theory to a new one is already under way, at least in Europe and outside the USA; he even brushes a general sketch of it. Moreover, he argues that the reason why the US researchers have not shown much interest and thus have contributed nothing to the new theory is the creationists' counter-productive pressure; indeed the US biologists and

---

[iii] Staune has a whole chapter titled 'Seeking Biology's Einstein – Urgent!'

scientists have been pushed to look at the situation in black and white: either Darwin (a theory of random, purposeless evolution) or God (direct ID creation).

To close this section, let me mention briefly the general lines of the new theory of evolution as Staune envisions it:

1. The main elements of Neo-Darwinism, which will remain the reference theory for explaining most aspects of microevolution (within a given species) and adaptation.

2. Some elements of Neo-Lamarckism, which will be necessary in order to explain why some organisms (from bacteria to animals) seem to adapt quickly in certain situations, at a rate that is much faster than Neo-Darwinism would have it.

3. New elements, synthesised from the ideas of Conway-Morris and De Duve (channeling, quantum bias), Denton (archetypal forms), Gould et al. (accelerated evolution that looks like jumps) and additional contributions (some of which Staune mentions but I have chosen to leave out[iv]).

theistic evolution

Although Staune is eminently concerned with the philosophical and religious implications of the new paradigms of science, and though he does spend a few pages to underscore the Catholic Church's acceptance of evolution and its theories, he surprisingly does not give any attention to the so-called theistic evolution views that many (at least in the Anglo-Saxon world) scientists, philosophers and theologians (Robert Russell, Holmes Rolston, John Polkinghorne, John F. Haught, Keith Ward etc.) have discussed in recent years. Let me try to briefly cover this topic.

It is interesting to relate that the first academic to be informed by Darwin about his full new theory immediately accepted it by giving it a theistic interpretation. Indeed, in 1857 Darwin sent a sneak-preview version of his *Origins* to Asa Gray, the Harvard professor, who was already an illustrious botanist. Gray was a firm Christian believer, a devoted Presbyterian, so he saw in Darwin's theory a beautiful description of how God brought about creatures in the world. He and Darwin ended up exchanging very

---

[iv] For instance, Vincent Fleury's radical thesis by which genetic factors (and thus mutations) play essentially no role at all in Evolution.

interesting and meaningful letters[v]; from those we have come to peep into Darwin's beliefs and their evolution. Gray also became a strong defender of the new theory and published his essays in a book titled *Darwiniana*.

Pierre Teilhard de Chardin (1881–1955)

Perhaps the most illustrious and fascinating theistic evolutionist of the past century is the French Jesuit paleontologist and philosopher Pierre Teilhard

de Chardin, who was early on more famous for his Omega Point and Noosphere, but was more and more often mentioned in the context of evolution. Indeed, when he wrote his famous book *The Human Phenomenon*, a grand narrative of the creation story up to and including man, his evolutionary outlook and nonliteral interpretation of the Bible offended the Church authorities, and the book was not allowed to be published until much later. (He has been rehabilitated since. In fact he had great influence on the Church's recent acceptance of evolution.)

Teilhard (as he is known for short) spent many years as a paleontologist, both in the laboratory (at the famous Muséum National d'Histoire Naturelle, in Paris) studying mammals, and in the field, doing excavations in the prehistoric painted caves in the northwest of Spain for a short while and then spending many years in China, where he participated in the discovery of, among other things, the famous Peking Man.

Teilhard was teleological through and through. He believed in 'orthogenesis', the idea that evolution unfolds in an upward, directional and purposeful way. In fact, he did not see biological evolution as limited to nature on earth but rather as a cosmic phenomenon, which englobes earth

---

[v] In a recent little article (*Milestones*, Online 11 (2007), http://www.templeton.org/milestones/), Denis R. Alexander tells us that Darwin exchanged correspondence with nearly 2,000 people, and that 14,500 of his letters have been collected by the Cambridge University Library Darwin Archive; about half of those have been published, making up 15 volumes. Two hundred of the letters were written to and by clergymen, thus giving us a good 'fossil record' of the evolution of his views. With Asa Gray, Darwin exchanged some 300 letters over 27 years.

itself, the solar system, and the universe at large, until a final Omega Point where complexity and conscious reach maximum levels and so the whole universe becomes conscious of itself and thus transcendent.

Theistic evolution has received some renewed attention lately, especially when the controversy over ID seemed to pit evolution proponents against theists. Some contemporary theologians and thinkers have tried to insist that the choice is not limited to evolution and ID; in fact another acceptable interpretation has existed essentially since day 1, as we saw with Asa Gray.

First, one must emphasise that the theistic evolution camp fully accepts and upholds the principles of evolution, including the important role played by chance. For example, Keith Ward, the renowned Oxford theologian, states matter-of-factly that evolution is 'now shown to be a major feature of the universe as we know it . . . '[38] And John Haught, a prominent Catholic American theologian, adds: 'A theology after Darwin also argues that divine Providence influences the world in a persuasive rather than a coercive way'[39].

One of today's most eminent proponents of theistic evolution is Robert J. Russell, a physicist who has been for the past two decades or so at the forefront of the development of the Science-Religion discourse as an academic field; in particular, he has been running the Center for Theology and the Natural Sciences in Berkeley, California. Russell's views have been described as 'an updated version of Asa Gray's thinking'[40], for he has proposed that evolution is perhaps guided by God by making use of indeterminacy (for us) at the quantum level. This then gives the appearance of randomness to mutations and transformational processes that lead to evolution. In this way, God drives the upward evolution of life towards higher forms.

Another illustrious contemporary thinker, who subscribes to theistic evolution, albeit from another scientific perspective, is Holmes Rolston, III[vi]. A philosopher, he is mostly known for his contributions in the field of environmental ethics and the science and religion dialogue. Regarding evolution, Rolston questions the centrality of natural selection as the prime agent of evolution. Instead, he promotes the idea of self-organization (which we found was put forward by Kauffman and Ho), which in his view brings about a definite upward move.

[vi] Among the honours he has received, Holmes Rolston, III was awarded the 2003 Templeton Prize for progress toward research or discoveries about spiritual realities; he has also been invited to give the Gifford Lectures at the University of Edinburgh in 1997–1998, a definite recognition of scholarship.

As to the (Christian) theologians who have addressed the question of evolution and tried to make it compatible with their world view and dogmas, it is interesting to note that they have tried to explore the meaning of and reason for the existence of 'the obvious randomness' (Haught's words[41]) as the primary factor in the evolutionary process. John Haught discards the idea that randomness is only a reflection of our human ignorance of the divine action in the world. Instead, he insists that contingency 'must be accepted as a fact of nature' and that indeed it is not only consistent with but also reflects divine love: 'An infinite love, if we think about it seriously, would manifest itself in the creation of a universe free of any rigid determinism (either natural or divine) that would keep it from arriving at its own independence, autonomy and self-coherence'[42]. He further finds in this view a possible solution to the problem of evil: 'Maybe the entirety of evolution, including all the suffering and contingency that seem to render it absurd, are quite consistent, after all, with the idea of a Providence that cares for the internal growth and emergent independence of the world'[43]. I shall discuss the (major) topic of divine action in the world in the next chapter, albeit briefly, for this major question would need to be discussed in a separate major treatment.

John Polkinghorne[vii] has presented theistic views on evolution similar to those of Haught. In a paper appropriately titled 'The Inbuilt Potentiality of Creation'[44], he writes (as I quoted at the top of this chapter) that God has brought into being a creation that could 'make itself'. He even also adopts Arthur Peacocke's 'striking' phrase, 'an Improvisor of unsurpassed ingenuity'[45]. Like Haught, Polkinghorne sees randomness as 'an expression of God's loving gift of freedom'[46], and finds in it an explanation of evil (as an example he cites mutations which sometimes turn cells cancerous) as 'the necessary cost of the evolution of complexity'. Finally, Polkinghorne addresses the important question of divine action; he insists that God's action cannot be disentangled from nature's or from man's: 'One cannot say "nature did this, human will did that, and God did the third thing"'[47].

The last important thinker I must bring up on theistic evolution is Keith Ward, the Oxford theologian, who also finds it crucial to address the question of randomness in nature. He too believes that chance is necessary for the existence of freedom: 'Chance, or indeterminacy – that is the lack of sufficient

---

[vii] John Polkinghorne is another prominent thinker; a physicist who decided to devote his life to the Church, he has written extensively on the question of science and religion and received the Templeton Prize in 2002.

causality – is a necessary, though by no means a sufficient, condition for the emergence of free and responsible choice on the part of agents generated by the evolutionary process'[48]. Furthermore, he insists that events may be random at their individual level but the process must be 'morally good overall'[49]. And to the main counter-argument of the atheists that the process of evolution is too random and exhibits no arrow or plan of creation, Ward responds that mutations are selected by the environment[50], which itself is determined by the laws of physics that were designed by God, therefore the whole process could be divinely pre-directed.

I shall close this section with Ward: 'One main argument for belief in theistic evolution is that it offers the prospect of framing such a view, which takes the findings of modern science and the testimony of the world's ancient religious and philosophical traditions with equal seriousness'[51].

human evolution

Darwin's *Origins* had barely one sentence on the relevance of his theory to human evolution. He did not want to make man the centre of his already controversial theory, so he focused only on animals, but he did promise that his theory would shed much light on our origins. Quickly the question was taken up by others, and Darwin tackled it fully in *The Descent of Man* in 1871. At that point, he had no fossil evidence of any kind to support his view of man's evolution.

Man was always considered special. Not only are we the only conscious and intelligent species, unlike all other species there exists only one variety of us. Take any two human beings, as diverse as you can find, say African Pygmies and Greenland Eskimos, you will find that their DNAs are very closely related: the difference between those two individuals would only be about 15 percent greater than the difference between two people from the same village[52]. But this, as Francisco Ayala insists[53], is the only evidence that we have a common ancestor and at some point groups among the descendants migrated and gradually populated the globe.

But man had also long been noticed to greatly resemble gorillas and other apes. For example, the name 'orangutan' comes from the aborigines' *orang*, meaning 'man', and *utan*, meaning 'forest'. Furthermore, modern genetics has shown[54] that our DNA differs from that of chimpanzees by only about 1 per cent, and that the difference is slightly greater for humans and gorillas and keeps on increasing for orangutans, baboons and more distant species. Ayala summarises the situation well: 'In some *biological* respects we are very similar to apes, but in other biological respects we are very different, and these differences provide a valid foundation for a religious view of humans as special creatures of God'[55] (emphasis in the text).

The DNA similarity/difference between humans and apes allows geneticists to determine the approximate time in the past when the split occurred between the human lineage and the apes. Indeed, by knowing the rate of occurrence of some crucial mutations, one can calculate how long it would have taken to get a certain difference between two daughter-species. One recent study finds this difference between 5.4 million years and 6.3 million years[56].

The first old human fossils to be uncovered came as early as the 1830s, including the famous Neanderthal man (1856). The research continued in the following decades; we have seen that Teilhard was a prime participant in that programme and in some of its discoveries, which came up

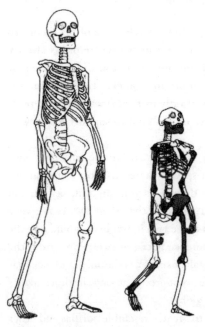

with skeletons that were often a few hundred thousand years old. The million-year barrier was spectacularly smashed in 1924 when small pieces of a skull found in South Africa were dated at between 2 million years and 3 million years. At least two decades passed before *Australopithecus* remains started to be found, although few – if any – scientists consider *Australopithecus* to be (fully) human.

The fossils of the 'hominid'[57] that became known as Lucy were discovered in November 1974 in the Afar region of Ethiopia. About 40 percent of the skeleton was recovered; its age was estimated at 3.2 million years. It was an extremely important discovery, for until the past few years when similar findings accelerated, it was the oldest hominid remains to be ever found.

How 'human' was Lucy? More accurately, how close to humanity was the Lucy species? That's the big question. How far back does the human species go? Difficult to say, exactly. To simplify, Lucy had mixed characteristics of humans as well as chimpanzees. She was only 1.10 m tall and weighed 29 kg; her brain was small (about one-third of the modern human's) and her arms were long; all this related her more to chimps. But she walked upright, and she and her tribe could climb trees and swing through them, judging from the length of their arms and shoulder blades. She was classified in the *Australopithecus afarensis* species. Most paleontologists tend to consider Australopithecines as pre-humans, or transitional forms between apes and humans.

In 2006, the remains of a fossilised child were uncovered in Ethiopia's Dikika region. It had been found in 2000, but it took more than five years to carefully remove the bones from the solidified mud. The remarkable near-complete skeleton was dated at about 3.3 million years; its age at death (most likely in a flood) was estimated (by analysis of the jaw by CT scans) at about three years. She was nicknamed 'Lucy's baby', metaphorically speaking. 'In my opinion, *afarensis* is a very good transitional species . . . a mixture of ape-like and human-like features'[58], Dr. Zeresenay Alemseged, the leader of the discovery team, told the BBC News.

In the past few years, dozens of such skeletons (usually partial) have been discovered, including the 2.5 million years old *Australopithecus garhi*, regarded as a link between Lucy and modern man. Most importantly, its discoverers have advanced circumstantial evidence that *A. garhi* was able to make tools. This would represent the real start of humans taking off from their animal ancestry. Sometime between 2.5 million years and 1.5 million years ago *Homo habilis* ('skilled human') had emerged (gradually); he had smaller teeth and a larger brain (about two-thirds of today's size). The reader will have noticed that the name has suddenly changed from *Australopithecus* to *Homo*; the crucial question is of course to what extent these species were human and whether that transformation was very gradual (as standard evolutionary theory says) or occurred in jumps (as new, controversial theses like Anne Dambricourt Malassé's propose).

Homo habilis became *Homo erectus* then *Homo sapiens*, literally the 'wise human', about 200,000 years ago. Indeed, the usage of tools allowed *H. habilis* to make hunting weapons, which then brought him bountiful meat. The protein-rich diet fuelled the brain and allowed him to develop faster and function with much greater capacity (our modern brains consume about one-fifth of calories we engorge). Changes in the environment then forced groups to migrate, either in search of places more abundant in food, or by following animals to either scavenge or hunt directly (such as the saber-tooth tiger who always left behind large amounts of meat that it could not cut and eat).

Finally, some 50,000–100,000 years ago, what anthropologists have characterised as the 'great leap forward' occurred: Human culture appeared in the form of burials, clothe-making and cave paintings. This represents a profound change: the development of symbolic thought, including religious ideas. The great religious traditions of the world have called this human: Adam.

The essential point of contention among scientists and thinkers is the stage of evolution at which one declares the species as 'human': homo habilis (about 2 million years ago), homo sapiens (200,000 years ago) or modern, cultured man (50,000–100,000 years ago).

On 2 October 2009, as I was putting the finishing touches to this book, the most extraordinary fossil skeleton, a 4.4-million-year old hominid, *Ardipithecus ramidus*, was presented to the world, via the prestigious journal *Science*. It was a remarkable discovery, as 110 pieces of the skeleton (about 45 per cent) were found, allowing researchers to reconstruct many of its features and attributes. It was not the oldest hominid fossil ever found; two previous discoveries of 6 million and 6–7 million years had been made, but from those two, only a skull piece and a leg bone had been found. Ardi, as the new hominid was nicknamed, lived in Ethiopia, which was then a forest land; as expected, she had a typically small brain (300 to 350 cm^3, similar to a female chimpanzee), was 120-cm tall and weighed about 50 kg. She did bring some important surprises – and with that misunderstandings and controversies.

The major surprise was that Ardi seemed quite different from chimpanzees, even though she had arms so long as to reach her knees in standing position. She walked upright but climbed easily and spent much time on trees. Her teeth seemed to indicate a lifestyle that was different from that of chimps. All in all, she was ape-like enough, but challenged some of the previous ideas of how humans got evolved from the chimps and other primates; now it seems more like the chimps evolved from Ardi, our ancestor.

I was surprised to note that for most of the day *al-Jazeera* TV, the Arab news network, was running, within the info strip at the bottom of the screen, the following title: 'Scientists refute Darwin's theory of human evolution with latest findings'. Worse, *al-Jazeera*'s website ran a summary piece titled 'Ardi gives a blow to Darwin's theory'[59], where one could read the following:

American scientists gave new evidence that Darwin's theory of evolution was wrong . . . humans did not evolve from ancestors resembling chimpanzees, hence shattering the old assumptions that mankind evolved from apes . . . Dr. Zaghloul al-Naggar [. . .] commented that westerners had started to go back to their senses after dealing with the origin of humans from a standpoint of materialism and denial of religions. In a phone conversation with *al-Jazeera*, he said that this important discovery deals a big blow to Darwin's theory. He added that the talk of 4.4 million years was an exaggeration, expecting the age of humans on Earth to be no more than about 400,000 years . . .

Islamic views

Adam is the central issue for Muslims with regard to evolution – at least nowadays. Contemporary religious scholars find it so impossible to conceive a pre-Adam species or even a possible multiplicity of Adams and lineages that have ended up disappearing (like Neanderthals, Java men etc.) that they're willing to reject the theory of evolution wholesale for that reason. Muslims did not always reject evolution; as we shall see, in previous times, up to the modern era (early twentieth century), they by and large accepted biological evolution and even welcomed it, as long as it did not present itself in purely materialistic, atheistic garb, even though the question of human evolution did often constitute a sore point. Nowadays, however, the rejection of Darwinism (in any form) is nearly unanimous.

As a case in point, let us consider the views of Mohammed S. R. al-Bouti, one of the most famous and respected Muslim scholars of the past three or four decades[60]. He is a conservative traditionalist, but he tries to adopt a rational approach of logical argumentation and scientific justification as much as he can (though he fails, in my opinion, due to his overly apologetic and subjective argumentation). One of his most important books is *Kubra al-Yaqiniyyat al-Kawniyyah* (*The Greatest Cosmic Certainties*)[61], which has gone through numerous editions and printings; in it he presents the reader with a rational defense of orthodox Islamic dogmas. In this book, he devotes a whole chapter to the question of evolution, and in some of its sections he even launches into a 'scientific' critique of Darwinism. Most interestingly, he starts with man; here are the first lines of that chapter:

The Muslim should be familiar with the following truths about Man and his reality; s/he must then make that a certainty and build upon it the meaning of his/her faith in Allah:

a. Man is the best and most noble of all creatures;

b. Man was created – in terms of his [physical] nature – from clay, and he has multiplied from the first man, Adam, peace be upon him[62];

c. Man was created from the very beginning in the best appearance and in
perfect shape; he has not evolved during his history in a way that would
have moved him gradually from one species to another[63].

A few pages later he warns the (Muslim) reader: 'And be careful not to pay
attention to what some Sufis say, that Adam (peace be upon him) whose
creation is related in the Qur'an was preceded by many other Adams'[64].

A number of observers have remarked that one can easily read some
human evolution in the various episodes of Adam's story in the Qur'an;
others have noted that some verses can easily be understood to support the
general biological evolutionary scenario. This issue is presented briefly in
the sidebar.

Evolution in the Qur'an?

There are a number of verses in the Qur'an that can be (and have been)
interpreted as lending support to evolution, both the human part of it and
the general biological one. Here are a few:

1. Human Evolution

- 'And He created you in stages . . . ' (71:14).
- 'He it is Who created you from clay, then He decreed a term/era . . . '
  (6:2).
- 'Thy Lord said unto the angels: I am to create a hominid (*bashar*) out of
  clay. And when I fashioned him and breathed into him of My Spirit, then
  do prostrate before him . . . ' (38: 71–72).

2. Biological Evolution

- 'Say (O Muhammad): Travel in the land and see how He originated
  creation, then Allah bringeth forth the later growth. Lo! Allah is Able to
  do all things' (29:20).
- 'Do the unbelievers not know that the heavens and the earth were of one
  piece, then We parted them, *and We made every living thing of water*? Will
  they then not believe?' (21:30).
- 'Allah hath created every animal out of water. Of them is (a kind) that
  goeth upon its belly and (a kind) that goeth upon two legs and (a kind)
  that goeth upon four. Allah createth what He wills. Lo! Allah is Able to
  do all things' (24:45).

---

al-Bouti then launches into an attack on Darwinism, a critique he tries
to base on scientific and rational arguments. First, he says, 'The reality

we observe strongly contradicts what Darwin called "the law of selection and survival of the most suitable"'. He adds: 'Why did nature produce the weak species if it is doomed?'; 'why didn't the higher apes acquire mental capabilities like man did?'; and 'studies of old mummified Egyptian plants and animals have shown that they do not differ from today's and that they have not changed during the past centuries . . . '[65] It quickly becomes clear that al-Bouti's understanding of Darwin's theory is flawed and that his approach is biased and pre-directed.

Unfortunately, al-Bouti is a very typical example of today's Islamic discourse on evolution. In the introductory section of this chapter, I gave several examples of important contemporary Islamic figures (e.g. al-Naggar) that have spoken, written and sometimes issued fatwas against evolution. There are, however, a few serious voices, which have emerged in recent times to present much more reasoned and knowledgeable views on evolution, and I shall review these at some length further down, but they are very rare. Before that, I would like to show that the Muslims' attitude towards evolution was not always this negative and rigid, both during the classical era and in more modern times (in the decades that followed the publication of Darwin's theory).

classical-era Islamic views

It may come as a surprise to readers, both Muslims and westerners, that scholars and thinkers (mostly philosophers) of the golden Islamic era discussed evolutionary ideas at some length and often ended up adopting them without objections. One should be careful to note, however, that the concept of evolution was not really used in those times, the terms used then were 'change', 'stage' (*tawr*, in Arabic, and *atwar*, plural) and 'transformation'. In fact, in his study of Arab reactions to Darwinian ideas from 1860 to 1930, Adel A. Ziadat, though he is not mainly concerned with the classical era, makes the following remarks[66]: 'Arab thinkers of the Middle Ages, who took the idea of evolution from the ancient Greeks, gave great consideration to the ideas of organic evolution and transformationism in the plant and animal kingdom [. . .] Arab writings in some ways approached those of Charles Darwin'. He adds: 'A number of influential Arab thinkers of modern times [. . .] denied the fact that the theory of evolution was a discovery of Darwin and Wallace. Others indicated that what Darwin explained was a part of Arab elaborations on the whole notion of transmutation'[67].

Similarly, Ayub Khan Ommaya[68] mentions al-Jahiz (776–868)[viii], who in his most famous work *The Book of Animals* (*Kitab al-Hayawan*), gave descriptions of avians who showed clear signs of evolution and adaptation during their migrations. According to Ommaya, al-Jahiz contributed to the concepts of evolution. He adds that Ibn Miskawayh (in his book *al-Fauz al-Asghar – The Smallest Accomplishment*) 'provided the modern concept of human origins as part of an evolutionary process'[69].

But what did exactly the classical Muslim thinkers say about evolution? Did they really 'approach' Darwin's ideas or even precede them?

At this point one should recall the idea of the Great Chain of Being that had prevailed in ancient times. According to this idea, all existence can be filed into a vertical hierarchical structure that starts with rocks at the bottom, as these only have 'existence', then plants, which have 'existence and life', then animals, which also have motion and appetite, then man, who has 'spirit' in addition to the previous attributes, then the angels and finally God. There are subdivisions within this classification; for example, rocks and minerals are ranked by their 'quality', with gold at the highest level, silver lying underneath it, lead lying below, etc. Now the crucial issue is that this great chain can be seen as a ladder, that is, one can have transformations from one species to another. One thing is certain that alchemists believed that minerals could be transmuted, and they spent centuries pursuing that goal; it is less clear whether scientists and philosophers believed that such transformations (which would amount to evolution) applied to living species as well. So one must keep this in mind when reading the views of the classical Muslim thinkers: Did they simply express ideas of great chain of being or did they actually imagine some possible evolutionary schemes?

Mahfuz A. Azzam has done a doctoral thesis on the classical philosophers' views on evolution; he has published his work as *Mabda' at-Tatawwur al-Hayawiyy lada Falasifat al-Islam* (*The Principle of Biological Evolution in the Works of the Classical Muslim Philosophers*)[70]. The most important views he mentions are those of al-Farabi, Ikhwan as-Safa, Ibn Miskawayh and Ibn Khaldun.

al-Farabi's view can be summarised in the following quote:

> Beings start with the least perfect one, then they rise gradually, until each has reached its most perfect form.

[viii] al-Jahiz was one of the earliest intellectuals of the Islamic tradition, a multi-disciplinary polymath who was both an early rational philosopher and one of the great literary figures of the Arab-Muslim tradition.

The classification of these beings starts with the lowliest, then the better and better ones, until one reaches the best; the lowest one is the common matter, then minerals, then plants, then non-speaking animals, and there is no better than the speaking animal (man)[71].

The views of Ikhwan as-Safa are much more detailed and complex. In their encyclopedic 'Epistles', one reads the following:

(1) One may believe either that creatures were brought out by the Maker all at once, or imagine that they were produced gradually, one by one till the last one, over times and epochs, or one may even think that some came at once and some came gradually . . . However, one cannot find any observational evidence for the all-at-once creation scenario, whereas one can find much evidence with just one observation to make the case that beings were created gradually[72]. (2) The last rung for plants is attached to (or a continuation of) the first rung for animals, and the last rung for animals is attached to the first rung of humans, and the last rung of humans is attached to the first rung of angels[73]. (3) If you look carefully, you will find that the fundamental reasons and principles behind the existence of creatures is the instinct of survival and the abhorrence of disappearance[74]. (4) A plant can transform into an animal, as can be read in the quote given at the top of this chapter from their '*Rasa'il*'. (5) Some have argued against the role of nature and thus attributed all its actions to the Maker, whether these actions were good or bad, while some have attributed the good actions to God and the bad ones to 'nature' (including sometimes to the stars) or to 'habits' or to 'devils' . . . [75]

The authors go on to argue that God has created a 'global spirit' that acts in nature, making it active and creative. Finally, the Ikhwan as-Safa take the evolution of species as a given observation and infer from it that the universe must be of finite age[76].

Ibn Khaldun, the fourteenth-century North African Muslim philosopher, historian, and polymath, was also close to the evolutionary paradigm:

Look at the world of creation, how it started from minerals, then plants, then animals, in a beautiful way of gradation and connection . . . where the meaning of connection in these creatures is that the end of the horizon for each is ready in a strange way to become the first in the line of what comes after it; the animal kingdom is vast and diverse in its species, such that it reached in gradual formation the human being, who distinguishes himself by mind and vision; to it the animal kingdom has risen from apes, which possess conscience and feeling but not vision and thought, and that is the start of the human world; this is what we observe[77].

One should note that the similarity between man and apes, in both their physiologies and lifestyles, was also noted and described at length by several classical-era authors, as Azzam describes[78] in his thesis.

Finally, Azzam quotes this amazing poem by Jalal ad-Din Rumi, the famous thirteenth-century Sufi master:

> Man first appeared at the level of inanimate matter,
> Then it moved to the level of plants,
> And lived years and years a plant among the plants,
> Not remembering a thing from its earlier inanimate life.
> And when it moved from plant to animal,
> It did not remember anything from its plant life,
> Except the longing it felt for plants,
> Especially when spring comes and beautiful flowers bloom,
> Like the longing of children to their mothers,
> They don't know the reason for the longing to their breasts.
> Then the Creator pulled Man – as you know – from its animal state,
> To his human state,
> And so Man moved from one natural state,
> To another natural state,
> Until he became wise, knowledgeable, and strong as he is now,
> But he does not remember anything from his earlier states,
> And he will change again from his current state[79].

modern-era Islamic views

Readers may be surprised to learn that a vigorous debate took place among Arab intellectuals around the theory of evolution beginning in 1876. (One should recall that Darwin's *Origins of Species* was published in 1859, and *Descent of Man* appeared in 1871.) As Ziadat explains in his excellent review, the debate, which was conducted through books and cultural magazine articles, focused more on philosophical, religious and social implications than on the scientific aspects of the theory. Still, it is important to emphasise the fact that the positions covered a wide spectrum, ranging from simplistic rejection to acceptance in toto. Indeed, some of the Arab secularists saw in Darwinism the embodiment of the modern, scientific spirit of the times and the way to pull the Arab world from its backward and irrational mindset.

The first response to Darwinism came from a non-Arab Muslim, Jamal Eddine al-Afghani, the leader of the reformist pan-Islamic movement, whom we have encountered in Chapter 1. In 1881, while in India, al-Afghani published his *Refutation of the Materialists* in Persian; it was soon translated into Arabic, and from that to other languages. His book was mainly a reaction to the modernist Islamic revival (read revamping) movement of Sir Syed Ahmad Khan, whose school of thought had derogatively come to be called *naycheri*, a corruption of 'naturalist', the belief that explanations of the world must rely solely on natural, i.e. material assumptions (hence rejecting any type of miracle, large or small). Several authors have remarked[80] that it appears from his book that al-Afghani had not actually read Darwin's works; his description of the theory of evolution was often mistaken and confused. For example, he wrote the following :

[Darwin] wrote a book stating that man descends from the monkey, and that in the course of successive centuries as a result of external impulses he changed until he reached the stage of the orangutan. From that form he rose to the earliest human degree, which was the race of cannibals and other Negroes. Then some men rose and reached a position on a higher plane than that of the Negroes, the plane of the Caucasian man[81].

(A very wrong account of Darwinism indeed!) In one passage, he sarcastically comments on Darwin's acceptance of Lamarck's 'inheritance of acquired characteristics': 'Is this wretch deaf to the fact that the Arabs and Jews for several thousand years have practiced circumcision, and despite this until now not one of them has been born circumcised?'[82]

Later in his career, however, al-Afghani seems to have moderated his views. Indeed, as Ziadat reports, when the Muslim leader was asked to comment on a poem by Abu al-'Ala al-Ma'arri (eleventh century) in which animals are said to have been generated from inorganic matter, al-Afghani replied that al-Ma'arri meant what the other Arab philosophers meant when they proposed that mud could transform into plant and plant into animal[83]; he added: 'If the doctrine of evolution is based on these premises, then the Arab scientists preceded Darwin'[84]. He did not then voice any criticism of either the Arab philosophers or Darwin; on the contrary he seemed to accept the general macroevolution stated in those views.

It is interesting to note that as soon as more careful readings and discussions of Darwinism were made, reactions became much more welcoming, even from the Muslim religious scholars.

Abu al-Majid M. R. al-Isfahani, was a Shiite scholar who wrote a book in 1914 with a title that seemed to imply a rejection of Darwinism: *Naqd Falsafat Darwin* (*Critique of Darwin's Philosophy*). The Iraqi scholar, however, did not actually contradict the theory of evolution itself, but was rather mainly concerned with the status of man. He realised and stressed that the religious texts only provided metaphorical creation stories, but he still insisted that the origin of man from an animal form cannot and must not be accepted. His fundamental principle was that materialists could only conjecture on questions of creation and origin, while religious people could rely on the absolute truths that could be found in their texts.

Hussein al-Jisr was a Sunni Lebanese scholar who in 1887 published a book[85], which later received the Ottoman Sultan's prize. In it he conducted a meticulous and clear discussion of Darwinism from a positive, confident Islamic standpoint. He strongly believed that Islam always supported truths in all matters and encouraged progressive thought, including naturalism as long as it did not deny the central principle of a creator; this then led him to accept all evolutionary processes and even what we would today call 'methodological materialism'. Indeed, he cited several Qur'an verses to show their concordance with evolution and even insisted that the Qur'an pointed to the creation of life from inanimate matter.

al-Jisr also subscribed to the Averroesian double principle of (1) harmony between philosophy/science and revelation, and (2) the necessity to conduct *ta'wil* (hermeneutics) on religious texts (the Qur'an) when their literal reading leads to a clash with well-established truths of nature. He further insisted that some aspects of nature could not (in his view) be explained by materialism, for example, the characteristics inherent in some physical laws or phenomena, such as why the gravitational or the electromagnetic force followed an inverse-square law (in today's parlance) or why energy changed from one form to another (e.g. heat to light). Finally, true to his times, he described the universe as either a clock or a steam engine, and like Paley, inferred the existence of a clock- or engine-Maker.

Mustafa H. al-Mansuri was a Muslim Egyptian scholar who was convinced by Darwin's theory, he saw it as a new scientific and philosophical paradigm that established the capability of natural laws to explain phenomena pertaining to life, species, evolution and man. In his writings (ca. 1914) he declared that no financial remuneration, no matter how large, could possibly reward

Darwin for the revolution in knowledge that he had produced. al-Mansuri also wanted to show harmony between science and religion, and insisted that Darwin's theory had been falsely identified with atheistic materialism.

In today's vocabulary, al-Mansuri and other Muslim intellectuals of that time can be described as theistic evolutionists.

Finally, Ismail Mazhar, an Egyptian Muslim writer, was the first to translate *Origins of Species*; he did that in stages: the first five chapters in 1918, four more in 1928, the complete work appearing in Arabic only in 1964. He presented his strongly supportive views of evolution in the long preface to his translation as well as in a book on the topic that he published in 1924. Mazhar was a positivist at heart, but he refused to impose a materialistic interpretation of evolution and considered religion as a useful 'second stage' (in Auguste Comte's taxonomy) in the evolution of societies, highlighting Islam's strife for knowledge and for the eradication of ignorance.

Other modern Muslim responses have been recently related by Abdul Majid[ix] in his article 'The Muslim Responses to Evolution'[86]. According to their positions, he classifies scholars into three categories: (1) the literalists who view evolution as totally contradictory and incompatible with the Islamic teachings; (2) the modernists, who insist that evolution must be accepted totally; (3) the 'moderates' who consider that some but not all aspects of the theory can be accommodated by Islam.

In the first group, Abdul Majid puts Shihab-ud-Din Nadvi[x], Wahiduddin Khan[xi], and Harun Yahya, and one may add the group of perennial

---

[ix] Abdul Majid was, at least then, assistant professor in the department of zoology of the Government Postgraduate College, Mansehra, Pakistan.

[x] Shihab-ud-Din Nadvi was an Indian Muslim scholar (he died in 2002). The landing on the moon challenged him to address issues related to Islam, science, man, creation, space, evolution etc. He wrote dozens of books on such topics in Urdu and became very popular in the Indian subcontinent. On the subject of evolution, he has written *The Creation of Adam and the Evolutionary Theory*.

[xi] Khan, Wahiduddin (1925–) is an Indian Muslim scholar who focused on modern thought, including the sciences; his main concern was to show the relevance and importance of Islam for the modern age. On the subject of evolution, he has written *God Arising: Evidence of God in Nature and in Science*, where he says: 'For a hundred years this theory held sway over human thought. But then further investigations revealed that it had loopholes. It did not fully fit in the framework of creation. In certain fundamental ways, it clashed with the order of the universe as a whole'. (Cited by Abdul Majid, op. cit.)

philosophers that we have encountered in Chapter 4, namely Seyyed Hossein Nasr, Martin Lings and Frithjof Shuon. In the second category, Abdul Majid mentions Ghulam Ahmad Pervez[xii] and his student Abdul Wadood, who wrote the book *The Phenomena of Nature in the Qur'an and Sunnah*, in which the evolutionary stages are 'derived' from the Holy Book. In the third group, Abdul Majid puts a number of thinkers, including Mohammad Iqbal, Hossein al-Jisr, al-Asfahani, Inayatullah Mashriqi, Ahman Afzal, Israr Ahmed and Absar Ahmad[87]. It is interesting to note that several of these authors tried to distinguish between the factual parts of the theory, parts which they accepted and accommodated into the Islamic world view, and aspects that were more hypothetical or model-dependent, which they felt free to discard. In particular, some distinguished between Darwinian evolution and non-Darwinian evolution, an important distinction that I have made in this chapter.

As we can see, there was a rather wide spectrum of positions vis-à-vis Darwin among Muslim intellectuals during that early period (late nineteenth and early twentieth century), though Ziadat considers those reactions to be by and large an acceptance of the theory as 'the wisdom and will of God'; in his view, 'Muslim writers . . . provided a religious sanction to Darwin's science'[88]. Ziadat concludes: 'If one uses the phrases "evolution occurred by God's control" and "universe was created for a purpose" and "materialism is a neutral thought", one finds total support among Arab religious thinkers, Muslim and Christian'[89]. He further comments: 'Were there any differences between Muslim and Christian Arab religious thinkers, concerning Darwin's theory of evolution? The answer is not difficult to find. While both were open to Darwinism, this study suggests that Muslims were more ready to accept Darwin's evolution than were the Christian Arabs'[90].

contemporary Islamic views

The late nineteenth- and early twentieth-century period in the Arab world is known as the *Nahda* (Renaissance). It did not last or blossom into a

---

[xii] Pervez, Ghulam Ahmad (1903–86) was a controversial Pakistani Islamic scholar. He specialized in the Qur'an and its sciences, but his rejection of some hadiths brought him the ire of the traditionalists. His views were often considered outside of the mainstream of Islamic thought, for instance his insistence on the reinterpretation of the concepts of 'angels' and 'jinns', as well as miracles.

permanent development movement for the Arab society, whether economical or cultural. Intellectually, however, that period was a very rich one, and was characterised by fascinating ideas, exchanges, and personalities on the scene. We should therefore not be too surprised to learn that Darwin's theory was well received by and large, even welcomed and integrated with the world view of Arabs, by the Muslim scholars.

By contrast the late twentieth and early twenty-first-century period in the Muslim world is characterised by the strong fundamentalist, literalist brand of religion that has taken hold. And that is why the overwhelming majority of Muslims today, including elites, whether educated in modern universities or religious ones, reject evolution altogether. It is indeed rare to come across pro-evolutionary views expressed by any Arab/Muslim intellectual[91].

One of the rare exceptions, however, is the Syrian intellectual Mohamad Shahrour, who has undertaken a very interesting Qur'anic approach to the question of human evolution[92]. He starts by declaring that the Arabic/ Qur'anic words *insan*, usually understood as 'man', and *bashar*, usually rendered as 'human being', must be distinguished; they refer to two very different stages of human evolution. Indeed, in reviewing the story of Adam in the Qur'an, he shows that each time the word *insan* ('man') is used, there is a clear connotation of 'comprehension' (mental capacity), 'abstract conception' (of metaphysical entities, in particular) and 'intelligence'. By contrast, the word *bashar* is used only in the context of its creation, well before it has evolved to *insan* and become mentally capable. One could simply say that Shahrour wants to identify the *bashar* stage with hominid (or even Homo) and *insan* with modern man. He finds support for this idea in the fact that the Qur'an refers to the 'breathing of God's Spirit' into the hominid/homo (the verses Q 38:71–72 cited in the previous sidebar); note that Shahrour finds a significant hint in the usage of the word *ja'il* (making) instead of *khaliq* (creating) in this verse. He then constructs a story of human evolution from the Qur'anic verses that read in total accordance with the modern theory:

- Hominids standing up (the bipedal posture): 'O man! What has beguiled you from your Lord, the Gracious one, Who created thee, Fashioned thee in due proportion, and gave thee a just bias; in whatever Form He wills, does He put thee together' (Q 82: 6–8).
- Development of language: 'He has created "insan"/man; He has taught him speech' (Q 55:3–4).
- Learning of burial: 'Then Allah sent a raven scratching up the ground, to show him how to hide his brother's naked corpse. He said: Woe unto me!

Am I not able to be as this raven and hide my brother's naked corpse? And he became full of regrets' (Q 5:31).

- Sacrifice/offering: 'And relate to them in truth the story of the two sons of Adam, when they both made an offering, but it was accepted from one of them and was not accepted from the other' (Q 5:27).
- Spiritual development: '(The other son) said: Allah only accepts the sacrifice of those who are righteous' (Q 5: 28).
- Clothing/covering oneself: 'O ye Children of Adam! We have shown you clothing to cover your private parts, and clothing as an adornment to you' (Q 7: 26).
- Discovery of fire: 'He Who has produced for you fire from the green tree, so that with it you kindle (your own fires)' (Q 36:80).
- Finally, Revelation, the making of man, God's vice-regent on earth, and the 'fall' of Adam and Eve from 'paradise'. I should note that what was meant by paradise in Adam's story has often led to some debate among scholars. Shahrour (and others) has (have) offered convincing argumentation to explain that here the paradise in question is earthly; for example, Adam and Eve ate from the forbidden tree because they sought immortality, which means they were mortals, which is an earthly, not heavenly, characteristic of existence.

Shahrour summarises the whole story by emphasising two things: (1) there were many hominid/homo creatures before Adam; (2) God then 'selected' Adam and breathed into him from His Spirit; this later action is the transformative act that produced the jump from the animal state to the human one[93]. Francisco Ayala has referred to this ape-to-human transformation as 'the mystery of how a particular ape lineage became a hominid lineage, from which emerged, after only a few million years, humans able to think and love, who have developed complex societies and who uphold ethical, aesthetic, and religious values'[94].

Muslim creationism

Creationism is the belief that humans and animals were created roughly in the forms they presently have, and have undergone little or no evolution. In some extreme forms of creationism, it is claimed that everything in the universe, from plants and animals to planets and the universe itself, were created all at once at some specific time (a few thousand years) in the past.

Creationism as a movement is strong in the USA and in the Muslim world; it is virtually non-existent elsewhere. As I've mentioned in the introductory section of this chapter, however, what is most troubling is that unlike the situation in the USA, where creationism is widespread in the general (religious) public and only a tiny fraction of the highly educated segment of society holds such views, in the Muslim world creationism is quite equally present among the educated elite as among the less educated public.

One can find strong anti-evolutionary views in the writings of important Muslim thinkers (e.g. Seyyed Hossein Nasr), as I will briefly show further down, but one major creationist Muslim movement is the Turkish group known as Bilim Arastirma Vakfi (BAV), i.e. Foundation for Research and Science, which was created in 1991 by Harun Yahya, the pen name of Adnan Oktar, whom we met briefly in Chapter 7. With over 150 books published under his name in over a dozen languages, well-produced but cheaply sold magazines, audio-visual material that is often distributed for free and a website that contains tons of free material, Yahya (and his group) target(s) a very wide audience. Indeed, his message, though explicitly referring to Allah, the Creator, and his wisdom, is often soft and made to appeal educated and modern readers; indeed, his fame and success is as high among Muslims in the West[xiii] as it is among Muslims in Islamic countries.

His books and articles are blunt. Reviewing just a few titles gives a good idea of what is in store: *The Evolution Deceit*; *The Scientific Collapse of Darwinism and Its Ideological Background*; *The Disasters Darwinism Brought to Humanity*; *New Fossil Discovery Sinks Evolutionary Theories*; *Thermodynamics Falsifies Evolution* etc. Keep in mind that Yahya has no science background at all; the biographical note at the end of his articles says, he 'studied arts at Istanbul's Mimar Sinan University and philosophy at Istanbul University'.

Let me give a few samples of Tahya's anti-evolutionary writings; the following are taken from one of his most popular titles, *Allah Is Known Through Reason*[95]:

> Darwin's fanciful ideas were seized upon and promoted by certain ideological and political circles and the theory became very popular. [. . .] When Darwin put forward his assumptions, the disciplines of genetics, microbiology, and

---

[xiii] In a study conducted among Muslim students in Amsterdam in 2004–05, Danielle Koning found wide rejection of substantial parts – if not all – of Darwin's theory of evolution; students often referred to Harun Yahya's website and publications; they were very familiar with many of the creationist arguments against evolution ('Anti-evolutionism among Muslim students', *ISIM Review* 18, Autumn 2006, pp. 48–9).

biochemistry did not yet exist. If they had, Darwin might easily have recognized that his theory was totally unscientific and thus would not have attempted to advance such meaningless claims: the information determining species already exists in the genes and it is impossible for natural selection to produce new species by altering genes[96].

When terrestrial strata and the fossil record are examined, it is seen that living organisms appeared simultaneously. [. . .] The vast mosaic of living organisms, made up of such great numbers of complex creatures, emerged so suddenly that this miraculous event is referred to as the 'Cambrian Explosion' in scientific literature[97]. As may be seen, the fossil record indicates that living things did not evolve from primitive to advanced forms, but instead emerged all of a sudden and in a perfect state. [. . .] Not a single transitional form verifying the alleged evolutionary 'progression' of vertebrates – from fish to amphibians, reptiles, birds, and mammals – has ever been found. Every living species appears instantaneously and in its current form, perfect and complete, in the fossil record. In other words, living beings did not come into existence through evolution. They were created[98].

Now, upon reading some of Harun Yahya's material, anyone familiar with debated on evolution quickly realises that most of the arguments are recycled from American creationist literature on the subject. Indeed, as some authors[99] have detailed, there is a strong collaboration between Yahya's BAV and the main American creationist organization, Institute for Creation Research (ICR); in addition to official contacts and requests for assistance[xiv], books produced by ICR were translated to Turkish and distributed to public school teachers[100]. In fact Yahya himself praises the American movements of creationism and ID: 'Some trends developing in the USA such as "Creationism" or "Intelligent Design" prove by scientific evidence that all living things were created by Allah'[101].

[xiv] Arnould (God vs. Darwin, p. 140) cites the American creationist magazine *Acts and Facts*, which relates contacts made by the Turkish education minister with ICR in the mid-eighties requesting assistance for his attempts to remove the theory of evolution from the official curriculum; also, John Morris and Duane Gish, the US leaders of the creationist movement, visited BAV in 1992, and several speakers at the 'End of Evolution and Fact of Creation' conference in Istanbul in 1998 were American. See also Edis, Taner, *All Illusion of Harmony: Science and Religion in Islam* (Amherst, NY: Prometheus Book, 2007).

Before I address some of Yahya's main 'refutations' of evolution, I would like to highlight one grave methodological ill that I have found prevalent in his writings: quoting out of context and using only partial statements made by established biologists and paleontologists to convey a meaning that is often very different from the one originally intended by the authors. Here are examples of that unacceptable approach:

- When in 2002 the nearly complete cranium and parts of the jaws of a 6–7 million years old hominid (later named as Toumai) was found in Chad[102], it constituted a stunning discovery, not because it 'rocked the very foundations of the theory of evolution'[103] (as Yahya writes) but because it challenged the prevailing paradigm that man came from a hominid lineage that existed in Ethiopia (not Chad) and started to evolve towards the human form some 5 million years ago. Yahya becomes more strongly misleading when he quotes Henry Gee, the senior editor of *Nature* who in a short article in *The Guardian*[104], correctly argued that the new skull 'shows, once and for all, that the old idea of a "missing link" is bunk . . . ' He added: 'It should be quite plain that the very idea of the missing link, always shaky, is now completely untenable'. But Yahya, misinterpreting these words for his readers, immediately adds: 'In brief, the drawings of the "evolutionary ladder that stretches from ape to man" that we do frequently encounter in newspapers and magazines have no scientific value at all'. Did Yahya (or his writers) read Gee's article? If yes, then he/they should have clearly understood that Gee sees the ladder as clearer now than ever.
- In a section discussing his favourite topic of the Cambrian Explosion, Yahya refers to a short article[105] by Richard Fortey in the 2001 issue of prestigious journal *Science*. He quotes Fortey in the following words: 'The beginning of the Cambrian period, some 545 million years ago, saw the sudden appearance in the fossil record of almost all the main types of animals (phyla) that still dominate the biota today'. Yahya then explains to his readers that 'the same article notes that for such complex and distinct living groups to be explained according to the theory of evolution, very rich fossil beds showing a gradual development process should have been found, but this has not yet proved possible'[106]. The reader is led to conclude that Fortey considers the Cambrian Explosion as a serious challenge to the theory of evolution. But when one goes back to the article in question, one finds that Fortey is only weighing two possible explanations for the Cambrian Explosion: the very rapid (over 10 million

years) evolutionary activity (like Gould proposed) or the 'phylogenetic fuse – an extended period of evolutionary genesis that left little or no fossil record'. How such statements can be cited as supporting 'The Collapse of Darwinism' (the title of Yahya's article) says much about the methods used by the creationists.

Now, as far as the arguments that Harun Yahya uses to 'disprove' evolution, I may cite the following as the main ones because they are repeatedly made in his various books and articles:

1. 'The fossil record refutes evolution'; 'no transitional forms have yet been uncovered. All the fossils unearthed in excavations showed that contrary to the beliefs of evolutionists, life appeared on earth all of a sudden and fully – formed'[107]. My earlier sidebar on 'missing links', or transitional forms, showed (visually) several examples which counter this claim.

2. 'Thermodynamics falsifies evolution'. He writes: 'Evolutionary theory ignores this fundamental law of physics. The mechanism offered by evolution totally contradicts the second law [of thermodynamics]'[108]. This argument, actually an old favorite of creationists in the USA, insists that the second law, sometimes simplistically referred to as the law of disorder, stipulates that disorder (entropy) must always increase, and thus nothing complex can naturally arise out of simple and disordered matter. What every physics student actually learns is that entropy (disorder) increases globally, but order can certainly appear in some parts of the system. Indeed, we always see that complexity arises naturally around us; for instance, crystals (like snow flakes) form out of simple disordered water molecules, heavier nuclei form by fusion from small, light ones and so on. When conditions exist, for example, gravitational attraction, thermal energy or quantum tunneling, complexity develops; there is nothing new or shocking in this. The more complexity is needed, the more special are the conditions required and the longer time one will have to wait, but the process is certainly not declared to be impossible by physics.

3. 'Probability is too low for proteins to form naturally'; he writes: 'An average-sized protein molecule is composed of 288 amino acids of which there are twelve different types. These can be arranged in 10^{300} different ways. Of all these possible sequences, only one forms the desired protein

molecule. The rest of them are amino-acid chains that are either totally useless or else potentially harmful to living things. In other words, the probability of the formation of only one protein molecule is "1 out of 10^{300}". The probability of this "1" to occur is practically impossible. (In mathematics, probabilities smaller than 1 over 10^{50} are accepted as "zero probability"). Furthermore, a protein molecule of 288 amino acids is rather a modest one compared with some giant protein molecules consisting of thousands of amino acids'[109]. Yahya makes a conceptual and mathematical mistake in performing this calculation: He assumes that all the amino acids come together at once when forming proteins; in that case, of course, the probability would be ridiculously low; in reality, however, the process takes place step by step. This argument is as false as claiming that the uranium nucleus (which has 238 nucleons in its most common form) could never form because the probability that 92 protons and 146 neutrons fuse together at once is (similarly) negligibly small; in fact, we know that protons fuse (in three steps) to form helium-4, then heavier and heavier nuclei form by fusion.

The creationist campaign of Harun Yahya and his group seems to have accelerated lately. In 2007, the group produced a lavishly illustrated *Atlas of Creation*, a creationist book in two volumes of about 800 pages each, one that would normally cost over \$100; thousands of copies of this book were mailed free of charge to high schools in France and Switzerland[110]. As I was writing this, I checked his website and found an insistent advertisement offering 450 books, 120 VCDs and 100 DVDs for free – including the DHL shipping!

On 17 February 2007, *Gulf News* (the leading English newspaper in the Gulf region) published a long story with pictures and a sidebar, all under the title 'Debating the Origin of Life'[111]; the subtitle of the news piece stunned me: 'Darwin's Theory Will Be Removed from Public School Curriculums Next Year'! The story explicitly referred to Harun Yahya and his group's efforts in the UAE, organising exhibitions and distributing free documentation to students and teachers. The article reported the latest success of the group, quoting the Senior supervisor of biology at the curriculum development centre in the ministry of education who announced the removal of evolution from the curriculum of Grade 12 (the only grade at which Darwin's theory had been taught briefly until now). It was funny to see the biology supervisor justifying the decision by the need to cover new topics like 'recent advances in DNA technologies'.

I fired a letter to the editors, making the following points:

- The theory of evolution is a scientific theory; it purports to explain the diversity of organisms and their features on the basis of a principle (evolution) and specific mechanisms (natural selection etc.). It is not intrinsically atheistic; many Muslims find no contradiction between their faith and acceptance of evolution, which is strongly supported by scientific evidence. To take one conservative viewpoint (Muslim, Christian or others) and present it as 'the religious view' is simply wrong.
- Those who claim that the Qur'an contradicts the evolutionary scenario are simply making a simplistic and literalistic reading of the Holy Book. This is similar to claiming that the Qur'an rules out the Copernican theory because it says 'the Sun rises' and 'the Sun sets'. Today we know that one must make a more intelligent reading of such verses. Likewise, we must learn to understand the concept of creation in a less literalistic way.
- People may be surprised to learn that many famous Muslim scholars of the golden era of the Islamic Civilization, scholars like al-Farabi, al-Jahiz, Ikhwan al-Safa and Ibn Khaldun, all noted the 'gradation', or even 'evolution', of organisms in nature. How much have we regressed!
- To claim that evolution is 'only a theory' is to show woeful ignorance about both evolution and the nature of scientific theories.
- To justify removing the subject from the curriculum in order to make room for discussions of latest scientific advances like DNA technology is similar to justifying the removal of the Copernican theory (of a sun-centred solar system) from the syllabus in order to make room for a discussion of advances in satellite technology. It is a self-contradictory and dishonest argument at the best.

The editors published parts of my letter.

Yahya is the boldest Muslim campaigner against evolution and for creationism. There are, however, other less visible, but perhaps stronger anti-evolutionists in the Islam–science domain today. One is the famous Seyyed H. Nasr, who has always opposed the theory of evolution on philosophical and theological grounds, even though sometimes he tries to look at it from a scientific perspective. In a 2006 article devoted to the question[112], he starts with the following words: 'I have studied not only physics but also geology and paleontology at Harvard, and so it is with this background that I reject the ordinary understanding of the Darwinian theory of evolution even on scientific grounds'. (It should be pointed out that Nasr studied physics as an undergraduate at Harvard some 50 years ago.) He further writes: 'A triangle is a triangle, and nothing evolves into a triangle; until a triangle becomes a triangle, it is not a triangle. So if we have three loose lines that gradually meet, even if there is one micron of separation, that is not a triangle. Only a triangle is a triangle. And life forms also have a finality of their own'. He also uses another argument, which he calls 'logical criticism': 'How could something greater come out of something lesser?' And further: 'A lot of the pictures and paintings used to demonstrate evolution are a hoax. Yes, 95% of our neurocells are similar to those of monkeys, but this does not prove anything'.

But does Nasr have any critique of substance to offer, other than these uninformed statements? The most he can come up with is referring to some European critics of evolution, non-Darwinian biologists (not any of those we have encountered), who have expressed different viewpoints, though Nasr doesn't tell us whether they, as I suspect, are non-Darwinian *evolutionists* (as opposed to *creationists* or Intelligent Designers like Behe, whom he mentions approvingly).

Regarding his own position with regard to evolution, first Nasr insists that only God, the Life-Giver (*al-Muḥyi*, which is one of His 99 attributes) can turn inanimate matter into something alive, and secondly, he accepts microevolution but strongly rejects macroevolution. Finally, he considers theistic evolution as 'worse than the Darwinian idea' because: 'it is no longer scientific evolution'; 'it . . . will not satisfy the agnostic or atheistic biologists'; and 'it ties the Hands of God through a process that we believe we know, but we really do not know'.

Another Muslim thinker who has been a participant in Islam–science debates is Osman Bakar[xv], a disciple of Nasr who also strongly opposes evolution[113]. He sees this theory as fundamentally materialistic in its bases and outlook; for him, Darwinism seeks to deny nature's evident dependence on Allah, its creator.

Finally, a contemporary Egyptian religious scholar, Abdes-Sabour Chahine, who is quite famous in the Islamic cultural landscape, has also

---

[xv] Osman Bakar (1946–) is a philosopher of science; he has published extensively on topics relating to the history and philosophy of Islamic science; since 2000 he has been at Georgetown University, Washington DC, holding the Malaysia Chair of Islam in Southeast Asia.

presented his views on the topic of human evolution. In a book titled *Abi Adam* (*My Father, Adam*), first published in Cairo in 1998, Chahine tells us that he had been developing his thesis for at least three decades[114]. His thesis, as it turns out, is twofold: (1) there was some long development of humans over millions of years, though they did not evolve from any kind of animal species; (2) there is no macroevolution, no species evolves from or produces another. He makes it clear and insists that he does not accept Darwin in any way. His intention, he says, is to rid the Islamic culture of two viruses: (1) all the 'Israelite' (old and new) stories that have been pushed into the Muslims' understanding of various historical issues; and (2) the literalist approach to the religious texts.

Chahine's book is full of errors, essentially on every scientific point; for example, he gives geological ages in the tens of billions of years (the pre-Cambrian period alone is given at 71.125 billion years). He shows a very wrong understanding of the theory of evolution and is happy to quote disparate pieces of writing, the main one from 1956 that claims to falsify Darwin's theory. He writes the following:

> There is no need to expound on the fact that all current scientific efforts aim at rejecting Darwin and his claims . . . and we can simply say that after all the critique that it has been subjected to, Darwin's theory has become a very weak claim; it has no bearing on the question of Man's origin, even though it may have contributed much in biology and anthropology[115].

He further adds: 'The idea of "creative evolution" – and we say "idea" and not "theory" – has thus fallen. . . . In its place, the truth of "separate creation", which religion (Islam) decrees, has triumphed: Man has always been human, and apes have always been apes'[116]. Still the book was met with a firestorm of criticism and calls for a ban; he was taken to four courts, including the higher court of appeals; and one reviewer wrote, 'Chahine commits such errors and sins toward his religion that if someone else had done that, Chahine himself would have called for his stoning with trash . . .'[117]

The rest of the book is then an attempt to show that man goes back to millions of years instead of the usual thousands or tens of thousands that traditional creationists claim. Chahine makes the same distinction, as Shahrour before him, between *insan* and *bashar*; he accepts the existence of many adams (humans) before the famous Adam of the religious story; that the paradise from which Adam was expelled was an earthly one (perhaps in east Africa); and many parts of the creation story must be read with a metaphorical sense and a hermeneutical approach.

conclusions

The first major idea that must be reiterated and re-emphasised here is that the process of evolution is an established fact of nature. I have devoted a long section of this chapter in an attempt to provide succinct but multiple lines of evidence of observational support for evolution. Even I did not cover all areas of substantiation; for example, I left out evidence from geographical distribution, from speciation, and other lines. No one can ignore or reject the facts of evolution (on any grounds) and expect to be taken seriously; this applies equally to the special area of human evolution and to the general field of biology. One cannot repeat that Islam (or any other religion) does not contradict science while rejecting a whole part of natural science, dismissing it as 'a hoax'.

The second important idea I have tried to highlight is that the theory of evolution, like any other theory of science, can take a variety of versions; the one that will be accepted in the end will fit the data (observations) best. So far, Darwinian evolution, which is essentially based on random mutations and natural selection (plus some secondary effects like sexual selection and genetic drift), remains the dominant theory among biologists. But I have also pointed to the growing amount of work being produced here and there that shows that a newer theory of evolution may have to be produced, by adding to the standard theory elements like self-organization, a bit of Neo-Lamarckism, and some downplaying of natural selection.

Here too, as in all of science, one must clearly identify the metaphysical bases of the theory and therefore not confuse the scientific aspects (methodology, results etc.) from the interpretations. And that is why I devoted a whole section to Theistic Evolution, which accepts all of the scientific parts of the theory but proposes a nonmaterialistic interpretation. Clearly, evolution does not automatically imply atheism or even materialism.

The second part of this chapter was devoted to the Islamic position with regard to evolution. With the review I presented various thinkers' views, going back to the medieval era and reaching up to the present, and tried to emphasise the fact that attitudes towards Darwin's theory in the Muslim world were much more welcoming and accommodating – at least in a theistic evolution conception – early on; antievolutionary stands grew and became largely dominant in the second half of the twentieth century with the appearance and expansion of fundamentalism. Today, intelligent proevolution voices (even in a theistic version) are extremely rare. Antievolutionists lump together wrong understandings of the evolution theory, knee-jerk reactions against any human evolution (claiming a special status for man in the cosmos) and arguments against the materialistic standpoints of Darwinists. Creationism has grown and spread in the Muslim world, whether in the popular form presented and disseminated by Harun Yahya or in the pseudo-intellectual presentations made by the perennial philosophers or the *I'jaz* advocates. The arguments and methods that the Muslim creationists have adopted are largely those of the American creationists, and one can only hope that the recent defeats that the latter have suffered in the US courts and media will start occurring in the Muslim world (though there has been no recent sign of slowdown, much less retreat).

It is incumbent upon enlightened Muslim scientists and thinkers to step up and on the one hand oppose this movement, which as I have shown does not even adopt serious methodology and arguments, and on the other hand show that a theologically acceptable version of evolution exists, one which can be adopted and further developed on the basis of the rich Islamic tradition. As Mehdi Golshani has written, 'The belief in an evolutionary mechanism for the emergence of species does not negate the idea of Divine creation'[118].

Finally, evolution is highly important in the science–religion/Islam debates, for it is there that one sees the clear difference between those who adopt a simplistic, literalistic reading of the scriptures (in all areas of life and thought) and those who accept the application of hermeneutics and the principle of multiple, multilayered reading of the Texts.

part iii

outlook

IO

Islam and science . . . tomorrow

In a Parisian hotel, 12 Muslim academics have gathered to discuss topics at
the interface between science and Islam. Several have come from France, but
others have come from as far as Canada and the United Arab Emirates. Six of
them are physicists, with a dominance of astrophysics, one is a mathematician, one is a sociologist, two are philosophers, and two are experts in Sufism.
The strong presence of physicists is not surprising; as Robert Griffiths has
said, 'If we need an atheist for a debate, I go to the Philosophy department.
The Physics department isn't much use . . . '¹ Plus, contemporary science–
religion issues (in the West) are often strongly flavored with physics and
cosmology topics.

  These Muslim intellectuals have identified each other from their previous
efforts at addressing issues that arise from the academic practice of science
and philosophy. They have seen a growing number of western intellectuals
tackle epistemological and metaphysical (fundamental) questions pertaining
to science as well as modern issues which press Muslims to reflect on topics
like evolution, design, God's action in the world, the nature of Reality
etc. These thinkers have thus chosen one topic each; they research them
in between meetings and interact by email (exchanging material, questions
and ideas); they then gather every nine months or so to make semi-formal
presentations to the group, address questions that are posed to them and

hear suggestions for further exploration. Among the topics chosen were the unifications principles of Islam (*tawheed*) and modern physics (GUT, TOE)[i]; Islam, Quantum Mechanics (QM) and the nature of Reality; Islam, design and the anthropic principle; Islam and the limits of knowledge (Gödel) etc. The outcomes of this project are multifold: an academic proceedings volume; numerous media appearances (interviews and shows); a series of articles in newspapers and magazines; and several open-public forums, where general presentations are made and questions entertained for the benefit of the general public.

It is interesting that this project, meant to benefit the Muslim world by developing a new and modern Islam–science group of specialists, is organized and hosted by a Parisian institution[2] and funded by an American foundation[3]. The project does, however, in my view represent one important road to the future of Islam and science, not the least because it interfaces with the West and tries to establish solid methodologies, unlike much of what has come to pass for an Islamic discourse on modern science.

It would indeed be an understatement to say that the Islamic discourse on – and engagement with – modern science is in need of revamping. One problem that Muslim academics will need to publicly address is *I'jaz*, as I have explained in this book. Another big question is the metaphysical foundation or envelope of modern science, the materialism or naturalism, in both the methodological and the ideological versions. In the 1980s, Muslims did try to formulate an Islamic position (or actually positions), but the main ones (Nasr, al-Faruqi and Sardar) found themselves either in dead ends (the first two) or running out of steam and being overrun by the development of science itself (the third one). It is time to revisit this question, in the framework of education. Finally, a host of important topics seem to have been barely touched by Muslim thinkers, and they are to be tackled, lest we leave young Muslim minds lost and confused; some of these topics include miracles; divine action; the nature of time, creation Reality etc.

In this short chapter I wish to touch upon these topics – and invite other thinkers (and perhaps myself too) to explore them more fully in the future. I also wish to identify other issues that are of an educational and

---

[i] GUT = Grand Unified Theories; TOE = Theory Of Everything.

social nature, which I consider as vital for the future of the Islam–science discourse.

miracles and science

One of the questions I asked the professors and students I recently surveyed on science and religion issues (see Appendix C) was 'Do you believe in miracles?' I will present the results to this question shortly; what I wish to stress here is that a good number of the non-Muslim professors wrote 'Define' next to the word miracles or underlined it and put a question mark over it. None of the Muslim respondents, professors or students, did any such thing.

Indeed, the definition of miracles is crucial to the discussion, and we shall see that much of the debate – and the positions adopted by various theologians, thinkers and scientists – rests on how the concept is defined. Let us here, for a starter, define miracles as 'phenomena that seem to contradict nature's laws or course' (something that would imply divine action or intervention), not phenomena that cannot be explained by science today. For example, if you let go of an object in your hand and, instead of dropping to the ground, it hangs in the air or even moves upward, the well-established course of nature (here, downward motion under gravity) will have been 'violated'. The idea of 'violation of the laws of nature' was how David Hume defined miracles, and this then led him to declare them impossible; needless to say, this reasoning was attacked by numerous thinkers[4] in more recent times, essentially judging it as circular thinking.

But it is not nearly as simple as phenomena that go against the course of nature. What about spontaneous remission, the sudden shrinking and disappearance of a well-developed and sometimes advanced cancer tumor; does that contradict the course of nature; is that a miracle? Probably not, for one may still explain it away as some physical (natural) process we have yet to understand. Then what about Jesus curing a man of his blindness by simply rubbing some mud over his eyes and asking him to wash it in some water source; does that contradict the laws of nature; is that a miracle (as it has always traditionally been declared)? Then how does one declare an event to be a miracle? This too is a gray area; the Catholic Church has various committees, one on top of the other, to filter claims of miracles; for example, out of some 7,000 submissions to the committee at Lourdes (the famous French city that is visited by some 6 million people each year for its presumed production of miracles), only 67 have been declared as miracles in the past 150 years (20 in the first 100 years, and 46 in the last 50 years)[5].

Today the question of miracles constitutes perhaps the clearest bone of contention between science and religion, and it will still be for some time, until scientists have developed a better understanding of many such occurrences (especially in the area of medicine) and theologians have developed a reasonable standpoint on this concept. In September 2002, *Zygon* (the premiere academic journal of religion and science issues) devoted a whole issue to miracles; in December 2004, a poll[6] among 1,100 physicians in the USA found that 74 percent of them believe in miracles[ii]; in September 2006, the French popular science magazine *Science et Vie* (known for its rationalist approach) published a special issue fully devoted to the question. Surprisingly enough, *Science et Vie* both recognised the existence of countless unexplained (though not necessarily unexplainable) cases and acknowledged the important, yet often downplayed role of the mind. The magazine, however, gingerly suggested[7] several leads or investigation tracks – if not yet explanations – for the 'miraculous' healings that doctors witness much more often than often known or admitted: (a) powers of the immune system; (b) induction of death (apoptosis) of tumor cells; and (c) the role of the mind.

The role of the mind is stressed by Prof. Edouard Zarifian[iii], who insisted that medicine has always relied first and foremost on the relationship that the patient and the doctor (or healer, or shaman) establish together, and that only recently and in the more materialistic societies has medicine turned into a mechanical exercise[8]. For instance, he mentions that in France 4.8 medicines are prescribed on average for a patient, compared to 0.8 in Northern Europe. But the most important manifestation of the mind's role in healing is the famous placebo effect, where a patient is given a bland (non-active) pill, yet one observes a surprising rate of self-healing: about 30 per cent over all pathologies of body functions, 20–50 percent for migraines, about 50 per cent in metastatic bone pain and 45–75 per cent in headaches of various sorts. Perrine Vennetier, the *Science et Vie* writer, then suggests that the healing miracles of Jesus could be explained by this placebo

---

[ii] In fact, 55 percent of the doctors said 'they have seen treatment results in their patients that they would consider miraculous' available at http://www.worldnetdaily.com/news/article.asp?ARTICLE_ID=42061.

[iii] French emeritus professor of medical psychology and psychiatry; he died in February 2007.

effect, especially because in those cases faith must have been extraordinarily high and thus the effect of the mind proportionally strong. She says: 'We must hold knowledge but also faith. As in the magical healings, "belief" in that "man-medicine" plays a role'[9].

So science seems to be both admitting the existence of some nonmechanical, mind aspects to the healing process and suggesting possible routes for fuller understanding and explanation. But this hardly exhausts the spectrum of miracles that traditional religions (Christianity, Islam etc.) report and ask the faithful to accept. These miracles range from Moses parting the Red Sea[iv] to Jesus healing the leper and the blind, and coming back to life (the Resurrection) three days after having been crucified, to Muhammad journeying from Mecca to Jerusalem, ascending to heaven and coming back to Mecca in one night. Some of the 'miracles' can be interpreted as allegorical, at least by nonliteral religious commentators; for instance, the multiplication of food by Jesus, allowing a whole crowd to eat from a basket of bread and fish (a story which has a very similar counterpart in Muhammad's case) is often interpreted allegorically. Likewise perhaps, Muhammad's Night Journey and Ascension; indeed, my proposed multi-level reading approach to the Qur'an can allow people to adopt different views and understandings of the same verses. Other miracles, however, seem to indeed violate the laws of nature, such as Jesus walking on water, his turning of water into wine and his resurrection.

Christian theologians (e.g. Terrence Nichols and Keith Ward) have proposed interesting ideas in addressing the question of miracles. First, Nichols insists that 'the evidence for miracles, ancient and modern, is respectable and deserves attention'; he views them as events that are 'consistent with, but transcend, natural processes'[10]. This theologian suggests two approaches for dealing with miracles: (a) the phenomenon may be an extreme, singular case of natural processes, akin to black holes (with gravity) and superconductivity (with electricity); (b) the event can only be explained by divine intervention, and for this he invokes either the indeterminacies of QM or the chaos theory. Nichols speculates that 'in some extreme circumstances,

---

[iv] One should recall how the Bible explains the parting of the Red Sea: 'The Lord drove the sea back by a strong east wind and made the sea dry land' (Exodus 14:21); we shall see later that one of the main proposals of God's action in the world (whether in miracles or in 'small' interventions) is through non-linear (chaos) processes and changes in the microscopic conditions, which produce macroscopic ('butterfly') effects.

such as the presence of great faith, the laws of nature, while not changed, behave differently from the way they do in ordinary contexts'.

Keith Ward takes a similar stand. He suggests that 'laws of nature . . . are best seen not as exceptionless rules but as context-dependent realizations of natural powers'[11]. But he leaves open the possibility that miracles may not 'fall under formulable scientific laws'; he adds that 'there is every reason for a theist to think that there are higher principles than laws of nature'.

In Islam, the existence and nature of miracles is a question on which schools of thought differ. One of the most common positions declares that only the Qur'an and possibly a few events in Muhammad's life (e.g. the Night Journey and Ascension and the 'splitting of the Moon' discussed in the book's introduction) constitute miracles, though the latter events could be explained spiritually, allegorically or even naturally (in the case of moon splitting). Another frequent position is the belief that only prophets, being inspired, supported and possibly empowered by God, could produce miracles, but not mere humans. A third position one encounters, especially among Sufi-inclined people, is that miracles are reserved to prophets, but 'saints' (awliya') are given 'gifts' by God (karamat), divine largesses that allow the saint to minister largesses onto others; it should be noted, however, that saint stories abound with the most astounding unnatural events and feats that can only be defined as miracles.

In the survey that I conducted, the question 'do you believe in miracles?' received the following responses from Muslim professors and students (the results were very similar among the two groups):

- Yes, I have no problem with that: 66 per cent.
- Only prophets have miracles, none such could occur today: 35 per cent.
- Yes, but I still need to find an explanation to them: 20 per cent.
- No, science contradicts miracles: 4 per cent.

(The results total more than 100 per cent, as some respondents chose more than one answer.)

By contrast, 44 per cent of non-Muslim professors chose the last answer; the rest were split between the first choice and the third one.

Absar Ahmed, a contemporary Pakistani philosopher, has reviewed the Muslim positions on miracles and presented his own views as well[12]. He starts by dismissing Sir Syed Ahmad Khan's denial of miracles; the knighted Indian Muslim reformer of late nineteenth century had in fact argued that the laws of nature were 'a practical promise of God that something will happen so, and if we say it can happen otherwise we are accusing Him of going against His promise, and this is inconceivable'[13]. Khan insisted that his rejection of miracles was not because they were irrational, but because God (in the Qur'an) tells us otherwise. Absar Ahmed rejects this position, for in his view God must not be limited; He can certainly rise above the laws of nature.

The Pakistani philosopher goes on to distinguish between 'natural miracles', which are described as portents or signs (*ayat*) in the Qur'an, and 'supernatural miracles', the historical ones (by the various prophets), and the revelation itself (the process and the Book), which are called *ayat bayyinat* by the Qur'an. In fact, the Qur'an, like the Bible, never uses the term 'miracle'; instead, it is always 'signs' by which God draws people's attention or makes some point. The first category, which he characterises as 'micro-miracles', can be regarded as God's action in the world, i.e. in the ordinary affairs of life; he explains those as being similar to normal actions, as God through the mind–body connection is present at all times. The 'macro-miracles', however, he considers to be totally beyond any naturalist interpretation; these are not, he insists, 'a matter of cognition, [. . .] a matter of sense-based limited rationality'.

Finally, Mehdi Golshani adopts the viewpoint of the respected Schii scholar Murtada Mutahhari (1919–1979) and considers the divine 'patterns' (laws) of nature as invariable; however, miracles may occur and should not be viewed as exceptions but rather as part of the more extensive pattern, the whole of which we may not have uncovered yet[14].

The question of miracles is an obvious area of dialogue between science and theology. It is hoped that other serious Muslim explorations and positions can be made and presented to the world, joining those of the Christian thinkers, who have been struggling with the topic in recent times.

science and prayers

One large area of interface between science and religion that I have not – for total lack of expertise – touched upon in this book is medicine. In particular, the role and effect of prayer upon one's health has long been believed to be important and positive. Yet this needs to be explored, both conceptually and experimentally. And indeed there have been some studies in the West that have produced equivocal results. I have not found, however, any Muslim studies of either kind, and would like to encourage scientists and students to delve into this field and produce some Islamic works.

In the aforementioned special issue on miracles, the ultra-rational magazine *Science et Vie* acknowledges the role of 'meditation, prayers, yoga, hypnosis and support groups' in the 'longer survival of patients'[15]. The evidence for this general claim is highly mixed, however, especially with the most recent and most extensive study finding no or negative effects of prayers.

The physician who has studied this topic most attentively and carefully is Herbert Benson[v]. He describes his interest in the power of prayers in a 1997 essay[16]; he starts with his observations of the placebo effect, which he renames 'remembered wellness', and the impact of relaxation. He quickly infers from the placebo effect that the mind has uncanny abilities to trigger the release of medicinal chemicals to heal some sicknesses. He also stresses the 'magnificent long-term preventive and restorative properties' of relaxation exercises 'for people who elicited it regularly'[17]. He notes the special effectiveness of the technique for hypertension, chronic pain, insomnia, infertility, premenstrual syndrome, headaches, cardiac arrhythmias etc. Then he describes the various relaxation exercises that patients used and concludes the following: 'Suddenly, I came to a startling conclusion: This is prayer!' Benson then relates his (limited) experiment with patients who, in 80 per cent of cases, 'chose to say a prayer to elicit the calm of the relaxation response'. He further notes that '25 percent of those who elicited the relaxation response experienced an increase in spirituality, whether or not they intended to [. . . and] experienced fewer medical symptoms than those who did not experience spirituality'.

Benson then researched the topic and found that out of 1,066 medical journal articles, only 12 (1.1 percent) assessed religious factors. Still he found some confirmations of his own observations, for instance, a study in 1995 had shown that 'people over the age of fifty-five who had open-heart surgery were three times more likely to survive if they reported receiving solace and comfort from religious beliefs'. And the general literature review seemed to indicate that, indeed, 'belief in God lowered death rates and increased health'.

[v] Herbert Benson, MD, has been chief of the division of behavior medicine at Deaconess Beth Israel Medical Center, founding president of the Mind/Body Medical Institute, and Mind/Body Medical Institute Associate Professor of Medicine, Harvard Medical School. Dr. Benson is the author of, among other books, Timeless Healing: The Power and Biology of Belief' (with Marg Stark).

On the basis of these observations and preliminary conclusions, Herbert Benson applied for and received a large grant ($2.4 million) for the study of the effects of prayer on illness from the John Templeton Foundation. His was not the first such study; indeed, it had been preceded by several in the recent past, but those had been deemed to be flawed; so the new, extensive and rigorous one was supposed to establish firm results and conclusions on the question. In all 1,802 patients were monitored after coronary bypass surgery in six hospitals; they were divided into three groups: one was prayed for and told about it, another was prayed for but not told, and the third group was not prayed for at all. The prayers were said by members of three congregations, with only first names of patients being given to the church participants, the latter being asked to pray in whatever way they liked but used the phrase 'for a successful surgery with a quick, healthy recovery and no complications'[18]. The results were quite unexpected: those who were prayed for and told about it had higher rates of complications, like abnormal heart rhythms, perhaps due to the anxiety generated by the heightened expectations; one should note that the increase in complications was only by 8 percent (from 51 to 59 percent). Benson and his team were quick to stress that these results should not be regarded as the last word on the subject. Indeed, many have commented that it was rather surprising that all those prayers didn't overcome the anxiety and the physiological complications, and the way the experiment was set up and conducted exhibited a variety of flaws and opened the door to a multitude of uncertain conclusions: Do prayers have any value (with God) when they are made by strangers who have no knowledge of and no relation to the patient? Is God's answer to the prayers supposed to always be to the particular requests (such as that specified phrase) or is the answer perhaps what God Himself wishes to do for the patient? Is the absence of 'a quick, healthy recovery and no complications' evidence for or even indication of having been ignored? Is God's response supposed to have a specified time interval after which one must assume it is not coming?

Indeed, however skeptical and dismissive one may be, this experiment cannot be considered as the last word on the subject, and I hope Muslim doctors, psychologists, statisticians and religious scholars get into the field and conduct their own systematic experiments. I cannot, however, stress enough the need for such experiments to be highly rigorous, lest they open the door for even more confusion and irrational beliefs, such as those I decried regarding Qur'anic healing and such in the Introduction of this book.

divine action

Both the previous issues of miracles and praying for God's help in healing patients are directly related to the crucial topic of divine action. This is probably the most important area that theologians and specialists of science–religion questions must address: How does God act in the world – whether in extraordinary (miraculous) situations or in everyday circumstances, such as sickness? In fact, one must first ask: Does God act in the world, or does He let things work out naturally? (In principle only deists believe that God created the world and then withdrew to observe only; theists believe that God does act . . . somehow.) And if God acts indeed, does He do so only through the normal processes of nature or, at least sometimes, by some direct interventions, going beyond the natural processes?

In the survey I conducted at my university, I asked the students and professors the following: 'If you believe in God, which of the following statements do you subscribe to?' The choices and the results are as follows:

- God acts on our lives directly, including in natural phenomena and catastrophes: 77 per cent.
- God acts in the world, but only according to the laws of nature: 15 per cent.
- God exists but does not act in the world; He lets things work out (according to the laws of nature): 8 per cent.

Here too the results from the students were statistically close to those of the professors. I should also note that 26 per cent of non-Muslim professors chose 'I do not believe in God', while only about 2 per cent of Muslim professors and students chose that option. So it becomes clear that as the belief in God has been overwhelmingly accepted by Muslims, the choice regarding His (inter)action with (in) the world reduces mostly to: does He act directly or does He act (only) through the processes of nature.

This issue has focused many thinkers' attention lately. The latest opus has come from Keith Ward, the Oxford theologian, who published *Divine Action: Examining God's Role in an Open and Emergent Universe* in 2007. Two years earlier, John Polkinghorne, the physicist–theologian, published *Science and Providence: God's Interaction with the World*. Numerous scholarly articles[vi] have

---

[vi] A search for 'Divine Action' in the title of academic papers through the university's electronic library database turned up 34 references.

also been written lately; among them, to cite just a few published recently, are as follows: R. J. Berry's 'Divine Action: Expected and Unexpected'[19]; Ross L. Stein's 'The Action of God in the World – A synthesis of Process Thought in Science and Theology'[20]; Alvin Plantinga's 'Divine Action in the World'[21]; Edward L. Schoen's 'Divine Action and Modern Science'[22]; Michael Epperson's 'Divine Action and Modern Science'[23]; Larry Chapp's 'Divine Action and Modern Science'[24], and many others.

Theologians and scholars of science and religion have suggested a variety of ways to understand (if not explain) God's action in the world. Many, such as Polkinghorne, Russell and Gregersen, have focused squarely on physical processes: QM; chaos; both effects in tandem. Others, e.g. Donald MacKay, have rejected the mechanical framework and suggested an 'artistic' type of divine influence. Others, particularly those subscribing to 'process theology', like Arthur Peacocke, have suggested some kind of holistic information flow, through which God acts on the whole world, with a trickle-down effect, then leading to results at specific points. Let us review these propositions briefly.

From early on, observers noted that the intrinsic indeterminism of QM could be a doorway for God's action in nature, since one would normally assume that God (the omniscient and omnipotent) is able to set the outcome of the 'wave function collapse process' to one particular choice among those that the physics of the situation allows. (Note, however, that Peacocke does not even accept such absolute omniscience and omnipotence; he speaks of God's 'self-limited omniscience'[25] and 'self-limited omnipotence'[26] and declares, 'God has made the world in such a way that God does not know definitively, but only probabilistically'[27]; one then wonders what is the difference between such a 'God' and a good physicist.) If God then can determine the outcomes of any quantum mechanical process, which will always appear indeterministic to us (according to standard Quantum Theory), then He can 'steer' events in a way that will be transparent to us. This invisible action or intervention is not necessarily a source of unease, despite what non-theists will say (that there would thus never be any evidence for this, hence the need to reject it, at least on the basis of Occam's razor). There is, however, one problem with this proposal: Quantum mechanical indeterminism exists only at the microscopic level; and as macroscopic entities and events are a superposition of zillions of microscopic quantum objects, one would either have to postulate a divine action on each of them, thus leading to some macroscopically observable action or conclude that the superposition will kill the little microscopic effect, as it is known that quantum mechanical interference is destroyed by superposition.

The second, more fruitful proposal of physical divine action are the non-linear processes that lead to chaos: tiny effects in the initial conditions of a system, whether microscopic or macroscopic lead to hugely amplified results. Here again, as tiny interventions and changes are essentially impossible to notice, God could take such an approach, which is not intrinsically indeterministic, for His actions. This would be the perfect explanation for the parting of the Red Sea by the 'strong east wind' (the Bible's words), although this would also be the perfect ground for believing in God's intervention in natural catastrophes, which many lay people believe are God's punishing acts, a viewpoint which more liberal and humanistic people abhor.

Finally, one can also suggest a combination of the two effects, whereby a divine projection of the quantum wave function to a specific outcome at the microscopic level with a non-linear (chaos type) amplification can lead to the macroscopic effect being sought.

Peacocke objects to these approaches, which he calls 'God of the (to us) uncloseable gaps'[28]. He finds such actions fundamentally unacceptable, for they 'would imply that these processes without such interventions were inadequate to effect God's creative intentions if they continued to operate in the, usually probabilitistic, way God originally made them and continues to sustain them in existence'[29].

My response to Peacocke is through my favourite analogy with the computer programmer who designs a software, say a game, which can run throughout on preset values that determine the various developments, but which also allows the player to enter different values at specific points or use certain joker cards according to one's needs and wishes; such cards could at the start be given to the player (who would be born with such capacities) or be won during the game (through good deeds). Now, in our world/game, one would have to add an active element (God) who, instead of the automa-tised software code, would decide upon each request (e.g. prayer) whether the wish will be granted, or whether the game shall proceed naturally. The whole game, with the preset laws, initial conditions and limitations (some wishes are by default rejected), but also with the (inter)active parts, must be looked at together. There is then no contradiction between the laws as they continue to operate and the interventions which occur as part of the game.

Finally, MacKay (followed by Berry[30]) has suggested a double analogy to describe his view of divine action: firstly one should recall the wave-particle complementarity of quantum objects and apply it to macroscopic subjects, and secondly, one should envision a more suggestive, artistic influence that God can exert upon various subjects when He wishes to produce a certain effect. In this proposition, humans in particular (but other creatures and objects as well) would have two complementary ways of interacting: a physical one and a nonmaterial one, with both being sustained by God, which MacKay describes as more of a Cosmic Artist than a Cosmic Mechanic.

On the Muslim side, there have been very few, if any, proposals for the explanation of God's action in the world[31]. (It is a very sensitive issue, and one runs the risk of diverting too much from the orthodoxy and thus being labelled a heretic.)

One recent article that addressed the topic, albeit not from a modern or contemporary viewpoint, is Rahim Acar's 'Avicenna's Position Concerning the Basis of the Divine Creative Action'[32]. In it the author considers Avicenna's views and concludes that the great eleventh-century Muslim philosopher and polymath regarded God's creative action as similar to the action of natural things. He built on his principle that the universe is necessary and concluded that creation (initial or continuous) is more of a natural than a voluntary action.

A similar article, though a much wider one in scope, is Abdelhakim al-Khalifi's 'Divine Action between Necessity and Choice'[33]. This author addressed the same subject as Acar but enlarged his investigation from the classical philosophers (al-Farabi and Avicenna) to the important theological schools of Mu'tazilism and Ash'arism. In particular, he contrasts the Ash'arites' (orthodox) views that God's action is totally free and unconstrained with the Mu'tazilites' (rational theology) position that God's act of creation was free (contrary to Avicenna's view) but that God has constrained himself by the principles of pushing only towards goodness and rewarding/punishing for following/disobeying divine directives to that effect.

Before I close this section, let me mention and comment on two recent articles by Muslim intellectuals (from the aforementioned Parisian group), who wanted to explore Islamic 'resonances' with QM: Abdelhaq M. Hamza's 'A Reflection on Quantum Mechanics: An Islamic Perspective' and Eric Younes Geoffroy's 'Les voies d'accès à la Réalité dans le soufisme' ('Channels of Access to Reality in Sufism').

Hamza is a physicist. In his essay, he first presents a wide overview of modern physics, with its foundations, successes and failures, hoping 'to show that modern science, with Quantum theory as its backbone, points to a higher science, the "science of al-Tawhid . . . '" He then branches into QM somewhat briefly, enough to declare that Einstein's dream of a deterministic theory is still alive, for the 'paradox' of a deterministic Schrödinger equation and an indeterministic measurement outcome could only be resolved by a future theory. Hamza then insists that 'the Islamic stand point [. . .] is that of Certainty (*yaqeen*) and Determinism (*tahqeeq*). Nothing is left to probabilities in Islam. The whole creation is subject to "*Koun Fa Yakoun*" (Qur'an 2:117)'. He adds, 'As opposed to physics, which has endured the transition from determinism to indeterminism, Metaphysics has always been deterministic. It succeeds wherever and whenever physics fails. [. . .] The divine decree that underlies the absolute metaphysical determinism replaces the probabilistic indeterminism of modern physics'. I do not think physics and metaphysics can be so compared, and I do not share this deterministic dream.

Hamza's outlook may be contrasted with Polkinghorne's views. The physicist and Anglican priest insists on indeterminism as an ontological reality and necessity on theological grounds[34]. His indeterminism is not only in the quantum world but also 'at all levels', including the limits of non-linear processes at which initial conditions become crucial. Polkinghorne argues for 'an ontologically open world at all levels', which would make it much more widely subject to God's action. This bold proposition is supported by Niels H. Gregersen and Robert J. Russell; the latter writes: 'I place my wager with John and Niels: indeterminism is there in the world at all levels'[35].

Geoffroy is an expert in Sufism. In his essay he tries to find parallels between the concept of Reality in QM and in Sufism. He is impressed by the Veiled Reality interpretation of QM of Bernard d'Espagnat, who argues (quite convincingly) that there is a level of Reality at the quantum level that is beyond our measurement capability but allows wave-particles to keep connections such as those exhibited by the correlation experiments of Alain Aspect[36], and Nicholas Gisin and coworkers[37], those correlations that Einstein, Podolsky and Rosen had declared to be a paradox that would constitute the death knell of the QM theory. (It did not.) My main issue with Geoffroy's reading of 'parallels' between the two visions (QM and Sufism) is first and foremost that the 'deeper level of Reality' postulated by quantum physicists to explain those connections is totally inaccessible to us, whereas Sufis insist and place their whole doctrine on the possibility of acceding to that Reality (capitalised for its divine nature), which is all-encompassing, thus fundamentally different from d'Espagnat's views.

educational and social issues

There remain a few important nontechnical issues that we must address if we wish to see the field of science and Islam develop and thrive seriously tomorrow. Such issues include the need to teach the philosophy of science (in a reasonable and enriching way), the need to revisit and present the history of science, including the Islamic contribution and the rise of modern science, in much more rigorous way, the need to engage in a serious dialogue with the Muslim theologians and scholars and convince them that science today has much to say on topics that they have monopolised for too long, the need to educate the public on science issues that are rather closely related to the religious domain, and finally the need to link with non-Muslim thinkers who have developed expertise in the field of science and religion.

Let me elaborate a little on these issues.

First, the philosophy and history of science. It is very sad and quite astounding that practically no philosophy of science is taught in schools and universities of the Arab-Muslim world, except perhaps in departments (e.g. philosophy) which cannot ignore such a topic. But the fact remains that none of the names and schools of thought that I have presented and engaged in Chapters 3 and 4 are known to any Arab-Muslim students. This explains why the few Muslims who have contributed to the important debates on the philosophy of science (its metaphysical bases, its methods etc.) live outside of the Muslim world, or at least outside of the Arab world: Nasr (in the USA), Sardar (in the UK), Golshani (in Iran) etc.

This is an extremely serious situation, for this in my view is the essential reason for the existence of a very skewed understanding of the relation between science and religion generally, and Islam in particular, among most Muslims. This is also the reason why thousands of Muslim scientists are in fact technicians who are competent in some narrow area but have no knowledge or understanding of the bigger picture; this is also why they very often adopt traditional, or even irrational views on most issues, from Qur'anic healing to I'jaz. This further explains the fact that in the (limited) survey that I conducted, I found that Muslim professors were no less, and sometimes more, orthodox (rejecting evolution, for example) than students and the general public. I strongly believe that the main remedy to many of the ills that plague the Arab-Muslim world in the realm of science is the introduction of serious courses on the philosophy of science.

Likewise, for the history of science, which is barely taught (often in just a few introductory pages of science textbooks) in schools and universities of the region. Let me relate one anecdote to illustrate the effect of this state of affairs.

In December 2006 I participated in a scientific conference that tried to address in a rigorous way Islamic topics where astronomy was the prime field of relevance; for example, the start of Islamic months, such as Ramadan, for which one needs to ascertain the first sighting of the thin new lunar crescent; the construction of an Islamic calendar (based on the Islamic months); the determination of prayer times, particularly in high-latitude regions (for which calculations are difficult to perform accurately and where the prayer times themselves are difficult to define); the scholarly review of the Muslims' contributions to astronomy in history etc.

One of the speakers boldly claimed that the Copernican (heliocentric) theory had been proposed by Muslim astronomers; he specifically and rather casually mentioned al-Biruni and Ibn al-Shatir. When he finished his talk, I asked for the floor and pointed out the historical fallacies of his claims and explained the confusion he had made between the heliocentric proposition Copernicus made – which he had definitely *not* taken from the Muslim astronomers – and the technical aspects of Copernicus' (erroneous) geometrical orbits (still then based on circular epicycles and deferents), which he probably took from the then-sophisticated research of the Muslims. I strongly emphasised the fact that the first idea was certainly of Copernicus, and that it was so major, original and correct that we now refer to it as the Copernican Revolution, and stressed the fact that the second part, the one Copernicus probably took from the Muslims, turned out to be totally wrong and was discarded 60 years later (400 years ago now) by Kepler. I was then stunned to hear a participant, a full professor of physics no less, get up from his seat and shout at the top of his lungs that, I quote, 'Copernicus stole his theory from Ibn al-Shatir, and there is historical research which proves this . . . ' I then got up and said, 'I challenge Prof. X to dig up that important piece of "historical research" and to email it to all of us'. I did not wait for the two professors (the presenter and his defender) to send me/us any such piece of 'historical research', for I knew there was no such work; instead, I researched the topic a little, and within a week had written a short article rebutting those claims and sent it to the two professors and the conference organizers.

Similarly, many Muslim writers often tend to simplify complex historical and scientific issues in a manner that leads to gross distortions and further miscomprehensions. I will give just one glaring example: In his recent (2006) 500-page book titled *L'Islam fondateur de la Science* (*Islam, the founder of Science*), Nas E. Boutamina titles his sixth chapter as 'The Era of Plagiarism and Looting', where over more than 100 pages he presents in tabular form dozens and dozens of claims of plagiarism of Muslim works by Renaissance authors and scientists; according to him[38], Copernicus and Kepler plagiarised their magnum opuses from al-Biruni, al-Farghani, Az-Zarqali, al-Bitruji and Ibn Yunus; Tycho Brahe would have plagiarised from al-Buzjani and so on . . .

What are we to make of these cases? Are these extreme isolated occurrences that we must not make too much of?

It is my experience that Muslim academics often display faulty methodologies and make very serious factual errors about science and its history. Secondly, I wanted to highlight through the above story a serious problem we Muslims tend to have in dealing with our history, from its scientific component to its theological and political elements. On this issue, writers – except for experts who specialise in this (e.g. F. Jamil Ragep and George Saliba) – tend to be extremists who either describe the classical era as an extraordinary golden age of major discoveries, important innovations and methodical explorations or dismiss the whole historical narrative we find in our books as a large myth[39] constructed to make Muslims feel good about their heritage and thus believe that their culture and religion have indeed been – and thus can still be – capable of producing a great civilization, complete with sciences and technology.

What is the cause of this strong tendency? I believe that despite many efforts at unearthing and presenting to the public (Muslim and western) the wealth of scientific works that the golden-era Muslim scholars produced, there is still serious ignorance as to what was done exactly. Most of the discourse on the Islamic civilization has remained superficial and ill-informed. Hence, there is need to teach the (whole) history of science rigorously and vigorously in the general curriculum.

The next important issue is the need to engage the Islamic scholars in a serious dialogue and convince them that scientists have much to say on topics that have for too long remained the monopoly of the religious scholars and their discourse. While there is no doubt in people's minds that human knowledge evolves and grows, it is often understood that religions, especially Islam, are (is) absolute, immutable and transcendent principles, which are set in rigid frames of reference. But we know today that religions – and Islam is no exception – cannot afford to adopt a stationary attitude, lest they find themselves clashing with and overrun by modern knowledge, and religious principles appear more and more quaint and obsolete.

A few Muslim thinkers have recently made similar points. For example, Professor Faheem Ashraf has written the following:

> This enhanced knowledge of the universe has improved our understanding of the Quran and Ahadith. So science, though it has provisional theories has become a key requirement in understanding Religion, as it explained several things like existence of other worlds and fused status of heaven and earth, etc which our old exegetes were unable to explain. Thus science is not only a requirement of material comfort but also a key requisite for appreciation of religion . . . *science has provided us a universal language with which we can interact with the people of other religions and cultures*[40]. (Emphasis in the text.)

Speaking specifically of Islam, Ziauddin Sardar goes even further[41]; he believes that Islam must not just enrich itself with science but also reform itself on the basis of the dynamic that modern science has created in society and history. Sardar cites the Arab Human Development Report that was produced by the United Nations' Development Program in 2003, which called for the reactivation of *ijtihad* (the old, largely shut, religious tradition of intellectual effort and innovative religious research of Islam) and 'the protection of the right to differ'. Sardar reminds us that *ijtihad* was the driving force of the scientific spirit that pushed the Muslim civilization forward. He concludes: 'It is now widely thought that science itself can play an important role in reopening the gates of Ijtihad. So the revival of science in Muslim societies and the reform of Islam itself can proceed hand in hand. Similar thoughts are being echoed by the Organization of the Islamic Conference's standing commission on scientific and technological cooperation. The commission has argued that substantial increases in scientific expenditure and original work would not only improve Muslim societies, but would have a catalytic effect on Islamic thought'[42].

The last recommendation that I would like to make is the increase in collaboration with western thinkers who have developed considerable expertise on the various issues surrounding modern science and its developments, its philosophical and theological implications and the response of religious (mainly Christian) thought in this realm. Many names come to mind in this regard, among them Denis Alexander (in the UK), Philip Clayton (in the USA), Jean Staune (in France) and Keith Ward (in the UK), to name just a few. Clayton, for example, has proposed[43] the organization of an international program he would call 'Science and the Spiritual Quest in the Abrahamic Traditions', along the lines of the highly successful multi-tradition program[vii] 'Science and the Spiritual Quest' that ran between 1995 and 2003. Staune has, through the Université Interdisciplinaire de Paris, of which he is the Secretary General, managed the international science and religion in Islam project that was led by Bruno Abdelhaq Guiderdoni. James F. Moore, a theologian, has called[44] for an interfaith dialogue on science and religion issues; in particular he identified the design argument as an important issue of common interest that has come back to foreground lately; he has also called for collaboration on practical issues (lifestyle, praxis, social impact etc.). And last but not the least, the John Templeton Foundation has funded a number of programs with the aim of helping Muslim intellectuals develop expertise and scholarship in the area of Islam and science.

The international framework and scholarly community exists and is ready to interact with us; it is up to Muslim thinkers to take steps to collaborate and benefit for the sake of our people's intellectual future.

[vii] Clayton (previous reference) describes the success of the programme through the following outcomes: 'Between 1995 and 2003 SSQ held 16 private three-day workshops on two continents, involving 123 new scientists in constructive dialogue at the intersections of science and spirituality. The program organized 17 public events in nine countries on four continents. Taken together, these events reached close to 12,000 audience members firsthand and many millions more through the media – some 250 million, according to the official estimates of one media research firm. Six books covering the research output of SSQ have been published on four different continents or are currently in production. The group's website (www.ssq.net) lists four full-length video products and contains a massive amount of supplementary material'.

epilogue: a conversation with
my students

Students at the American University of Sharjah (AUS) come from over
60 nationalities (according to the admissions office). There are no statistics
on their religious affiliations, as they are not asked any question about that
in the documents they fill at the time of enrolment. But from my general
observations, I would say that roughly 80 per cent of them are Muslim.
However, they show a wide spectrum of religious attitudes concerning
dogmas and practice; for example, about half the girls cover their heads and
dress conservatively, while the other half adopt western styles. By regional
standards, students at AUS tend to be more liberal than typical Muslim
youngsters; they also tend to have higher academic potentials and achieve
higher levels of learning.

The survey on Religion and Science I conducted at AUS in the fall 2007
semester generated quite a bit of interest. A number of faculty members who
filled the questionnaire later asked me about the general results. Many of
my students inquired about my project – and my own views. A few of them
wanted to discuss the topic more widely and in more depth, so I arranged
for a discussion session in my office.

Marzieh is an Iranian student of Engineering; she is bright, studious and
highly interested in cultural and religious issues, not to mention scientific
ones. She is rather conservative in her outlook, but intellectually (perhaps
due to her Shiite background) she tends to be quite open-minded.

Rami is a Palestinian student of Journalism; he has become fascinated by the astronomical topics he has been learning about and may end up specialising in science reporting, something that is rare in this part of the world.

Aya is an Egyptian student majoring in International Studies; she is an A-student who likes to completely master the material in every subject she studies.

Mohammad is a Pakistani student of Engineering; he is bright, witty and highly attracted to philosophical discourse as well as scientific and cosmic topics. He too looks and sounds quite conservative (he once participated and won a prize in an Islamic competition), but appearances are often deceiving, for he displays progressive attitudes and viewpoints and is equally comfortable quoting Einstein and 'The X-Files', al-Ghazzali and Pink Floyd.

The students were first and foremost eager to know the results of the survey. I briefed them quickly, pointing out some of the main, striking outcomes. They then asked me for my general views; I said I would tell them as much as they wanted to know, but I preferred to first ask them for their views.

I started by asking them: Do you think there is, in general, some problem(s) between Islam and Modern Science?

Rami was the first to jump in; he said: 'I don't think there is any problem between Islam and Modern Science. The problem is with us. Our ignorance is what hinders us from bridging the gap between the two and realizing that the Holy Qur'an speaks of many miracles of Modern Science'. I jumped right back: 'Hold that (*I'jaz*) idea, I will certainly want to come back to it in a minute!' He added: 'Sure, but there are some issues that can pose problems, like cloning, which is some humans' attempt (and failure) to become God, and also the tampering with one's natural form (by plastic surgery), which is the way God created you'. I said: 'Hmm, those are interesting "practical" issues, but I was thinking more of conceptual ones'.

Marzieh spoke: 'Somehow yes, I feel there are some conflicts. Some Islamic beliefs are not compatible with modern sciences, the whole philosophy of modern science is problematic, the rejection of God and of anything sacred . . . ' Aya concurred: 'There are some problems between Islam and Modern Science, because science in general is much younger than religion, and therefore, it still needs much time and exploration for it to reach a fuller understanding of the world. But in fact if one looks at things in the right way, one will see that Islam and Modern Science are two faces of the same coin; they complement each other'.

So I looked at Mohammad, knowing his predilection for philosophical issues. He said: 'I don't think there is any sort of problem between Islam and Science. Problems are usually created by people who confuse the nature of science with the nature of religion. For example, those who try to prove to the general public that the Qur'an is divine *because* it displays scientific facts that were discovered only recently . . . '

I then offered some thoughts of my own. I first stressed the need to define what one means by 'Islam': Are we talking about the Qur'an, in that case one must distinguish between the verses, the principles and the details that can be found in there, but also all the interpretations that have become practically attached to the Text and an 'integral' part of its understanding; or do we mean by Islam the fundamentals of the religion (Qur'an and Hadiths); or are we talking about the whole Tradition? On the other hand, one must also define the term science: Do we mean the general knowledge of the world we have acquired, or the method used to understand nature? Are we aware of the bases upon which the whole enterprise is built? Depending on which definitions of Islam and science one adopts, for example, if one buys the whole materialistic approach of modern science, then conflicts may arise or harmony can be found . . .

Then I said: 'Now, I am quite impatient to come back to that question of modern scientific information in the Qur'an, but before I do that, I would like to ask Marzieh how she personally tries to reconcile the two (since, unlike the others, she had said that there were some conflicts between Islam and science)'. Marzieh said: 'I tell myself that theories are the products of human thought and are therefore not necessarily the final, absolute answer'. Rami added: 'Education is the key. Enlightening our minds with what God has said in the Holy Book allows us to advance further . . . We must realize our limits and avoid the repercussions that result from our overly free actions'. Aya commented: 'I personally believe in what is stated in the Qur'an and in the teachings of the Prophet. I believe they contain the truth, and my past experiences have proven me right. Now, when I am faced with any new scientific theory, I try to look at it objectively first (understand it with no judgment), then analyze it and see if there are any similarities between this theory and the Islamic logic. If proof is given along with this theory, and there is no clash between it and Islam, I accept it. Otherwise, I try to see where the theory has gone wrong and try to find a solution. In due time, science (with its theories) will converge toward the Islamic view'. And Mohammad simply said: 'I don't attempt to reconcile the two, simply because I do not consider the Qur'an to be a science textbook'.

I commented that the reconciliation between the two depends strongly on the reading (literal vs. interpretative) that one adopts for the religious texts. The more literal the person is, the more problems she/he will find in harmonising science with Islam.

Then I went back to that *I'jaz* issue: 'So tell me, how much science do you think there is in the Qur'an?' Rami said: 'Are you familiar [with] Harun Yahya? He has done extensive work trying to extract scientific facts from the Qur'an!' Mohammad was getting agitated, but I said: 'Before I let others speak, I should point out that Harun Yahya has done some of that but much less than others like Zaghloul al-Naggar. Make sure you read my book when it comes out'. Aya also believed in the scientific *I'jaz* of the Qur'an and said: 'In many verses, a host of scientific facts were revealed. I can give many examples . . . '

Mohammad then said: 'I think that the science which is in the Qur'an is only of the illustrative kind. There are examples like the stages of the embryo, the water cycle, etc. I use the word "illustrative" because I believe that the verses regarding such scientific facts were not revealed so that only those with scientific knowledge could understand them; they could be understood by the people who were contemporary to the Prophet and those who lived in earlier times. Most people, including those who insist on the "miraculousness" of those verses are unaware that most if not all of those facts were actually known, verified, [and] proven by Greek scientists and mathematicians long before the Qur'an was revealed. Moreover, many who try to match superficial interpretations of the Qur'an with scientific facts often end up misinterpreting the verses'. He further added, 'If I see that a scientific fact is in line with an exoteric interpretation of a Qur'anic verse, it is well and good; in the opposite case, I tell myself that (a) this must be a temporary scientific finding, and the future may reveal otherwise; (b) there may be a useful esoteric interpretation; [and] (c) my understanding of the entire matter may be flawed, and I need to change my approach entirely'.

I challenged him: 'And if all that fails?' He replied: 'If all that fails, then I have to question my beliefs and assess whether I have to rely on the Qur'an to such an extent . . . ' I was impressed with Mohammad's honesty and intellectual integrity.

Then I turned to Marzieh, waiting for her views. She said: 'If by science we mean modern and technical science such as engineering, then I don't think we can find them exactly in the Qur'an; however, some of the bases of these sciences are there. I also believe that the purpose behind their inclusion (by God) in the Qur'an is not exactly for teaching us, but rather to show us (and everyone) God's power, since at the time of the revelation no human knew about them and therefore this Book came from a source with absolute knowledge. But I should add that having these sciences in the Qur'an does not increase or decrease its worth'.

I then decided to explore one of the – perhaps the most – explosive topics between Islam and science. I asked: 'Do you believe that Islam contradicts Darwin's theory?'

Rami said: 'Yes. The Qur'an clearly states that Adam was the first human to inhabit the earth. It's possible that apes of the past had a larger body structure but we can't say for sure. Both may have changed in size through the course of time but never did one evolve into the other . . . ' Aya abounded in the same sense: 'According to Darwin's theory, humans originated from the process of evolution, a claim which contradicts the Islamic perception of creation. Humans are superior to animals; they shouldn't be classified under the same species. They are superior physically, psychologically and intellectually. Therefore, as creation and evolution are totally contradictory, I don't think there's a way of reconciling the two views. Creation simply implies bringing into existence something from nothing, while evolution implies that something came into existence from another thing already existing before it. Fundamentally, they are different'. I turned to Marzieh and asked: 'What do you think? Can the two be reconciled?' She said: 'I don't know the theory well enough, but within my knowledge and understanding, I don't think there is any way to reconcile Islam with Darwinism'. Mohammad was eager to bring a more hopeful view: 'It is the interpretation of the Qur'an by certain individuals that contradicts Darwin's theory of evolution. The Qur'an does not deny the process of evolution anywhere, although it does propose the concept of Adam and Eve. There is much research addressing this very question (e.g. the so-called "Mitochondrial Eve"), and I would not be surprised if in the future, we might determine that there were really an Adam and an Eve! Till such a time, we must recognize evolution as an ongoing and necessary process. One interesting idea that I came across was a statement by a scholar that God chose a particular pair of a male and female homo sapiens and breathed the *ruh* (soul) into them thereby distinguishing them from their other counterparts (He also compared eating the forbidden fruit to their becoming aware of their transition from mere "animals" to real human beings who have a higher level of understanding). But no one knows for sure what or how things happened . . . '

The students were eager to hear my views on the subject. I said: 'Well, it is quite impossible for me to fully explain my stand on the question, for it took me tens of pages to cover the topic in my book, but I'll give you the gist of it. First, we have no choice but to accept any strongly proven fact, and evolution is a fact of nature that has been established by many different methods. Darwin's theory is not the final one on the subject, just like Newton's theory of gravity was followed by Einstein's, but it remains a very solid one, except in some minor aspects. Now, the problem can arise from two directions: (1) a materialistic reading of the theory of evolution, ruling out any divine plan or purpose behind the process, and (2) a literal reading of some verses, especially those pertaining to the creation of humans. A less rigid reading and understanding of both the Qur'anic text and the scientific theory erases much, if not all, of the apparent conflict'.

I had only one or two more questions for my students – before I started presenting them with my personal philosophy on the whole subject. I asked them: 'Do you think that science in general leads people to more belief or to more atheism?'

Marzieh said: 'I think that at first science leads to atheism but with more knowledge and honest thinking, it leads to belief. I have found this in the general development of modern science and in the stories of contemporary scientists'. Rami concurred: 'It leads to more belief, as students of science begin to unravel the marvel and might that only the Almighty is capable of. After all, how could this intricate universe have been created without a Supreme Being behind it all?!' Aya agreed: 'I believe that science in general leads people to more belief and helps enlighten nations. There are many examples of people in the West and in the East whose intellectual journeys took them from doubt or unbelief to faith and religion'. Mohammad was searching for a relevant quote: 'What was that famous utterance of Einstein about science without religion and religion without science?' I helped him out: '"Science without religion is lame, religion without science is blind"'. Mohammad smiled and commented further: 'It all boils down to whether a scientific theory appears to be in conjunction with one's beliefs or is opposed to them. In the first case, a person's beliefs are fortified whereas in the second case, a person is forced to question his/her beliefs and/or reinterpret them. But historical clashes between the Church and scientists have created a false impression of antagonism, and as a result, in certain cases science led people to think that if they are convinced by science then they cannot accept religion . . . '

I thought I should point the students towards the whole discussion on the design argument, the fine-tuning of the universe, the anthropic principle and the natural theology program that have become a staple of the religion–science discussions in recent times. They were largely unaware of all that, and they welcomed the invitation to explore those ideas later.

Finally, I asked: 'Do you think the educational curriculum in the Arab/Islamic world needs to be modified in order to bring about a greater harmony between modern science and Islam? What should be changed?'

Mohammad replied: 'The curriculum should encourage people to question everything. People may fear that this may divert young minds away from Islam, but I personally believe it would bring them closer to the truth and may even bring out facts in the Qur'an that people have been unaware of till today. As the famous tagline of the series *The X-Files* says, the truth is out there. We just have to be willing to reach out for it'. Aya agreed with Mohammad but added a religious aspect to the envisioned curriculum: 'New, revised curriculums should integrate the study of science with Islam. Students should be taught all the theories of science and helped to consider whether they challenge Islam or support it. They should be given all the information they need in order to reach their own conclusions; ultimately, the goal should be to prepare students to propose better theories in the future, to correct the ones they believe are false'. Rami stressed the importance of education: 'First and foremost, we must educate; only this way will Islam and science not clash'. Marzieh then added an important new idea: 'We must introduce or at least greatly enhance the teaching of the history and philosophy of science for all students at the high school level. This will train students to think broadly and critically from a younger age. The process would then be completed at the university level, alongside the scientific and technical courses'. I smiled and said: 'You know, Marzieh, this is one of the main recommendations I give at the end of my book; you seem to be 25 years ahead of me . . .'

It was a delightful discussion, showing bright students who were struggling to harmonise all that they had learned and come to believe or accept; clearly they were not being helped much in that endeavor. Muslim scientists, philosophers, thinkers and educators have a tremendous responsibility in this regard. These attempts (small conversations, surveys, books) need to be widened in scope and increased in strength and frequency.

Let me now finish with a few concluding thoughts, which represent my views as I presented them to the students. These also constitute a brief summary of some of the main ideas I have developed in various parts of this book.

There is no question that the Qur'an urges man to reflect and seek the truth, wherever it may be. But one must keep an open mind and a wide viewing angle. The more microscopic one is in his/her approach to Islam (or any religion), the more one is liable to run into difficulties. Science must also be looked at macroscopically, that is, as a search for truths and general understanding, not for facts or explanations, which can always be improved upon, sometimes with drastically novel theories.

With that view on science and religion, it becomes much easier to find common grounds between the two; they both become forces that push humans to find truths, one by religious intuition, and the other by its own methods (plural, as I explained in Chapter 3). These two forces must act in tandem, not in opposite directions. One must not, however, push this complementarity and mutual support to an extreme and start identifying the two by, for example, finding scientific facts in the religious texts.

And I should also underscore one of the recommendations that I made in my last chapter, namely, religion (represented by theologians and scholars) must try to take full account of the scientific discoveries and methods; likewise, science must try to benefit from philosophy and religion in searching for meaning in its quests and in strengthening its bases (as I showed, for instance, with regard to cosmology).

Today the religious thought in Islam suffers from two main problems: (1) the refusal of the *ulemas* (religious scholars) to grant the (non-religious) specialists the right to propose ideas on any issue that is perceived as religious in any way; (2) the superficial character of the scientific culture of Muslims in general, and its virtual inexistence in the *ulamas'* education, particularly in the Arab world. This is a serious educational and social problem, and I have addressed some aspects of it briefly in Chapter 10.

How can we help develop the discourse on Islam and science? This depends on one's goals: Is one seeking a remedy to the malady which exists in this regard in the Muslim society today, or is one interested in producing a discourse which can contribute to the worldly, multicultural effort to reconcile science and religion more generally and globally? For the latter's goal, we must collaborate with our western (non-Muslim) colleagues on topics such as those I have raised in Chapter 10. For the first objective, clearly a huge educational endeavor is needed; the whole manner in which science is treated and presented in the Arab/Muslim world must be reformed in all venues (schools, universities and media). Seminars must be organized and books must be written, and the whole effort must be led by intellectuals with moral, financial, cultural and scientific support by colleagues everywhere

(East and West). In particular, joint seminars with the *ulamas* and educators must be organized as often as possible; there is an urgent need to expose *ulamas* and teachers at all levels of various discussions that have been conducted at the interface of science, philosophy and religion.

There is much work to be done, and too few people are seriously involved in this endeavor. I truly believe that this is one of the main ways to generate a Muslim renaissance.

appendix a: collective article 'towards an open-minded science' (*le monde*, 23 feb. 2006)

If scientists abandon metaphysical and spiritual reflection, they will cut themselves off from society.

We are a group of scientists from the most diverse scientific and cultural backgrounds. We share the belief that religious or metaphysical ways of thinking should not, a priori, interfere in the ordinary practice of science. However, we also consider that it is legitimate, indeed necessary, to reflect, a posteriori, on the philosophical, ethical, and metaphysical implications of scientific discoveries and theories. Indeed, to fall short of doing so would be to isolate many scientists and science itself from a large proportion of society.

It is a debate which includes the most diverse opinions. Whereas Richard Dawkins has famously asserted that it has been possible to live as a perfectly fulfilled atheist since the publication of Darwin's *Origin of Species*, Arthur Eddington, on the other hand, has written that since 1926, the year of the synthesis in Quantum Mechanics, intelligent people can once again believe in the existence of God. But there is no shortage of biologists who affirm the compatibility of Darwinism with a belief in a creator, and physicists for whom quantum physics in no way diminishes the credibility of materialism.

Today the legitimacy of this debate in France as in the USA and other countries faces two types of confusion. These are linked to the intense media attention that has surrounded the so-called Intelligent Design movement. This movement transgresses the limits of science as it counts numerous creationists who deny some of the basic tenets of modern science. In addition, the movement has an unhealthy political agenda of modifying science education in American schools.

The first confusion is between creationists and those who completely accept the theory of evolution while submitting different hypotheses regarding its mechanisms, including the possibility of internal factors. The term creationist should only be used to describe those people who deny a common ancestor to all the main forms of life on earth or who deny that evolution led the original forms of life to present day beings. If we don't apply this rigour regarding the use of these terms, all Jewish, Muslim, Christian, or Deist scientists could be described as creationist because they believe in a creating principle. And the majority of the founders of modern science, including Newton, Galileo, Descartes also, for the same reason. We can see how this leads to a tremendous degree of confusion.

The second confusion is even easier to make as it concerns the use of the same term: 'design'. For example, one needs to distinguish between those who say that progress in astrophysics does not exclude the philosophical idea that the universe is designed and the Intelligent Design movement. Hence, in 1999, the American Association for the Advancement of Science (AAAS), one of the world's largest scientific organisations, which edits the journal *Science* organised a three-day conference on *Cosmic Questions*, which included a entire day dedicated to debating *Is the Universe designed?* Of course, none of the proponents of the Intelligent Design movement were part of this meeting. It was a meeting between professional astronomers. This area stems from research carried out in the 1980s that found that the universe appeared to be fine-tuned for the appearance of life. Moreover, the slightest modification in the constants of the universe would render it incapable of developing any form of complexity. This area of research, which concerns the Anthropic Principle, has given rise to numerous publications in peer review journals. For some scientists the fine-tuning of the universe lends renewed possibility to the hypothesis of the existence of a creator (without in any way providing proof thereof). Others have vehemently rejected such hypotheses. It is a good example of debates about the philosophical and metaphysical significance of major scientific discoveries. These sorts of debates take place within the context of the mainstream academic community and those involved should not be mistaken with those who deny the fundamental basis of science as creationists do. It is essential, on this point, to make clear that the acceptance of methodological materialism, which is at the basis of method used in most scientific disciplines (in the eyes of many of its specialists quantum physics is an exception to the rule), should not be presented as leading to, or validating, philosophical materialism.

We, therefore, wish to assert with force the following:

- To evoke the existence of a movement, such as the Intelligent Design movement, to discredit certain scientists who affirm that recent scientific discoveries lend more credibility (without ever providing proof) to nonmaterialistic philosophies is to create a confusion which should be condemned.
- Accusing some scientists, as this has been the case recently in France, of taking part in a campaign of 'spiritualist intrusion' into science is unethical and runs against the freedom to debate which must exist on the philosophical and metaphysical implications of recent discoveries in science. It is also an example of double standards as these same people do not accuse Richard Dawkins of being involved in 'materialist intrusions' into science.
- Acting in this way is to do a disservice to science. At a time when the young are lacking in motivation to take up scientific careers, and when science is subject to numerous criticisms, often abusive, or ill-informed, science needs to be as open as possible (among other things to the question of meaning) and should not seal itself off in a way which is characteristic of scientism.
- In France, the Interdisciplinary University of Paris (UIP), whose activities we have all participated in, has brought this debate into the public sphere during its ten years of existence. The UIP has done this in an open and rigorous manner and we think that this approach should be supported.

We hope, with this common declaration, to help the French public and the French media in particular to avoid the confusions mentioned above; to be interested in the richness of contemporary debates in the philosophical and metaphysical implications of scientific discoveries made in the twentieth century; to respect all the authors of this debate so long as they base their arguments on facts that are accepted by the entire scientific community.

Jacques ARSAC, Computer Scientist, Academy of Sciences
Mario BEAUREGARD, Neurologist, University of Montreal

Raymond CHIAO, Physicist, Professor at Berkeley University

Freeman DYSON, Physicist, Professor at the Institute of Advanced Studies at Princeton

Bernard D'ESPAGNAT, Physicist, Academy of Moral and Political Sciences

Nidhal GUESSOUM, Astrophysicist, American University of Sharjah, UAE

Stanley KLEIN, Physicist, Professor at Berkeley University

Jean KOVALEVSKY, Astronomer, Member of the French Academy of Sciences

Dominique LAPLANE, Neurologist, Professor at the University of Paris VI

Mario MOLINA, Nobel Prize for Chemistry, University of San Diego

Bill NEWSOME, Neurologist, Professor at Stanford University

Pierre PERRIER, Computer Scientist, French Academy of Sciences

Lothar SCHAFER, Physical Chemist, Professor at Arkansas University

Charles TOWNES, Nobel Prize for Physics, Berkeley University

TRINH XUAN Thuan, Astronomer, Professor at the University of Virginia

appendix b: the flaws in 'a new astronomical Qur'anic method for the determination of the greatest speed c' by Dr. Mansour Hassab-Elnaby

In addition to the methodological and conceptual errors that I noted in my review of this paper (in Chapter 5), I would like to highlight in some detail the technical errors made by the author.

The derivation starts with the Qur'anic verse, which is rendered as follows: 'God rules the cosmic affair from the heavens to the earth. Then this affair travels to Him (i.e. through the whole universe) in one day, where the measure is one thousand years of your reckoning' (32:5); I have already noted the bias inherent in his translation.

The author then draws a certain number of conclusions from this verse:

1. He states: 'The Qur'anic expression "of your reckoning" leaves no doubt as to our understanding of the year as the lunar year'; however, we must note that as it is of 'people's reckoning', it should be referring to 'synodic' (lunar phase) months, not sidereal (lunar orbit) months, which no one can infer from 'reckoning'.
2. He writes: 'This affair . . . crosses in *one day* a maximum distance in space equivalent to that which the moon passes during *one thousand lunar years* (i.e. 12,000 sidereal months) . . . ' This is a huge conceptual leap, as the verse speaks of time periods, not of distances at all.

3. An equation is then constructed:

> Distance crossed in vacuum by the universal cosmic affair in one sidereal day = length of 12,000 revolutions of the moon around the earth.

$$Ct = 12,000 \, L,$$

where

> C is the velocity of the cosmic affair; t is the time interval of one terrestrial sidereal day defined as the time of one rotation of the earth about its axis (relative to the stars), i.e. 23 hr, 56 min, 4.0906 sec = 86,164.0906 sec; L is the inertial distance which the moon covers in co-revolution around the earth during one sidereal month, i.e. L is the net length of the moon's orbit due to its own geocentric motion, without the interference of its spiral motion caused by the earth's revolution around the sun, i.e. L is the lunar orbit length excluding the effect of the solar gravitational field on the measured value.

Note the usage of 'impressive' scientific terms in defining L, which could have simply been explained as the 'circumference of the moon's orbit with respect to a fixed earth'.

4. Then L is calculated as $L = 2 \pi R$, R being the average moon-earth distance, and $V = L/T$ is the average orbital speed of the moon, T being the (sidereal) orbital period of the moon.

5. Then an angle α is thrown in, so the previous equation $V = L/T$ becomes $L = V \cos\alpha \, T$, α being introduced as the 'angle traveled by the earth–moon system around the sun during one sidereal month'.

The author justifies it with the following reasoning: 'Since the presence of the sun changes the geometrical properties of space and time, we must screen out its gravitational effect on the earth-moon system according to the validity condition of the second postulate of special relativity, i.e. we must only consider the lunar geocentric motion without the heliocentric motion of the earth-moon system. Thus a velocity component $V_0 = V \cos\alpha$ representing the net orbital velocity of the moon. is introduced for calculating the net length L of the lunar orbit assuming a stationary earth'. This makes no physical or geometrical sense at all.

6. Finally, combining the two equations

$$Ct = 12,000\,L$$

and $L = V \cos \alpha\, T$,

the author arrives at the value of C: 299,792.5 km/s, which he triumphantly compares to the standard value of 299,792.46 km/s.

appendix c: survey of 'science and religion' views at the American University of Sharjah, UAE – fall 2007

During the fall of 2007, I conducted a survey on 'Science & Religion' issues among professors and students at my university, the American University of Sharjah. About 100 students and 100 faculty members (about one-third of the faculty body) responded to the survey.

The survey first starts by identifying the respondents by categories:

1. Student vs. faculty member
2. Male vs. female
3. Ethnic background: Arab, westerner, Asian (non-Arab)
4. Religious background/affiliation; the question was phrased: 'Do you consider yourself . . . Muslim, Christian, other (specify if you wish), nonreligious, atheist?'

The survey stated clearly and was conducted in total anonymity; no names were collected whatsoever, and no attempt was made to link the responses to any individual. Due to the relatively small numbers of respondents, it was difficult to draw more conclusions other than by general categories (students vs. professors, Muslims vs. non-Muslims); it would have been interesting to compare male and female responses, for example, or science and engineering faculty members with language and humanities professors.

Below are the questions and the responses by category:

1. Ages (in years)

Muslim students	thousands (%)	tens of thousands (%)	hundreds of thousands	millions (%)	billions (%)	tens of billions (%)	don't know (%)
How old is the universe?			2	17	10	5	66
How old is the earth?	2	2	3	31	5	3	55
How old is the sun?		2	3	25	5	3	63
How old is humanity?	16	2	2	14	2	2	64

Muslim professors	thousands (%)	tens of thousands (%)	hundreds of thousands	millions (%)	billions (%)	tens of billions (%)	don't know (%)
How old is the universe?				7	9	12	72
How old is the earth?				19	28		54
How old is the sun?	2			14	28	2	54
How old is humanity?	10	10	10	12			60

non-Muslim professors	thousands (%)	tens of thousands (%)	hundreds of thousands	millions (%)	billions (%)	tens of billions (%)	don't know (%)
How old is the universe?		2		8	19	23	48
How old is the earth?		2	2	12	41		43
How old is the sun?		2		12	38	2	46
How old is humanity?	8	10	27	8	8		37

2. Miracles: Do you believe in miracles?

	Muslim students (%)	Muslim professors (%)	non-Muslim professors (%)
Yes, I have no problem with that	55	70	38
Only prophets have miracles, none such could occur today	36	29	
Yes, but I still need to find an explanation to them	17	15	24
No, science contradicts miracles	8	6	44

3. God's Action: If you believe in God, do you think that

	Muslim students (%)	Muslim professors (%)	non-Muslim professors (%)
God acts on our lives directly, including in natural phenomena and catastrophes	80	75	40
God acts in the world, but only according to the laws of nature	12	18	29
God exists but does not act in the world; He lets things work out (according to the laws of nature)	8	7	31

4. Evolution

	Muslim students (%)	Muslim professors (%)	non-Muslim professors (%)
Evolution is only an unproven theory and I don't believe in it	62	62	10
Evolution is correct, except for humans	28	22	16
Evolution is strongly confirmed by evidence	11	14	74
Humans did not evolve from any earlier species	66	49	15
Humans evolved only as humans, not from animals	24	41	8
It is now a fact that humans evolved from earlier species	10	10	77
Evolution is against religion; it should not be taught	32	12	0
Evolution should be taught but as 'just a theory'	58	70	36
Evolution should be taught as a strong theory	17	19	64

5. The Qur'an and Science

	Muslim students (%)	Muslim professors (%)
The Qur'an contains explicit statements that are now known to be scientific facts	90	80
The Qur'an deals with natural phenomena and alludes to scientific facts but only vaguely	8	11
One should not try to read any scientific content in the Qur'an	2	9

notes

prologue

1 See, for an accessible and brief account, Karen Armstrong, *A History of God: The 4,000-Year Quest of Judaism, Christianity, and Islam* (New York: Ballantine Books, 1993, pp. 193–5).

2 Mazliak, Paul, *Avicenne & Averroes* (Paris: Vuibert/ADAPT, 2004, p. 106).

3 Alain de Libera, *La philosophie Medievale* (Paris: Puf-Quadrige, 2004).

4 al-Aribi, Mohamed, *Ibn Rushd wa falasifat al-Islam* (Beirut: Dar al-Fikr al-Lubnani, 1992, p. 8).

5 See the list given in the appendix of '*Averroes – Discours Décisif*' (Paris: GF Flammarion, 1996).

6 I am, of course, referring here, though very quickly for now, to the famous and momentous rebuttal written by Ibn Rushd (Tahafut a-Tahafut, 'The Incoherence of the Incoherence') to that deadly attack by Abu Hamid al-Ghazzali (1058–1111) a century earlier against philosophers, *Tahafut al-Falasifa* ('*The Incoherence of Philosophers*').

7 In 'Averroes, precursor of the enlightenment?' (*Alif*, 16 (1996), p. 6–18), Charles Butterworth writes: 'As striking, even as unwarranted as it seems at first glance, some signs may be adduced to support the claim that an early version of the famous eighteenth-century European Enlightenment is to be found in medieval Islamic philosophy, particularly in the writings of the famous philosopher, physician, and sometime royal consultant, Muhammad Ibn Ahmed Ibn Rushd or Averroes'.

8 My contributions on the topic include *The Determination of Lunar Crescent Months and the Islamic Calendar* (in Arabic) (Algeria: Dar al-Oumma, 1993, and Beirut, Lebanon: Dar al-Taliaa, 1997); 'Le Croissant du Ramadan et les Astronomes', La Recherche, Janvier 1999; presentations at the 'experts meeting' organized on 'crescent observability and the Islamic calendar' in Rabat, Morocco in November 2006 by the Islamic Organization for Education, Science, and Culture; and the international conference on Applications of Astronomical Calculations to Islamic Issues organized in December 2006 in the UAE.

introduction

1 al-Ghazzali: *Ihya Ulum Id-Din*, in *The Bounty of Allah* (compiled and translated by Aneela Khalid Arshed; New York: Crossroad Publ. Co., 1999).

2 Carl Sagan: *The demon-haunted world: science as a candle in the dark*, (New York: Ballantine Books, 1997, p. 26).

3 http://archive.gulfnews.com/articles/07/05/28/10128282.html

4 Dubai's 'Internet City' hosts Microsoft, Cisco, HP and other giant IT companies; 'Media City' encompasses international and Arab media powerhouses, such as Reuters, CNN, CNBC, Associated Press, MBC, al-Arabiyya and others; and 'Knowledge Village' hosts a large number of universities and colleges, including The British University in Dubai, Middlesex University, Dubai campus and University of Wollongong in Dubai.

5 In Dubai, Emirati nationals make up only about 10 percent of the population; in the whole country, the fraction is about 15 percent. At the American University of Sharjah, where I work and reside, the 350 faculty members represent some 50 nationalities, and the 5,000 students represent over 70 countries.

6 'We will make first UAE astronaut', *Gulf News*, 17 November 2005; see also, 'Sharaf Travel wins space tourism ticketing deal with Virgin Galactic', *Gulf News*, 2 May 2007. http://archive.gulfnews.com/articles/07/05/02/10122330.html

7 Abu Dhabi University was opened in September 2003 and moved to a new campus in October 2006; the Paris-Sorbonne Abu Dhabi University opened its doors in October 2006; Massachusetts Institute of Technology (MIT) has partnered with Abu Dhabi's Energy Company to establish Masdar, the region's first-ever zero-carbon campus of research. http://web.mit.edu/newsoffice/2007/abu-dhabi.html

8 The conference's programme, paper abstracts and video recordings of many talks can be found at http://nooran.org/con8/4.htm and pages within the website.

9 In March 2008 the ninth international conference on the 'Scientific I'jaz (Miraculous Aspects) in the Qur'an and the Sunna' was organized in Setif, Algeria (under the patronage of the country's president and with an opening speech given by the prime minister). The 10th conference will be held in Turkey in 2010.

10 *al-Khaleej*, 13 April 2007 (no. 10190), p. 9.

11 A web search for the phrase 'Qur'an healing conference' (with quotation marks as specifiers and limiters) in Arabic turns up 13,600 pages, most of them media reports. Still, none of the papers presented at the conference can be located. The newspaper reports were, however, detailed enough to allow one to get a clear idea of what each presenter stated.

12 *al-Khaleej*, 13 April 2007 (no. 10190), p. 9. The article also carried a picture of the device.

13 *al-Khaleej*, 13 April 2007 (no. 10190), p. 9.

14 Fahd bin Abd ar-Rahman Ar-Rumi: *Khasa'is al-Qur'an al-Kareem* (*Characteristics of the Noble Qur'an*) (Riyadh, Saudi Arabia: Maktabat at-Tawbah, 10th edition, 2000).

15 Ar-Rumi: *Khasa'is al-Qur'an al-Kareem*, p. 110.

16 http://roqia.host.sk/5555.pps

17 As of 30 September 2009; and the number keeps increasing fast, month after month.

18 'The Hour (of Judgment) is nigh, and the moon did rend asunder. But if they see a Sign, they turn away, and say, "This is (but) transient magic"' (Q 54: 1–2).

19 http://zuserver2.star.ucl.ac.uk/~apod/apod/ap021029.html

20 For example, As-Sayyid al-Jamili, *al-I'jaz al-'Ilmiy fil Qur'an* (*Scientific Miraculousness of the Qur'an*) (Beirut, Lebanon: Dar al-Fikr al-'Arabi, 2002).

21 Majid al-'Arjawi, Abdul, *al-Barahin al- 'Ilmiyya 'ala Sihhat al-'Aqida al-Islamiyya* (*The Scientific Proofs to the Correctness of the Islamic Creed*) (Damascus, Syria: Dar Wahy al-Qalam, 2003).

22 Sigismondi, Constantino & Imponente, Giovanni: The observation of lunar impacts', *WGN*, *Journal of the International Meteor Organization* 28/2–3 (2000), pp. 54–7.

23 http://news.bbc.co.uk/2/hi/science/nature/1304985.stm

24 Snow, Peter, *Maori Legends About Historical Impact Disaster*, http://abob.libs.uga.edu/bobk/ccc/cc032301.html

25 The other two favourable conditions are as follows: During a total eclipse, or far from the dark limb around first or last quarter (Sigismondi & Imponente: 'The observation of lunar impacts', op. cit.).

26 Salam, for whom I shall present a sidebar profile in Chapter 4, received his Nobel Prize in Physics in 1979; he was originally from Pakistan but had spent practically all of his professional life in England and Italy; Zewail received his Nobel Prize in Chemistry in 1999; he is originally from Egypt but has spent his entire career in the USA.

one

1 Shihab-u-Deen M. al-Abshihi, *al-Mustatraf* ("The Compendium"), edited by Yahya Murad, Mu'assasat al-Mukhtar Publ., Cairo, 2006, p. 13.

2 al-Abshihi: *al-Mustatraf* (Cairo, Egypt: Mu'assasat al-Mukhtar Publ., 2006, p. 14).

3 For example, Q 7:172 and Q 30:30.

4 For instance: Omar Sulaiman al-Ashqar, *Nahwa Thaqafatin Islamiyyatin Asilah* ("For a genuine Islamic knowledge"), Dar al-Nafa'is, Amman (Jordan), 1997; Mustafa Muslim and Fathi Muhammad al-Zughbi, *Ath-Thaqafatu al-Islamiyyah: ta'rifuha, masadiruha, majalatuha, tahaddiyatuha* ("Islamic knowledge: definition, sources, fields, and challenges"), Mu'assasat al-Risalah, Dar al-Bashir, Sharjah (UAE), 2004; 'Azmi Taha al-Sayyid et al. *Ath-Thaqafatu al-Islamiyyah: mafhumuha, masadiruha, khasa'isuha, majalatuha* ("Islamic knowledge: concept, sources, characteristics, and fields"), Dar al-Manahij, Amman (Jordan) 4th edition, 2002.

5 al-Bukhari: Vol. 9, book 93, hadiths 489.

6 There are several hadiths to this effect; see the discussions in al-Seyyid Sabiq, *Islamic Dogmas* (Cairo: al-Fath li l-I'lam al-'Arabiy Publ., 1992, pp. 28–9).

7 Hadith related by al-Tirmidhi in his classical collection, see al-Seyyid Sabiq: *Islamic Dogmas*, pp. 23–7.

8 For instance, http://en.wikipedia.org/wiki/99_Names_of_God_in_the_Qur'an, http://www.ecotao.com/holism/theism/more_names.html, http://lexicorient.com/e.o/99_names_god.htm

9 For instance, http://nur313.wordpress.com/2006/07/22/al-asmaa-husna-the-99-beautiful-names-of-allah-swt/,http://www.muhajabah.com/islamicblog/archives/veiled4allah/005074.php, http://grocs.dmc.dc.umich.edu/gallery/buildingislam/99_Names_of_God_American_Moslem_Center, http://www.fortunecity.com/victorian/beardsley/250/call.htm, http://www.flickr.com/photos/43046235@N00/122799318/

10 See al-Seyyid Sabiq: *Islamic Dogmas*, pp. 28–9.

11 Saleh as-Saleh: 'Whom must we worship', http://www.sultan.org/articles/wmww.html

12 For instance, al-Bukhari: Vol. 9, book 93, hadiths 515, 525y and others.

13 al-Bukhari: Vol. 9, book 93, hadith 535.

14 This arises from the verse *'Some faces, that Day, looking to their Lord'* (Q 75:22–23) and a few hadiths, which state explicitly: 'You people will see your Lord as you will see this full moon, and you will have no trouble in seeing Him'; 'You will definitely see your Lord with your own eyes'; 'Do you have any difficulty in seeing the moon on a full moon night?' They said, 'No, O Allah's Apostle'. He said, 'Do you have any difficulty in seeing the sun

when there are no clouds?' They said, 'No, O Allah's Apostle'. He said, 'So you will see Him like that . . . ' (al-Bukhari: Vol. 9, book 93, hadiths 529, 530, 532c respectively).

15 Chittick, William, *Science of the Cosmos, Science of the Soul: The Pertinence of Islamic Cosmology in the Modern World* (Oxford, UK: Oneworld, 2007, p. 87).

16 Armstrong, Karen, *A History of God: the 4,000-Year Quest of Judaism, Christianity, and Islam* (New York: Ballantine Books, 1993, p. 190).

17 Yusuf, Imtiaz, 'Badiuzzaman Said Nursi's discourse on belief in Allah: a study of texts from Risale-I Nur collection', *The Muslim World*, 89/3–4 (Jul.–Oct. 1999), p. 336.

18 Armstrong: *A History of God* (p. 170).

19 Qur'an 3:7.

20 Slightly edited by me from the report of Armstrong: *A History of God* (p. 193) of Majid Fakhry's *A History of Islamic Philosophy* (1970, pp. 313–14).

21 For example, Mazliak, Paul, *Avicenne & Averroes* (Paris: Vuibert/ADAPT, 2004); Geoffroy, Marc, *Averroes, l'Islam et la raison* (Paris: GF Flammarion, 2000) and references in these books.

22 Armstrong: *A History of God* (pp. 193–5).

23 See for instance, his treatise *Kitab al-Tawheed* (*The Book of Tawheed*), where essentially nothing 'modern' or 'reforming' can be found.

24 Indeed, in recent years, books of bridge-building inclinations have appeared, among which I should particularly hail Richard Bulliet's *The Case for Islamo-Christian Civilization* (New York: Columbia University Press, 2002).

25 http://www.vatican.va/holy_father/benedict_xvi/speeches/2006/september/documents/hf_ben-xvi_spe_20060912_university-regensburg_en.html.   All quotations here are taken from this official version.

26 See, in particular, the 'Open Letter to His Holiness Pope XVI', signed by 100 Muslim scholars and dignitaries, published in *Islamica Magazine*, 18 (2006), p. 26; and the article by Nayed, Aref Ali, 'A Muslim's commentary on Benedict XVI's Regensburg lecture', *Islamica Magazine*, 18 (2006), p. 46.

27 See, for example, the article by Murad, Abdal Hakim, 'Benedict CVI and Islam', *Islamica Magazine*, 18 (2006), p. 35.

28 Sabiq: *Islamic Dogmas*, p. 139.

29 See, for instance, Dawkins, Richard's *The God Delusion* (London: Bantam Press, 2006).

30 The most recent survey and analysis of (US) scientists' religious beliefs was conducted by Edward Larson and Larry Witham in 1997; they repeated a study performed by James Leuba in 1916, when it was found that 'only 40 percent' of US scientists believed in a supreme Being; Leuba had related that low figure to the spreading of the scientific mindset and predicted that unbelief would continue to increase as science spreads. Larson and Witham presented their results for the prestigious science journal *Nature* (386 (3 April 1997), pp. 435–36): 'The result: about 40 percent of scientists still believe in a personal God and an afterlife. In both surveys, roughly 45 percent disbelieved and 15 percent were doubters (agnostic)'. In a follow-up study ('Leading scientists still reject God', *Nature* 394 (1998), p. 313) Larson and Withham found that 'among the top natural scientists, disbelief is greater than ever'. Indeed, the percentage of 'leading' natural scientists (here they selected only from members of the National Academy of Sciences), who believe in God seemed to be dropping steadily over the twentieth century.

31 Omar al-Ashqar: *For a genuine Islamic knowledge*, pp. 101–105.

32 Q 30:30: '*So set your face upright for religion in the right state – the nature in which Allah has made men; there is no altering of Allah's creation; that is the right way, but most people do not know*'. The hadith Qudsi: 'I created all my servants naturally aware (of Me) and obedient (to Me), but devils came to them and diverted them from their religion' (in the collection of

Muslim). The Prophetic statement: 'Every newborn comes to the world in the right *fitrah* (pure godly state), and it is his/her parents who make him/her a Jew, a Christian, or a non-monotheist' (in the collection of al-Bukhari).

33 Muslim and al-Zughbi: *Islamic knowledge: definition, sources, fields, and challenges.*

34 al-Sayyid et al.: *Islamic knowledge: concept, sources, characteristics, and fields.*

35 Witham, Larry, *The Measure of God: Our Century-Long Struggle to Reconcile Science and Religion* (San Francisco, CA: Harper, 2005, p. 138).

36 Yahya, Harun, 'Quantum physics and the discovery of divine wisdom', http://www.riseofislam.com/rise_of_faith_04.html

37 O'Barret, Lakhdar, 'Quantum Islam: does physics confirm the Qur'anic worldview?', *Al Jumuah* 16/11 (Dec. 2004–Jan. 2005), pp. 48–50. O'Barret essentially took some ideas from *Science, Sense & Soul: the Mystical-Physical Nature of Human Existence* by Casey Blood (Los Angeles, CA: Renaissance Books, 2001) – with due acknowledgement – and connected with some vague Islamic principles.

38 Nasr, Seyyed Hossein: *The Heart of Islam: Enduring Values for Humanity* (SanFrancisco, CA: Harper, 2004, p. 9).

39 *Muslim*, volume 7, book 72, number 837.

40 al-Bukhari: Vol. 9, book 93, hadiths 647 and 648. It is worth noting that these hadiths are classified in the chapter on *tawheed* (oneness of God).

41 Nasr: *The Heart of Islam*, p. 11.

42 Nasr, Seyyed Hossein, *Ideals and Realities in Islam* (London: Aquarian, 1966, p. 28).

43 Nasr: *The heart of Islam*, p. 15.

44 Nasr: *Ideals and Realities in Islam*, p. 19.

45 Nasr: *Ideals and Realities in Islam*, pp. 23–5.

46 Armstrong: *A History of God*, p. 151.

47 Huda, Qamar-ul, 'Knowledge of Allah and the Islamic view of other religions', *Theological Studies*, 64/2 (Jun. 2003), p. 278.

two

1 Massimo Campanini, 'Qur'an and science: a hermeneutical approach', Journal of Qur'anic Studies, 7/1 (2005), p. 48.

2 Georges C. Anawati, 'Philosophy, theology, and mysticism', in J. Schacht and C. E. Bosworth (eds), The legacy of Islam (Oxford; 2nd edition, 1979).

3 Sachiko Murata and William C. Chittick, 'The Koran' (Introduction to their *Vision of Islam*, Paragon House, 1994, pp. XIV–XIX).

4 Reza Aslan, *No god but god: the origins, evolution, and future of Islam*, (Random House, 2005, p. 157).

5 Suha Taji-Farouki, 'Introduction', in S. Taji-Farouki (ed) *Modern Muslim Intellectuals and the Qur'an* (London: Oxford University Press, 2004, p. 20).

6 Seyyed H. Nasr: *The heart of Islam: enduring values for humanity*, op. cit., p. 23.

7 Ar-Rumi: *Khasa'is al-Qur'an al-Kareem.*

8 Ar-Rumi: *Khasa'is al-Qur'an al-Kareem*, pp. 18+

9 Aslan: *No god but God*, p. 159.

10 Murata and Chittick: The Koran, p. XVII.

11 Kenneth Cragg: *Readings in the Qur'an* (Brighton, UK: Sussex Academic Press, 1988, p. 15).

12 Cragg: *Readings in the Qur'an*, p. 28.

13 Muzaffar Iqbal, *Islam and Science* (Hampshire & Burlington, UK: Ashgate Publ., 2002, p. 89).

14 Murata and Chittick: The Koran.

15 Murata and Chittick: The Koran.

16 Mohamed Talbi and Maurice Bucaille, *Réflexions sur le Coran* (Paris: Seghers, 1989, p. 13).

17 Muhammad Asad, 'Symbolism and allegory in the Qur'an', n.d., http://www.geocities.com/masad02/appendix1

18 Note that whether one puts the comma at the first bracket position or the second, the role and importance of scholars changes drastically; note further that realising the possibility of the two different punctuations and of the two different meanings is itself contingent on the sophistication of the reader. There are many places in the Qur'an where the reading and the meaning change with the punctuation; the one cited here by Averroes is particularly important in that it insists on the special comprehension of the religious texts by people of knowledge. In any case, the Qur'an is full of praise for 'the people of knowledge/understanding'.

19 Asad: Symbolism and allegory in the Qur'an, for all quotes herein.

20 Mohamed Talbi, from his book *Iyal Allah: afkar jadida fi 'alaqat al-Muslim bi nafsihi wa bi al-akhirin*, ('God's children: new ideas on the Muslim's relation with himself and with others'), Tunis, 1996, as cited by Ronald L. Nettler, 'Mohamed Talbi on understanding the Qur'an', in Suha Taji-Farouki (ed.) Modern Muslim Intellectuals and the Qur'an (London: Oxford University Press, 2004, p. 231).

21 Dale F. Eickelman, Islamic Religious Commentary and Lesson Circles: Is There a Copernican Revolution?, G. W. Most (ed.) Commentaries. (Kommentare, Göttingen, 1999, p. 140).

22 Rainer Nabielek, 'Muhammad Shahrur, ein 'Martin Luther' des Islam', Inamo, 6/23–24, autumn/winter (2000), pp. 73–74.

23 Taji-Farouki: Introduction, p. 16.

24 Andreas Christmann, The form is permanent, but the content moves': the Qur'anic text and its interpretation(s) in Mohamad Shahrour's al-Kitab wal-Qur'an', in Suha Taji-Farouki (ed), Modern Muslim Intellectuals and the Qur'an (London: Oxford University Press, 2004, p. 272).

25 Christmann: The form is permanent, but the content moves, pp. 282–83.

26 Ar-Rumi: *Khasa'is al-Qur'an al-Kareem*, pp. 69–71.

27 Mehdi Golshani, *The Holy Qur'an and the Sciences of Nature: A Theological Reflection* (New York: Global Scholarly Publications, 2003).

28 Ghaleb Hasan, *Nazariat al-'ilm fi-l-Qur'an* ('The Theory of Knowledge/Science in the Qur'an') (Beirut: Dar al-Hady, 2001).

29 Golshani: *The Holy Qur'an and the Sciences of Nature*.

30 Murtada Mutahhari, *Farizeh 'Ilm* (Tehran: Goftar-e Mah, 1922, p. 137), quoted by Golshani, op. cit., p. 21.

31 Campanini: Qur'an and Science, p. 54.

32 Muzaffar Iqbal, *Science and Islam* (Westwood, CT and London: Greenwood Press, 2007, p. 21).

33 All Bucaille quotations here are from the previously cited Talbi and Bucaille, *Réflexions sur le Coran*.

34 Hasan: *Nazariat al-'ilm fi-l-Qur'an*, pp. 65–66.

35 Also, 'they say: We follow the ways we found our fathers following. What! Even though their fathers were wholly unintelligent and had no guidance?' (2:170).

36 Hasan: *Nazariat al-'ilm fi-l-Qur'an*, p. 79.

37 Q 30:22: 'And among His Signs is the creation of the heavens and the earth, and the variations in your languages and your colors: verily in that are signs for people of knowledge'; Q 29:43: 'And such are the parables We set forth for mankind, but only those who have knowledge understand them'.

38 Golshani: *The Holy Qur'an and the sciences of nature*, pp. 242–4.

39 Iqbal: *Islam and science*, p. 4.

40 Iqbal: *Islam and science*, p. 6.

41 Muhammad Iqbal, *The Reconstruction of Religious Thought in Islam* (Sang-e-Meel, 2004 edition).

42 This includes Mehdi Golshani and Muzaffar Iqbal in the aforementioned books and Muhammad Abdus Salam in Ideals and Realities (Singapore: World Scientific, second edition, 1987).

43 For instance, 'Say: Consider what is in the heavens and the earth' (Q 10:101) and 'Say: Travel in the land and see how He originated creation' (Q 29:20).

44 Golshani: *The Holy Qur'an and the sciences of nature*, p. 154.

45 al-Battani, *Az-Zij as-Sabi* ('The Sabian Tables'), p. 6, cited by Muntasir Mujahed, *Usus al-Manhaj al-Qur'aniy fi Bahth al-'Ulum al-Tabi'iyyah* (The bases of the Qur'anic methodology in natural-science research) (Djeddah: ad-Dar as-Su'udiyya li n-Nashr wa t-Tawzi', second edition, 2004, p. 100).

46 Mujahed: *Usus al-Manhaj al-Qur'aniy*, p. 278.

47 Mujahed: *Usus al-Manhaj al-Qur'aniy*, p. 138.

48 Mujahed: *Usus al-Manhaj al-Qur'aniy*, p. 141.

49 Iqbal: *The reconstruction of religious thought in Islam*, p. 8.

50 Iqbal: *The reconstruction of religious thought in Islam*, p. 6.

51 Campanini: Qur'an and science, pp. 54–55.

52 Campanini cites Richard C. Martin and Martin R. Woodward, *Defenders of Reason in Islam* (Oxford, UK: Oneworld, 1997).

53 Campanini: Qur'an and Science, p. 53.

54 Campanini: Qur'an and Science, p. 53.

55 Hasan: *Nazariat al-'ilm fi-l-Qur'an*, pp. 90–91.

56 Mujahed: *Usus al-Manhaj al-Qur'aniy*, p. 18.

57 Mujahed: *Usus al-Manhaj al-Qur'aniy*, pp. 31–34.

58 Golshani: *The Holy Qur'an and the sciences of nature*, pp. 27–28.

59 Ibn Rushd, *Fasl al-Maqal fi ma bayn al-Shari'a wa al-hikma min al-Ittisal* ('The Definitive Discourse on the Harmony between Religion and Philosophy'), available at: http://www.muslimphilosophy.com/ir/fasl.htm.

60 Campanini: Qur'an and science, p. 57.

61 I should add that the esoteric (batini) meanings of the Qur'an are even more central in the Sufi approach to the Qur'an.

62 Mohamed Talbi and Maurice Bucaille, *Réflexions sur le Coran* (Paris: Seghers, 1989, p. 13).

63 Hasan Hanafi, 'Method of thematic interpretation of the Qur'an', in *Islam in the Modern World* (Cairo: Anglo-Egyptian Bookshop, 1995, vol. 1, p. 417), quoted by Campanini: Qur'an and science, p. 60.

64 Campanini: Qur'an and science, p. 59.

65 Campanini: Qur'an and science, p. 59.

66 Campanini: Qur'an and science, p. 60.

67 Mohammad Hashim Kamali, 'Islam, rationality and science', *Islam & Science*, June 2003.

68 Kamali: Islam, rationality and science.

69 Talbi and Bucaille: *Réflexions sur le Coran*, p. 66.

70 Talbi and Bucaille: *Réflexions sur le Coran*, p. 71.

three

1 Sardar, Ziauddin, 'Muslims and the philosophy of science' (first published in 1997), in Ehsan Masood (ed.), *How Do You Know?* (London: Pluto Press, 2006, p. 108).

2 Sardar, Ziauddin, *Thomas Kuhn and the Science Wars* (London: Icon books/Totem books, 2000, p. 59).

3 Sagan, Carl, *The Demon-Haunted World: Science as a Candle in the Dark* (New York: Ballantine Books, 1997, p. 305).

4 In 'Islamization of knowledge or westernization of Islam' (*Inquiry* 7 (1984), pp. 39–45), Sardar refers to Mohamad Naquib al-Attas ('The de-westernization of knowledge' in *Islam, Secularism and the Philosophy of the Future* (London: Mansell, 1985), where he 'produces one of the most devastating criticisms of western epistemology', arguing that 'the all-pervasive scepticism, which knows no ethical and value boundaries, of the western system of knowledge is the antithesis of Islamic epistemology'.

5 Okasha, Samir, *Philosophy of Science, a Very Short Introduction* (Oxford: Oxford University Press, 2002, p. 1).

6 Sardar, Ziauddin, 'Islamic Science: The Way Ahead' (the text of a conference lecture given in 1995), in Ehsan Masood (ed), *How Do You Know?* (London: Pluto Press, 2006, p. 161).

7 Chittick, William, *Science of the Cosmos, Science of the Soul: The Pertinence of Islamic Cosmology in the Modern World* (Oxford, UK: Oneworld, 2007, p. 24).

8 Chittick: *Science of the Cosmos, Science of the Soul*, p. 35.

9 Chittick: *Science of the Cosmos, Science of the Soul*, p. 119.

10 Chittick: *Science of the Cosmos, Science of the Soul*, p. 34.

11 Chittick: *Science of the Cosmos, Science of the Soul*, p. 34.

12 Sardar: 'Muslims and the Philosophy of Science', in Ehsan Masood (ed), *How Do You Know?* (London: Pluto Press, 2006, p. 108).

13 Sardar: 'Science wars: a postcolonial reading' (first published as 'Above, beyond and at the centre of the science wars: a postcolonial reading' in 2001), in Ehsan Masood (ed), *How Do You Know?* (London: Pluto Press, 2006, p. 193).

14 Sardar: 'Science wars: a postcolonial reading', in Ehsan Masood (ed), *How Do You Know?* (London: Pluto Press, 2006, p. 195).

15 Alexander, Denis, *Rebuilding the Matrix: Science and Faith in the 21st Century* (Grand Rapids, MI: Zondervan, 2001, p. 76).

16 Briffault, Robert, *The Making of Humanity* (London: G. Allen & Unwin, 1928).

17 Randall, J. H., *The Making of the Modern Mind* (New York: Columbia University Press, 1976, p. 208).

18 Bachta, Abdekader, *L'Esprit Scientifique et la Civilization Arabo-Musulmane* (*The Scientific Spirit and the Arab-Muslim Civilization*) (Paris: L'Harmattan, 2004).

19 *Maqalat fi l-mantiq wa l-'ilm al-tibbiy* ('Essays on Logic and in the Medical Science').

20 Bachta: *L'Esprit Scientifique et la Civilization Arabo-Musulmane*, p. 66.

21 Bachta: *L'Esprit Scientifique et la Civilization Arabo-Musulmane*, pp. 68–9.

22 Chittick: *Science of the Cosmos, Science of the Soul*, p. 116.

23 For an interesting discussion of how and when 'science' started to appear in humanity's history, see Carl Sagan's chapter titled 'The wind makes dust' in *The Demon-Haunted World: Science as a Candle in the Dark* (pp. 309+).

24 Hackett, Jeremiah, 'Roger Bacon: his life, career, and works', in Hackett, Roger Bacon and the Sciences. Also, the manuscript 'The alchemical manual Speculum Alchemiae' has been ascribed to him.

25 Alexander: *Rebuilding the Matrix*, p. 235.

26 Kuhn: *The Structure of Scientific Revolutions*.

27 Sardar: 'Science wars: a postcolonial reading', p. 203.

28 Sardar: *Thomas Kuhn and the Science Wars*, p. 8.

29 http://www.martininstitute.ox.ac.uk/JMI/People/fellows/Ravetz+Jerry.htm

30 Sagan: *The Demon-Haunted World*, p. 11.

31 Sardar: *Islamic Science: The Way Ahead*, p. 162.

32 Sardar: *Islamic Science: The Way Ahead*, p. 163.

33 Nasr, Seyyed Hossein, *The Encounter of Man and Nature, the Spiritual Crisis of Modern Man* (London: Allen & Unwin, 1968).

34 Sardar: *Islamic Science: The Way Ahead*, p. 174.

35 Sardar: *Islamic Science: The Way Ahead*, p. 174.

36 Sardar: *Islamic Science: The Way Ahead*, p. 174.

37 Sardar: *Islamic Science: The Way Ahead*, p. 176.

38 Sardar: *Thomas Kuhn and the Science Wars*, p. 62

39 Random House Unabridged Dictionary (Random House, 2006).

40 Ferguson, Kitty, *The Fire in the Equations: Science, Religion, and the Search for God* (Grand Rapids, MI: Erdmans Publ, 1994).

41 Golshani, Mehdi, *Min al-'Ilm al-'Ilmaniy ila al-'Ilm al-Diniy (From Secular Science to Theistic Science)* (Beirut: Dar al-Hady, 2003, p. 154), citing 'Maududi on science', trans. by Rafat, *Journal of Islamic Science*, 10/2 (1994), pp. 113–14.

42 Alexander: *Rebuilding the Matrix*, p. 273; Okasha: *Philosophy of Science*, p. 121.

43 Haq, S. Nomanul, 'Science, scientism, and the liberal arts', *Islam & Science*, December 2003.

44 The letter, signed by Mario Molina and Charles Townes, both Nobel Prize winners, and other illustrious scientists and thinkers, such as Freeman Dyson and Bernard d'Espagnat, is reprinted in Appendix A.

45 Alexander: *Rebuilding the Matrix*, p. 82.

46 Alexander: *Rebuilding the Matrix*, p. 83.

47 Alexander: *Rebuilding the Matrix*, p. 84.

48 Alexander: *Rebuilding the Matrix*, p. 85.

49 Golshani: *From Secular Science to Theistic Science*, p. 90.

50 Golshani, Mehdi, *Can Science Dispense with Religion?* (Tehran: Institute for Humanities and Cultural Studies, 1998).

51 Golshani, Mehdi, 'Theistic science' in *God for the Twenty First Century* (Pennsylvania: John Templeton Foundation, 2000); 'Seek knowledge even if it is in China' (Metanexus: Views, 2002, http://www.metanexus.net/magazine/ArticleDetail/tabid/68/id/7557/Default.aspx); 'Does science offer evidence of a transcendent reality and purpose?', *Islam & Science*, 1 (2003), pp. 45–65; 'Some important questions concerning the relationship between science and religion', *Islam & Science* 3.1 (2003), pp. 63–83; 'Philosophy of Science: A Qur'anic Perspective' *al-Tawhid*, II/1, p. 1405.

52 Golshani, Mehdi, *The Holy Qur'an and the Sciences of Nature* (1997); *From Physics to Metaphysics* (Tehran: Inst. for Humanities and Cultural Studies, 1998); *Secular and Religious Science* (in Farsi, translated to Arabic in 2003).

53 Alexander: *Rebuilding the matrix*, p. 248.

54 Alexander: *Rebuilding the Matrix*, p. 249.

55 Golshani, 'Seek knowledge even if it is in China', Metanexus: Views, 2002, http://www.metanexus.net/magazine/ArticleDetail/tabid/68/id/7557/Default.aspx

56 Golshani: *From Secular Science to Theistic Science*, p. 7.

57 Bagir, Zainal Abidin, 'Islam, science, and "Islamic Science"', in Z. A. Bagir (ed), *Science and Religion in a Post-Colonial World – Interfaith Perspectives* (Adelaide, Australia: ATF Press, 2005, p. 45).

58 See, for instance, the United Nations Development Program's reports on the state of human development in the Arab world: 2002 (available at http://www.undp.org/arabstates/ahdr2002.shtml) and 2003, 'Building a knowledge society' (available at http://www.undp.org/arabstates/ahdr2003.shtml).

59 See, for example, the American Association for the Advancement of Science news release of 18 February 2005 at http://www.aaas.org/news/releases/2005/0218global.shtml.

four

1 Seyyed Hossein Nasr, in a lecture given to MIT students in November 1991; the transcript, with some reproduction errors, is available at http://web.mit.edu/mitmsa/www/NewSite/libstuff/nasr/nasrspeech1.html. See also William Chittick's (supportive) commentary in *Science of the Cosmos, Science of the Soul: The Pertinence of Islamic Cosmology in the Modern World* (Oxford, UK: Oneworld, 2007, pp. 99–101).

2 For example, 'Truly fear Allah those among His Servants who have knowledge' (Q 35:28).

3 Quoted in Nasr: *Science and Civilization in Islam*.

4 Cited by Golshani, Min al-*'Ilm al-'Ilmani ila al-'Ilm al-Diniy* (*From Secular Science to Theistic Science*) (Beirut, Lebanon: Dar al-Hady, 2003, p. 126).

5 Golshani: *From Secular Science To Theistic Science*, p. 265.

6 Sardar, Ziauddin, 'Rescuing Islam's universities', in Ehsan Masood (ed), *How do you know?* (London: Pluto Press, 2006, p. 45).

7 Sardar, Ziauddin, 'Islamization of knowledge or westernization of Islam', *Inquiry* 7 (1984), pp. 39–45.

8 Anees, Munawar, 'Illuminating Ilm', in Z. Sardar (ed), *How We Know: Ilm and the Revival of Knowledge* (London: Grey Seal, 1991, p. 10).

9 In 'Dewey departs: ideas on classifying knowledge', in Ehsan Masood (ed), *How Do You Know?* (London: Pluto Press, 2006, p. 114), Ziauddin Sardar cites Franz Rosenthal in his paper titled 'Muslim definitions of knowledge' and in his book, *Knowledge Triumphant*.

10 Golshani: *From Secular Science to Theistic Science*, pp. 85–6.

11 Sardar: 'Muslims and philosophy of science', first published in the Routledge Encyclopedia of Philosophy (Routledge, 1997), and reproduced in Ehsan Masood (ed), *How Do You Know?* (London: Pluto Press, 2006, p. 112).

12 Abu Hamid al-Ghazzali in Prolegomena to the Sciences, cited by Golshani: *From Secular Science to Theistic Science*, p. 22.

13 Abu Hamid al-Ghazzali, Deliverance from Error and Mystical Union with the Almighty (A translation of *al-Munqidh min al-Dalal* by Muhammad Abulaylah) George F. McLean (ed) (Washington, DC: Council for Research in Values & Philosophy, 2002); available at http://www.ghazali.org/books/md/index.html

14 Ibn Taymiyya in the grand collection of his epistles, cited by Golshani: *From Secular Science to Theistic Science*, pp. 23–4.

15 al-Dassouqi, F. A., *al-Islam wal 'Ulum al-Tajribiyyah* (*Islam and Experimental Science*) (Beirut, Lebanon: al-Maktab al-Islamiy & Riyadh, Saudi Arabia: Maktabat al-Khani, 1987, p. 104).

16 al-Dassouqi: *Islam and Experimental Science*, p. 104.

17 al-Dassouqi: *Islam and Experimental Science*, p. 105.

18 Khaldun, Ibn, Al Muqaddima (Prolegomena), cited by Golshani: *From Secular Science to Theistic Science*, p. 24.

19 Nasr himself stated that he coined the term 'Islamic Science' in the aforementioned lecture given to MIT students in November 1991 (available at http://web.mit.edu/mitmsa/www/NewSite/libstuff/nasr/nasrspeech1.html).

20 http://web.mit.edu/mitmsa/www/NewSite/libstuff/nasr/nasrspeech1.html.

21 These are what S. P. Loo refers to as 'the four hosemen of Islamic science', *Int. J. Sci. Educ.*, 18/3 (1996), pp. 285–94.

22 Bagir: *Islam, Science, and 'Islamic Science'*, pp. 40–1.

23 Nasr, Seyyed Hossein, 'Islam and the problem of modern science,' in Ziauddin Sardar (ed), *An Early Crescent: The Future of Knowledge and the Environment* (London: Mansell, 1989, p. 132).

24 Nasr in Sardar (ed): *An Early Crescent*, p. 132.

25 Nasr, Seyyed Hossein, *Science and Civilization in Islam* (Cambridge: Harvard Univ. Press, 1987, p. 22).

26 Kalin, Ibrahim, 'The sacred versus the secular: Nasr on science', in L. E. Hahn, R. E. Auxier and L. W. Stone (eds), *Library of Living Philosophers: Seyyed Hossein Nasr* (Chicago, IL: Open Court Press, 2001, pp. 445–62).

27 Nasr: *Islam and Contemporary Society* (London: Longman, 1982).

28 Kalin: 'The sacred versus the secular: Nasr on Science', op. cit. Kalin comments that 'the distinction between reason and intellect on the one hand, and their unity at a higher level of consciousness on the other, are the two fundamental tenets of the traditional school'.

29 Kalin: 'The sacred versus the secular'.

30 Nasr: *Islam and Contemporary Society*, p. 180.

31 Sardar: 'Islamic science: the way ahead' in *How Do You Know?*, p. 180.

32 Sardar, Ziauddin: *Explorations in Islamic Science* (London: Mansell, 1989, p. 65).

33 Nasr, Seyyed Hossein, *Islamic Science: An Illustrated Study* (London: World of Islam Festival, 1976, p. 193) quoted in Sardar: *Explorations in Islamic Science*, p. 122.

34 Sardar: *Explorations in Islamic Science*, pp. 128–9.

35 Stenberg, Leif 'Seyyed Hossein and Ziauddin Sardar on Islam and science: marginalization or modernization of a religious tradition', *Social Epistemology*, 10/3–4 (1996), pp. 273–87.

36 In *Explorations in Islamic Science*, Sardar wrote: 'The Ijmali position is similar to that of al-Ghazzali; the essence of Ijamali thought is reconstruction, complexity, and interconnection'.

37 Sardar, Ziauddin, *An Early Crescent: The Future of Knowledge and the Environment in Islam* (London: Mansell, 1989, p. 27).

38 Sardar, Ziauddin, *Desperately Seeking Paradise: Journeys of a Skeptical Muslim* (London: Granta books, 2004, p. 201).

39 Sardar: *Desperately Seeking Paradise*, p. 197.

40 al-'Alwani, Taha Jabir, 'Islamization of knowledge: premises, challenges, and perspectives', *IslamOnline*, 2004, available at http://www.islamonline.net/english/Contemporary/2004/05/Article01.shtml.

41 al-'Alwani: 'Islamization of Knowledge'.

42 Sardar: 'Islamization of Knowledge or Westernization of Islam', pp. 39–45.

43 See, for instance, Ibrahim Abderrahman Rajab, 'The 'contemporary Muslim' and the question of Islamization of knowledge,' 2003, available at http://www.islamonline.net/muasir/arabic/shahadat3.asp; and Abdekadir Guellati, 'Islamic modernity and the intrinsic renewal of the religious discourse: Part 3 – the genesis of the Islamization of knowledge', 2003, available at http://www.islamonline.net/Arabic/contemporary/2003/09/article01_c.shtml

44 al-'Alwani, Taha Jabir, 'Islamization of attitudes and practices in science and technology', available at http://www.amse.net/Islamization/2(Introduction).htm

45 al-'Alwani: 'Islamization of attitudes and practices in science and technology'.

46 See, for example, Nasr Hamid Abu Zayd, 'The Islamization of sciences and arts leads to a salafi discourse', *The Left* magazine 1 (March 1990), pp. 82–3. See also Aziz

al-Azmeh, Aslamat al-Ma'rifah wa Jumuh al-La'aqlaniyya as-Siyasiyyah ('The Islamization of knowledge and the hordes of political irrationalism') in *al-'Usuliyyat al-Islamiyyah
fi 'Asrina ar-Rahin* (*Islamic Fundamentalisms in Our Time*) (Cairo: Qadaya Fikriyya, 1993);
Ali Harb, 'Islamiyyat al-Ma'rifa wa Talghim al-Mashru' al-Hadariy' ('The Islamization of
knowledge and the mine-laying of the civilizational project'), *al-Hayat* newspaper 13997
(12 July 2001), p. 10.

47   al-Azmeh: 'The Islamization of knowledge and the hordes of political irrationalism'.

48   Harb: 'The Islamization of knowledge and the mine-laying of the civilizational project'.

49   Sardar, Ziauddin, 'Islamization of knowledge or westernization of Islam?', *Inquiry* 7 (1984),
pp. 39–45.

50   Sardar, Ziauddin, 'Islamization of knowledge: a state-of-the-art report', in Z. Sardar (ed)
An Early Crescent: The Future of Knowledge and the Environment in Islam (London: Mansell,
1989).

51   Tibi, Bassam, 'Culture and knowledge: the politics of Islamization of knowledge as a
post-modern project', in 'The fundamental claim to de-westernization', *Theory, Culture
and Society* 12 (1995), pp. 1–24 (Thousand Oaks: Sage).

52   al-Marzouqi, Abu Yaroub, 'Islamization of knowledge: a contrary view, an intervention
with Dr. Louay Safi', *Islamization of Knowledge* 14 (Winter 1998), pp. 139–66.

53   al-'Alwani: 'Islamization of knowledge'.

54   Rajab: 'The "contemporary Muslim" and the question of Islamization of knowledge'.

55   Sardar: *Desperately Seeking Paradise*, pp. 200–1.

56   Sardar: 'Islamization of knowledge or westernization of Islam?'

57   'A Step towards reevaluating the attitude of a Muslim scientist', available at
http://www.amse.net/Islamization/1(Introduction).htm

58   al-'Alwani: 'Islamization of attitudes and practices in science and technology'.

59   al-'Alwani: 'Islamization of attitudes and practices in science and technology', for all the
quotes of the present paragraph.

60   Sardar: *Explorations in Islamic Science*, p. 112.

61   Sardar, 'Islamic science; the way ahead', in Ehsan Masood (ed), *How Do You Know?* (London:
Pluto Press, 2006, pp. 187–88).

62   Sardar: 'Islamic Science; the Way Ahead', p. 183.

63   This list, from Munawar Anees' 'What Islamic science is NOT', *MAAS Journal of Islamic
Science* 2/1 (1984), pp. 9–19, is reproduced by Sardar in 'Islamic Science; the way ahead',
How Do You Know? p. 182.

64   Sardar: *Muslims and Philosophy of Science*, op. cit., p. 113. See also Gregg De Young,
'Should science be limited? Some modern Islamic perspectives', *Monist*, 79/2 (April 1996),
00269662.

65   Hoodbhoy, Pervez, *Islam and Science: Religious Orthodoxy and the Battle for Rationality* (London
and New Jersey: Zed Books, 1991, p. 75).

66   Stenberg, Leif, *Islamization of Science: Four Muslim Positions Developing an Islamic Modernity*,
op. cit., pp. 83–93.

67   Davies, Merryl Wyn, *Knowing One Another: Shaping an Islamic Anthropology* (London:
Mansell, 1988).

68   Stenberg: *Islamization of Science*, p. 84.

69   Stenberg: *Islamization of Science*, p. 96.

70   Stenberg: *Islamization of Science*, p. 96.

71   Sardar: *Desperately Seeking Paradise*, p. 328.

72   Wishart, Adam, *One in Three* (London: Profile Books, 2006, p. 249).

73   Sardar: 'Muslims and philosophy of science', in *How Do You Know?* p. 113.

74 Jamison, Andrew, 'Western science in perspective and the search for alternatives', in Jean-Jacques Salomon et al. (eds), *The Uncertain Quest: Science, Technology and Development* (Tokyo: United Nations Press, 1994, pp. 131–67).

75 The Nobel Prize in Physics 1979 – Banquet Speech. Available at http://nobelprize.org/nobel_prizes/physics/laureates/1979/salam-speech.html

76 Lai, C. H. (ed), *Ideals and Realities: Selected Essays of Abdus Salam* (Singapore: World Scientific, 2nd ed., 1987, pp. 179–213).

77 Sardar: *Explorations in Islamic Science*, pp. 5 and 25.

78 Sardar: *Explorations in Islamic Science*, p. 25.

79 Salam: 'Islam and science', in *Ideals and Realities: Selected Essays of Abdus Salam*, pp. 179–213.

80 Salam: 'Islam and science'.

81 Salam: 'Islam and science'.

82 Sardar: *Islamic Science; The Way Ahead*, p. 179.

83 For instance, Pervez Hoodbhoy, 'Muslims and the West after September 11', *Free Inquiry* 22/2 (Spring 2002).

84 See, for instance, his recent article 'Science and the Islamic world – the quest for rapprochement', *Physics Today* (August 2007), p. 49.

85 Hoodbhoy: *Islam and Science*, p. 2.

86 Hoodbhoy: *Islam and Science*, p. 137.

87 Hoodbhoy: *Islam and Science*, pp. 77–80.

88 Hoodbhoy: *Islam and Science*, p. 137.

89 The Beatles, Eleanor Rigby.

90 Sardar: *Explorations in Islamic Science*, p. 134.

91 Sardar: *Explorations in Islamic Science*, p. 166.

92 Sardar: *Explorations in Islamic Science*, p. 45.

93 Sardar: *Explorations in Islamic Science*, p. 45.

94 Sardar: *Islamization of Knowledge or Westernization of Islam*, p. 40.

95 Sardar: *Islamization of Knowledge or Westernization of Islam*, p. 157.

96 Sardar, *Thomas Kuhn and the Science Wars* (London: Icon books & Pennsylvania: Totem books, 2000, p. 65).

97 S. O. Funtowicz & J. R. Ravetz, quoted in Sardar: *Thomas Kuhn and the Science Wars*, pp. 64–5.

98 Sardar: *Explorations in Islamic Science*, p. 158.

five

1 Quoted by Muhammad *'Amarah, al-Islam wa Qadaya al-'Asr (Islam and Contemporary Issues)* (Beirut, Lebanon: Dar al-Wahdah, 1980, p. 75).

2 Hassab-Elnaby, Mansour, 'A New Astronomical Qur'anic Method for The Determination Of The Greatest Speed c', available at http://www.islamicity.com/Science/960703A.SHTML

3 al-Naggar, Zaghloul, *Min ayat al-I'jaz al-'Ilmiy – As-Samaa' fi l-Qur'an al-Kareem (Scientifically Miraculous Cosmic Verses of the Glorious Qur'an)* (Beirut, Lebanon: Dar al-Ma'rifa, 4th edition, 2007).

4 al-Naggar: *Scientifically Miraculous Cosmic Verses of the Glorious Qur'an*, p. 162.

5 Ben Salem, Kamel, 'Une nouvelle approche pour l'estimation des âges de la terre et de l'univers,' available at: http://piges.2presse.com/Une-nouvelle-vision-de-l-evolution

6 al-Naggar: *Scientifically Miraculous Cosmic Verses of the Glorious Qur'an*, pp. 130–135.

7   'Say: Is it that ye deny Him Who created the earth in two Days? And do ye join equals with Him? He is the Lord of (all) the Worlds. He set on the (earth), mountains standing firm, high above it, and bestowed blessings on the earth, and measure therein all things to give them nourishment in due proportion, in four Days, in accordance with (the needs of) those who seek (Sustenance). Moreover He comprehended in His design the sky, and it had been (as) smoke: He said to it and to the earth: 'Come ye together, willingly or unwillingly'. They said: 'We do come (together), in willing obedience'.

8   al-Taftanazi: *Qur'anic I'jaz in Light of the Modern Scientific Discoveries*, p. 171.

9   Bennabi, Malek, *al-Dhahirah al-Qur'aniyah* (*The Qur'anic Phenomenon*) (Damascus, Syria: Dar al-Fikr, 6th edition, 2006, p. 291).

10   http://www.ikitab.com/index.php3?menu=describe&table=Book&cat=Books&id= 3730&associate=

11   Dahrouch, Sana, 'La Génétique dans l'honorable Tradition Prophétique', available at http://www.55a.net/firas/french/?page=show_det&id=179

12   Syed, Ibrahim B. 'Qur'an and science', available at http://www.irfi.org/articles/ articles_1_50/Qur'an_and_science.htm

13   Gulen, M. Fethullah, 'Connecting science and religion in the Qur'an', *Science & Spirit*, available at http://www.science-spirit.org/article_detail.php?article_id=49.

14   al-Jamili, As-Sayyid, *al-I'jaz al-'Ilmiy fil Qur'an* (*Scientific Miraculousness of the Qur'an*) (Beirut, Lebanon: Dar al-Fikr al-'Arabi, 1st edition, 2002).

15   http://www.ahl-alquran.com/arabic/printpage.php?main_id=402&doc_type=1

16   http://www.elnaggarzr.com/index.php?l=ar&id=96&p=2&cat=6

17   http://www.nooran.org/Q/14.htm; http://www.kaheel7.com/modules.php?name= News&file=article&sid=633

18   Sardar: *Desperately Seeking Paradise*, p. 329.

19   http://www.Qur'an.gov.ae/content/view/43/49/lang,english/

20   *Gulf News*, 28 September 2006: http://archive.gulfnews.com/articles/06/09/28/10070739 .html. Zaghloul al-Najjar was described as a scholar, 'a prominent figure in scientific miraculousness of the Qur'an', 'the chairman of the Committee of Scientific Miraculousness in the Holy Qur'an' and 'a member of the Supreme Council of Islamic Affairs in Egypt'.

21   al-Naggar, Zaghloul, *Min Ayat al-I'jaz al-'Ilmiy fi l-Qur'an* (*On the Scientifically Miraculous Verses of the Qur'an*) (Cairo, Egypt: Maktabat al-Shurouq al-Duwaliyya, 2003, p. 72).

22   For example, al-Taftanazi: *Qur'anic I'jaz in Light of the Modern Scientific Discoveries*, pp. 112–14.

23   Wielandt, Rotraud, 'Exegesis the Qur'an: early modern and contemporary', in J. D. McAuliffe (ed.) *The Encyclopaedia of the Qur'an* (Leiden-Boston: Brill, vol. 2, 2002, pp. 124–41).

24   Jalees Rehman is a cardiology fellow at Indiana University School of Medicine.

25   Rehman, Jalees, 'Searching for scientific facts in the Qur'an: Islamization of knowledge or a new form of scientism?', *Islam & Science* (Dec. 2003).

26   Rehman: 'Searching for scientific facts in the Qur'an'.

27   Khan, Waheed ad-Deen, *al-Islamu yatahadda* (*Islam Challenges*) (Cairo, Egypt: Dar al-Buhuth al-'Ilmiyya, 2nd edition, 1973, p. 141).

28   ar-Razi, Fakhr ad-Din, *At-tafsir al-Kabir* (*The Great Commentary*) (Beirut, Lebnon: Dar al-Fikr, 1993, vol. 14, p. 278).

29   Wielandt: *Exegesis the Qur'an*, op. cit.

30   al-Musili, Sami Ahmed, *al-I'jaz al-'Ilmiy fi l-Qur'an: ta'sil fikriy wa tarikh wa manhaj* (*Scientific Miraculousness of the Qur'an: Conceptual Grounding, History, and Methodology*) (Beirut, Lebnon: Dar al-Nafa'is, 2001).

31 Mir, Mustansir, 'Scientific exegesis of the Qur'an – a viable project?', *Islam & Science* (Summer 2004).

32 As related by Abdallah ibn Mas'ud and narrated by al-Haakim.

33 As related by Abu Ali and as-Sutiy.

34 For instance, it has been narrated that the Prophet said: 'Each prophet has been given "signs" (or miracles), and what I have been given is the Revelation (the Qur'an)'.

35 al-Ghazali, Ihya' *'Ulum al-Deen (The Revival of the Religious Sciences)* (Beirut, Lebnon: Dar al-Fikr, 1995 edition, vol. 1, p. 272).

36 al-Ghazali, *Jawahir al-Qur'an (Pearls of the Qur'an)* (Beirut, Lebnon: Dar Ihya 'Ulum ad-Din, 2nd edition, 1996, p. 23).

37 Dallal, Ahmad, 'Science and the Qur'an', in J. D. McAuliffe (ed), *The Encyclopaedia of the Qur'an* (Leiden-Boston: Brill, vol. 4, 2004, pp. 540–57).

38 Talbi, Mohamed and Bucaille, Maurice, *Réflexions sur le Coran (Reflections on the Qur'an)* (Paris: Seghers, 1989).

39 Talbi and Bucaille: *Réflexions sur le Coran*, pp. 173–4.

40 Moore, Keith 'A scientist's interpretation of references to embryology in the Qur'an', *The Journal of the Islamic Medical Association*, 18 (Jan–Jun 1986), pp. 15–16; available at http://www.islam101.com/science/embryo.html

41 Qur'an 23:13–14, 39:6, 32:9, 22:5.

42 Moore: 'A scientist's interpretation of references to embryology in the Qur'an'.

43 See, for instance, *Scientific Facts Revealed In The Glorious Qur'an* (Beirut, Lebnon: Dar al-Kutub al-'Ilmiyya, 2006) and *al-I'jaz al-'Ilmiy fi l-Sunna a-Nabawiyya (The Scientific Miracles in the Prophetic Sunna (Hadith)* (Egypt: Nahdet Misr Publ., 7th ed., 2005).

44 al-Naggar: *Scientifically Miraculous Cosmic Verses of the Glorious Qur'an* (pp. 20–22).

45 http://www.nooran.org/O/15/15-8.htm

46 http://www.nooran.org/A/a4.htm

47 Az-Zindani, Abdul-Majid, *al-Mu'jiza al-'Ilmiyya fi l-Qur'an wa l-Sunna (The Scientific Miracle in the Qur'an and in the Sunna)* Cairo, n.d.

48 See, for instance, Taner Edis, 'Qur'an-science: scientific miracles from the 7th century?', available at http://www2.truman.edu/~edis/writings/articles/Qur'an-science.html

49 For example, Morris, Henry M., *Biblical Creationism* (Arkansas: Master Books, 2000); Is the Big Bang Biblical: And 99 Other Questions With John Morris (Arkansas: Master Books, 2003).

50 Pervez Hoodbhoy reports that the respected Indian astrophysicist J. V. Narlikar has pointed out that 'during the time that the Steady State theory of creation of the universe was in vogue, abundant scriptural evidence was gathered by religious Hindus to show how this was in perfect accord with the Vedas. Alas! this theory was eventually discredited and replaced by the Big Bang theory of creation. Not the least discomforted, Hindu fundamentalists quickly found other Vedic passages, which were in perfect accord with the newer theory and again proudly acclaimed it as a triumph of ancient wisdom' (Hoodbhoy, 1991, pp. 66–7).

51 al-'Arifi, Saud bin Abdelaziz, Menhaj al-Istidlal bil-Muktashafat al-'Ilmiyya 'ala-n-Nubuyya wa-r-Rububiyya – Dirasa Naqdiyya ('The methodology of arguing for the prophecy (of Muhammad) and for the (existence of) the divine on the basis of scientific discoveries – a critical study'), Majallat Umm al-Qura li 'Ulum a-Shari'ah wa-l-Lugha al-'Arabiyya wa 'Adabiha, 19/43, Dhul Hijja 1428 (Dec. 2007), p. 287.

52 al-Naggar, op. cit., pp. 69–70.

53 al-Naggar: *Scientifically Miraculous Cosmic Verses of the Glorious Qur'an*, p. 100.

54 al-Naggar: *Scientifically Miraculous Cosmic Verses of the Glorious Qur'an*, p. 50.

55 al-Naggar: *Scientifically Miraculous Cosmic Verses of the Glorious Qur'an*, p. 50.

56 al-Naggar: *Scientifically Miraculous Cosmic Verses of the Glorious Qur'an*, p. 69.

57 al-Naggar: *Scientifically Miraculous Cosmic Verses of the Glorious Qur'an*, p. 51.

58 al-Naggar: *Scientifically Miraculous Cosmic Verses of the Glorious Qur'an*, p. 117.

59 al-Naggar: *Scientifically Miraculous Cosmic Verses of the Glorious Qur'an*, p. 116.

60 al-Naggar: *Scientifically Miraculous Cosmic Verses of the Glorious Qur'an*, pp. 211–32.

61 al-Taftanazi: *Qur'anic I'jaz in Light of the Modern Scientific Discoveries*, p. 216.

62 al-Taftanazi: *Qur'anic I'jaz in Light of the Modern Scientific Discoveries*, p. 303.

63 al-Taftanazi: *Qur'anic I'jaz in Light of the Modern Scientific Discoveries*, p. 296.

64 al-Shatibi, Ibrahim, *al-Muwafaqat (The Concordances)*, Muhammad al-Khidr Husayn al-Tunisi, al-Matba'a al-Salafiyya (ed), 1922, vol. 2, pp. 80–1.

65 As quoted by Bustami Mohamed Khir, 'The Qur'an and science: the debate on the validity of scientific interpretations', *Journal of Qur'anic Studies*, 2/2, (Dec. 2000), pp. 19–35.

66 al-Naggar, Zaghloul, *al-'I'jaz al-'Ilmiy fi s-Sunna n-Nabawiyya (The Scientific I'jaz in the Prophetic Tradition (Sunna))* (Egypt: Nahdet Misr Publ., vol. 1 (7th ed.) and vol. 2 (5th ed.), 2005).

67 Abu Sway, Mustapha, 'Modern science and the hermeneutics of the "scientific interpretation" of the Qur'an', available at www.fau.edu/~mabusway/ScienceQur'anandHermeneutics.pdf

68 Khir: *The Qur'an and Science*, p. 33.

69 http://www.geocities.com/masad02/021

70 http://www.tafsir.com/default.asp?sid=8&tid=19643

71 http://www.geocities.com/masad02/008

72 http://www.geocities.com/masad02/appendix4

73 Campanini, Massimo, 'Qur'an and science: A hermeneutical approach', *Journal of Qur'anic Studies* 7/1, 2005, p. 58.

six

1 Ferguson, Kitty, *The Fire in the Equations – Science, Religion, and the Search for God* (Grand Rapids, MI: Eerdmans, 1994, p. 86).

2 Iqbal, Muzaffar, *Science and Islam* (Westport, CT & London: Greenwood Press, 2007, p. 30).

3 Hassan al-Hilly, Mohammed-Ali, *The Universe and the Holy Qur'an* (Beirut: Dar al-Kotob al-Ilmiyah, 2006, p. 25).

4 al-Hilly: *The Universe and the Holy Qur'an*, p. 13.

5 al-Hilly: *The Universe and the Holy Qur'an*, p. 12.

6 al-Hilly: *The Universe and the Holy Qur'an*, p. 59.

7 Taleb, Hichem, *Bina' al-Kawn wa Maseer al-'Insan, naqd li-naDhariyat al-'infijar al-kabir, haqa'iq mudhhila fil-'ulum al-kaqniyya wa-diniyya (The Structure of the Universe and the Destiny of Man – A Critique of the Big Bang Theory; Amazing Facts in the Cosmic and Religious Sciences)* (Dar al-Ma'rifah, Beirut: Dar al-Ma'rifah, 2006, pp. 189–93).

8 Taleb: *The Structure of the Universe and the Destiny of Man*, pp. 201-03.

9 Taleb: *The Structure of the Universe and the Destiny of Man*, pp. 323+.

10 Taleb: *The Structure of the Universe and the Destiny of Man*, p. 332.

11 http://www.youtubeislam.com/view_video.php?viewkey=b1282ba735e5e3723ed7

12 al-Mumin, A. Abdelamir, *al-Kawn, naDhratun turathiyyah wa mu'asirah (The Universe, a Traditional and Modern Look)* (Beirut: Dar al-Hujja al-Bayda, 2006, p. 126).

13 al-Naimiy, Hamid M., *al-Kawn wa Asraruh fi Ayat al-Qur'an al-Kareem (The Universe and Its Secrets in the Holy Qur'an)* (Amman, Jordan: Maktabat al-Ra'id al-'Ilmiyya, & Beirut, Lebanon: ad-Dar al-'Arabiyya li l-'Ulum, 2000).

14   al-Naimiy: *The Universe and its Secrets in the Holy Qur'an*, p. 90.

15   al-Naimiy: *The Universe and its Secrets in the Holy Qur'an*, p. 369.

16   In English in the (Arabic) text.

17   al-Naimiy: *The Universe and its Secrets in the Holy Qur'an*, p. 362.

18   In English in the (Arabic) text.

19   al-Naimiy: *The Universe and its Secrets in the Holy Qur'an*, pp. 93–4.

20   al-Naimiy: *The Universe and its Secrets in the Holy Qur'an*, p. 387.

21   al-Naimiy: *The Universe and its Secrets in the Holy Qur'an*, p. 101.

22   al-Naimiy: *The Universe and its Secrets in the Holy Qur'an*, p. 97.

23   al-Naimiy: *The Universe and its Secrets in the Holy Qur'an*, p. 354.

24   al-Naimiy: *The Universe and its Secrets in the Holy Qur'an*, p. 388.

25   al-Naimiy: *The Universe and its Secrets in the Holy Qur'an*, p. 386.

26   al-Naimiy: *The Universe and its Secrets in the Holy Qur'an*, p. 373.

27   al-Naimiy: *The Universe and its Secrets in the Holy Qur'an*, p. 374.

28   al-Naimiy: *The Universe and its Secrets in the Holy Qur'an*, p. 378.

29   'He Who created the seven heavens one above another: No want of proportion wilt thou see in the Creation of (Allah) Most Gracious. So turn thy vision again: seest thou any flaw?' (Q 67:3) and 'Do you not see how Allah has created the seven heavens, one above another' (Q 71:15).

30   al-Naimiy: *The Universe and its Secrets in the Holy Qur'an*, p. 146.

31   Nasr, *Introduction to Islamic Cosmological Doctrines* (New York: SUNY Press, revised edition, 1993, p. 275); Iqbal: *Science and Islam*, p. 10.

32   Iqbal: *Science and Islam*, p. 82.

33   Nasr, *Sciences et Savoir en Islam (Science and Knowledge in Islam)* (Paris: Sinbad, 1979, p. 121).

34   Qur'an 2:255.

35   Qur'an 24:35.

36   Iqbal: *Science and Islam* (pp. 30–1, 80); Nasr: *Cosmological Doctrines* (p. xiv).

37   They ask thee about the (final) Hour: when will be its appointed time? Say: 'The knowledge thereof is with my Lord (alone); none but He can reveal as to when it will occur. Heavy were its burden through the heavens and the earth; only all of a sudden will it come to you'. They ask thee as if thou wert eager in search thereof; say: 'the knowledge is with Allah (alone), but most men know not'.

38   Nasr: *Cosmological Doctrines*, p. 4.

39   Nasr, Seyyed H., 'Islamic cosmology – some of its basic tenets and implications, yesterday and today', *J. Islamic Sci.* 14/1–2 (1998), pp. 99–114.

40   Nasr: *Cosmological Doctrines*, p. 277.

41   Nasr: *Sciences et Savoir en Islam*, p. 120.

42   Nasr: *Cosmological Doctrines*, p. 277.

43   Chittick, William, *Science of the Cosmos, Science of the Soul: The Pertinence of Islamic Cosmology in the Modern World* (Oxford, UK: Oneworld, 2007, p. 31).

44   Iqbal: *Science and Islam*, p. 30.

45   Chittick: *Science of the Cosmos, Science of the Soul*, p. 141.

46   Chittick: *Science of the Cosmos, Science of the Soul*, p. 122.

47   Chittick: *Science of the Cosmos, Science of the Soul*, pp. 131, 136.

48   Chittick: *Science of the Cosmos, Science of the Soul*, p. 21.

49   Armstrong: *A History of God*, p. 174.

50   Armstrong: *A History of God*, p. 175.

51   Armstrong: *A History of God*, p. 181.

52   Nasr: *Cosmological Doctrines*, pp. 275–6.

53   Nasr: *Cosmological Doctrines*, pp. 5, 13.

54 Chittick, William C., *Man the Macrocosm*, Fourth Annual Symposium Of Muhyiddin Ibn 'Arabi Society, Jesus College, Oxford, 1987.

55 There is much disagreement, of course, on when the decline of the Islamic civilization really started and how fast (or slowly) that occurred. The Mongol sack of Baghdad (in 1258) is often mentioned as a great turning point (or collapse) for the Islamic civilization, though many scholars have pointed out that astronomy and medicine, in particular, continued on, with major observatories (Maragha, Samarqand, Istanbul) and scholars (al-Tusi, Ibn al-Shatir) appeared much later.

56 Allow me to mention in passing my own modest effort in this regard: *Qissat al-Kawn* (*The Story of the Universe*') (Algiers: Dar al-Maarifa, 1998 (1st edition) and 2002 (2nd edition); Beirut: Dar al-Multaqa, 2006).

57 Khudr, Abdulrahman, *Man in the Universe, in the Qur'an and in Science* (Saudi Arabia: Saudi Publishing House, 1983, p. 139).

58 Khudr, Abdulrahman, *Man in the Universe, in the Qur'an and in Science* (Saudi Arabia: Saudi Publishing House, 1983, p. 139).

59 Guiderdoni, Bruno, *Proceedings of the Workshop 'Science and the Spiritual Quest'*, R. J. Russell et al. (eds), 2000, in press; also available on http://www.geocities.com/siriusalgeria/cosmology.htm

60 Guiderdoni, Bruno, 'The exploration of the cosmos: an endless quest?', draft paper from the 'God, Life and Cosmos: Theistic Perspectives' conference, 6–9 Nov. 2000, Islamabad, Pakistan; text available at http://www.kalam.org/papers/bruno.htm

61 Guiderdoni: The exploration of the cosmos: an endless quest?

62 See, for example, NASA's 'WMAP Introduction to Cosmology' at http://map.gsfc.nasa.gov/m_uni.html

63 http://www.astro.ucla.edu/~wright/CMB-MN-03/Munich-07Nov05-clean.pdf

64 Primack, Joel & Abrams, Nancy Ellen, *The View from the Center of the Universe* (London: Fourth Estate, 2006, p. 6).

65 Primack & Abrams: *The View from the Center of the Universe*, p. 78.

66 Primack & Abrams: *The View from the Center of the Universe*, p. 8.

67 Primack & Abrams: *The View from the Center of the Universe*, p. 11.

68 Chittick: *Science of the Cosmos, Science of the Soul*, p. 69

69 Primack & Abrams: *The View from the Center of the Universe*, p. 41.

70 Heller, Michael, 'Can the Universe Explain Itself?', in W. Löffler & P. Weingartner (eds), *Knowledge and Belief (Wissen und Glauben)*, Proceedings of the 26th International Wittgenstein Symposium, Kirchberg, Austria, August 2003 (Vienna, Austria: Osterreichischer Bundesverlag/Holder-Pichler-Tempsky, 2004, pp. 316–28).

71 McMullin, Ernan, 'Is Philosophy relevant to Cosmology?', *American Philosophical Quarterly* 18/3 (July 1981), pp. 177–89.

72 Hawking, Stephen, *A Brief History of Time* (New York: Bantam Books, 1988).

73 Hartle, J. B. & Hawking, S. W., 'Wave function of the Universe', *Physical Review D*, 28 (1983), p. 2960.

74 Primack & Abrams: *The View from the Center of the Universe*, p. 276.

75 Primack & Abrams: *The View from the Center of the Universe*, p. 276.

76 Primack & Abrams: *The View from the Center of the Universe*, pp. 276–7.

77 Nasr, Seyyed Hossein, 'The question of cosmogenesis – the cosmos as a subject of scientific study', *Islam & Science* 4 (Summer 2006), p. 43.

78 Peacocke, Arthur, *Pathways from Science Towards God* (Oxford, U.K.: Oneworld, 2001, p. 1).

79 Peacocke: *Pathways from Science Towards God*.

80 Primack & Abrams: *The View from the Center of the Universe*, p. 295.

81 Primack & Abrams: *The View from the Center of the Universe*, p. 16.

82 Lampman, Jane, *The Christian Science Monitor*, July 9, 1998.

83 Sacks, Gordy, 'Faith in the Universe', *California Wild* (Summer 1998); available at http://www.calacademy.org/calwild/sum98/skywatch.htm

seven

1 Rushd, Ibn, *al-Kashf 'an Manahij al-Adillah fi 'aqa'id ahl al-milla (Unveiling the Methods of Proof in the Dogmas of Muslim Schools)* (Beirut: Markaz Dirassat al-Wihdah al-'Arabiyya, 2nd edition, 2001).

2 Hume, David, in *Dialogues Concerning Natural Religion* cited by Michael Ruse, 'The argument from design: A brief history', in W. A. Dembski and M. Ruse (ed.), *Debating Design: from Darwin to DNA*, pp. 13–31 (New York: Cambridge Univ. Press, 2004, p. 17).

3 Case-Winters, Anna, 'The argument from design: what is at stake theologically?', *Zygon* 35/1 (March 2000), pp. 69–81.

4 Ruse: 'The argument from design: a brief history', pp. 13–31.

5 Case-Winters: 'The argument from design: what is at stake theologically?', p. 71.

6 As cited by Ruse: 'The argument from design: a brief history', p. 23.

7 Ruse: 'The argument from design: a brief history', p. 23.

8 'Thus did We show Abraham the kingdom of the heavens and the earth that he mighty be of those possessing certainty. When the night grew dark upon him he beheld a star; he said: This is my Lord. But when it set, he said: I love not things that set. And when he saw the moon rising, he exclaimed: This is my Lord. But when it set, he said: Unless my Lord guide me, I surely shall go astray. And when he saw the sun rising, he cried: This is my Lord! This is greater! And when it set, he exclaimed: O my people! I am free from all that ye associate (with Him). Lo! I have turned my face toward Him Who created the heavens and the earth, as one by nature upright, and I am not of the idolaters' (Q 6: 75–79).

9 Aftab, Macksood, 'Groundwork on Islamic Philosophy', *Meteorite* 2 (Summer 2000), pp. 26–42, citing Kindi without reference.

10 Fakhry, Majid, citing al-Baqillani from his *'Tamhid'* Cairo, 1947, p. 45.

11 al-Ghazzali, Abu Hamid, *al-hikmah fi makhluqat-i-Llah (The Wisdom in God's Creation/Creatures)* (Beirut: Dar Ihya' al-'Ulum, 1978).

12 al-Ghazzali: *The Wisdom in God's Creation/Creatures*, pp. 15–16 (my own translations).

13 al-Ghazzali: *The Wisdom in God's Creation/Creatures*, p. 57.

14 al-Ghazzali: *The Wisdom in God's Creation/Creatures*, p. 35.

15 See earlier note.

16 Cited by Ruse: 'The argument from design: a brief history', p. 17.

17 Denton, Michael, *Evolution, a Theory in Crisis* (Bethesday: Adler & Adler, 1985, p. 342.

18 al-Ghazzali: *The Wisdom in God's Creation/Creatures*, pp. 74, 79, 80, 83, 86, 89–93, 99, 105.

19 al-Ghazzali: *The Wisdom in God's Creation/Creatures*, p. 103.

20 al-Ghazzali: *The Wisdom in God's Creation/Creatures*, pp. 111–12.

21 Note: 'Art' is the translators' rendition of *Sinaa'ah* (technique, or industry in modern vocabulary).

22 Ibn Rushd: *al-Kashf*, my translation.

23 Hawking, Stephen, *A Brief History of Time* (New York: Bantam, 1988).

24   Geoffroy, Marc, *Averroès – L'Islam et la raison* (*Averroes – Islam and Reason*), Paris: GF Flammarion, 2000, p. 106.

25   Iqbal, Muhammad, *The Reconstruction of Religious Thought in Islam*, Lahore, Pakistan: Sang-e-Meel 2004 ed., p. 33.

26   Iqbal: *The Reconstruction of Religious Thought in Islam*, p. 34.

27   Iqbal: *The Reconstruction of Religious Thought in Islam*, p. 35.

28   Iqbal: *The Reconstruction of Religious Thought in Islam*, p. 34.

29   Iqbal: *The Reconstruction of Religious Thought in Islam*, p. 35.

30   Yahya, Harun, *Allah is Known Through Reason* (New Delhi, India: Goodword Books, 2003 (first published in 2000), pp. 10, 27, 28.

31   See, for example, Zaghloul El-Naggar, 'Cosmological references in the Noble Qur'ân', available at http://english.islamway.com/bindex.php?section=article&id=102

32   Zaghloul El-Naggar, 'The Cosmic', available at http://www.55a.net/firas/english/?page=show_det&id=63

33   William Dembski is an American mathematician and philosopher; he is with Michael Behe, an American biochemist, one of the two leaders of the Intelligent Design (ID) movement, at least on its scientific front. Behe and Dembski have published a number of books on ID, most notable being Behe's *Darwin's Black Box: The Biochemical Challenge to Evolution* (1996), *The Edge of Evolution: The Search for the Limits of Darwinism* (2007), and Dembski's *The Design Inference: Eliminating Chance Through Small Probabilities* (1998), *Intelligent Design: The Bridge Between Science and Theology* (1999), and *The Design Revolution: Answering the Toughest Questions about Intelligent Design* (2004).

34   Miller, Kenneth R., 'Falling over the edge', *Nature* 447 (28 June 2007), pp. 1055–6; Coyne, Jerry, 'The Great Mutator', *The New Republic* online (14 June 2007) available at http://www.powells.com/review/2007_06_14

35   http://www.riseofislam.com/rise_of_faith_05.html

36   For example, 'Under God or Under Darwin?', his debate with Nicholas Matzke, spokesman for the US-based National Center for Science Education, 'Evolution vs. Intelligent Design'.

37   http://www.thewhitepath.com/archives/2007/03/turkeys_first_intelligent_design_conference.php

38   Akyol, Mustafa, 'Why Muslims should support intelligent design' (14 September 2004) available at http://www.islamonline.net/english/Contemporary/2004/09/Article02.shtml

39   Chikhli, Mohamed, 'Americans lean more toward the theory of intelligent design' (in Arabic), *al-Moustaqbal* (6 November 2005) available at http://www.almustaqbal.com/nawafez.aspx?StoryID=149354; Jalabi, Khalis, 'The Darwinian earthquake and the theory of intelligent design' (in Arabic), *al-Raee* (21 January 2006) available at http://www.arraee.com/modules.php?name=News&file=article&sid=8989; 'Intelligent design vs. Darwinism: a constant conflict' (in English) by the Health and Science staff of IslamOnline (21 December 2005) available at http://www.islamonline.net/servlet/Satellite?c=Article_C&cid=1158321448362&pagename=Zone-English-HealthScience%2FHSELayout

40   http://www.islamonline.net/English/Science/2005/07/article05.shtml (18 July 2005).

41   Rahim, Rizwana, 'Darwin vs. intelligent design-1', Islamic Research Foundation International website (no date given) available at http://www.irfi.org/articles/articles_451_500/darwin_vs_intelligent_design_1.htm

42   http://www.thewhitepath.com/archives/2006/10/turkish_minister_supports_intelligent_design.php

43   Case-Winters: 'The argument from design: what is at stake theologically?', p. 76.

eight

1   John Wheeler, Foreword to John D. Barrow and Frank J. Tipler's *The Anthropic Cosmological Principle* (New York: Oxford University Press, 1986, p. vii).

2   Nasr, Seyyed Hossein, *The Heart of Islam* (SanFrancico, New York: Harper, 2004, p. 243).

3   Leslie, John, (*Universes*, Routledge, 1989). This metaphor has since been adopted and adapted by many authors, e.g. Sir Martin Rees (*Our Cosmic Habitat*, Princeton University Press, 2001) and Jean Staune (*Notre Existence a-t-elle un Sens?* Presses de la Renaissance, 2007), who also cites Hubert Reeves etc.

4   Davies, Paul, *The Goldilocks Enigma: Why is the universe just right for life?* (London: Allen Lane, 2006).

5   Jacques Demaret and Dominique Lambert, *Le Principe Anthropique: L'homme est-il le centre de l'Univers?* (*The Anthropic Principle: Is Man at the Center of the Universe?*) (Paris: Armand Colin, 1994, p. vii).

6   Quoted by Demaret and Lambert: *The Anthropic Principle*, p. vii.

7   Quoted by Demaret and Lambert: *The Anthropic Principle*, p. ix.

8   Michel Paty in Ciel et Espace Hors Série, 2006, p. 68.

9   Barrow and Tipler: *The Anthropic Cosmological Principle*.

10 Wheeler in Barrow and Tipler: *The Anthropic Cosmological Principle*, p. vii.

11 Pochet, T. et al., *Astronomy & Astrophysics*, 243 (1991), p. 1.

12 Barrow and Tipler: *The Anthropic Cosmological Principle*, p. 411.

13 Demaret and Lambert: *The Anthropic Principle*, p. 116.

14 Davies: *The Goldilocks Enigma*, pp. 164–6.

15 Davies: *The Goldilocks Enigma*, pp. 166–70.

16 Davies: *The Goldilocks Enigma*, p. 170.

17 Davies, Paul, *The Cosmic Blueprint: New Discoveries in Nature's Creative Ability to Order the Universe* (New York: Simon and Schuster, 1988, p. 203).

18 Penrose, in Quantum Gravity 2 (C.J. Isham, R. Penrose, D.W. Sciama, eds.) (Oxford, 1981, pp. 248–9).

19 Gonzalez, Guillermo and Richards, Jay, *The Privileged Planet* (Washington, DC: Regnery Publ., 2004).

20 Prantzos, Nikos, 'On the Galactic Habitable Zone' (2006) available at http://arxiv.org/abs/astro-ph/0612316v1

21 Gonzalez and Richards: *The Privileged Planet*, p. 319.

22 Dicke, Robert H., *Reviews of Modern Physics*, 29, 355 (1957).

23 Wallace, Alfred Russel, *Man's Place in the Universe* (pp. 256–7 in the 1912 edition).

24 Barrow and Tipler: *The Anthropic Cosmological Principle*, p. 16.

25 Demaret and Lambert: *The Anthropic Principle*, pp. 138–9.

26 Demaret and Lambert: *The Anthropic Principle*, p. 21.

27 Demaret and Lambert: *The Anthropic Principle*, p. 22.

28 Demaret and Lambert: *The Anthropic Principle*, p. 23.

29 Staune, Jean, *Notre existence a-t-elle un sens? Une enquete scientifique et philosophique* (Presses de la Renaissance, 2007, pp. 177–8).

30 See Polkinghorne, John, *Reason and Reality* (London: Society for Promoting Christian Knowledge, 1991); see also Polkinghorne, 'The Inbuilt Potentiality of Creation', in William A. Dembski and Michael Ruse (eds.), *Debating Design* (Cambridge: Cambridge University Press, 2004, pp. 246–60).

31 Havel, Vaclav, 'The Need for transcendence in the postmodern world', speech given at the Independence Hall, Philadelphia, on 4 July 1994, published in *The Futurist* 29 (July 1995), p. 46.

32 George F. Ellis, cited by Michael D Lemonick and J. Madeleine Nash, 'Cosmic Conundrum', *Time* 164/22 (29 Novembre 2004), p. 58.

33 Smolin, Lee, *The Life of the Cosmos* (Oxford University Press; New Ed edition, 1997).

34 Davies, Paul, 'A brief history of the multiverse', *New York Times* (12 April 2003).

35 Davies: *The Goldilocks Enigma*.

36 Bostrom, Nick, 'The simulation argument: why the probability that you are living in a matrix is quite high,' *Times Higher Education Supplement* (16 May 2003). See also http://www.simulation-argument.com/

37 Davies, in *The Goldilocks Enigma*, p. 194, refers to Manson, Neil (ed.), *God and Design* (London: Routledge, 2003).

38 Dyson, Freeman, *Infinite in All Directions*, Perennial Library (Harper & Row, 1989, pp. 296–7).

39 See, for example, Primack, Joel and Abrams, Nancy Ellen, *The View from the Center of the Universe* (London: Fourth Estate, 2006).

40 Gonzalez and Richards: *The Privileged Planet*, pp. 215–8.

41 Denton, Michael J., *Nature's Destiny: How the Laws of Biology Reveal Purpose in the Universe* (New York: Free Press, 1998, p. 243).

42 Vilenkin, Alexander, *Many Worlds in One* (New York: Hill and Wang, 2006), and 'The Principle of Mediocrity', http://www.edge.org/3rd_culture/vilenkin06/vilenkin06_index.html

43 Davies: *The Goldilocks Enigma*, p. 225.

44 Gonzalez & Richards: *The Privileged Planet*, p. 306.

45 Davies: *The Goldilocks Enigma*, p. 226.

46 Davies: *The Goldilocks Enigma*, p. 226.

47 Davies: *The Goldilocks Enigma*, p. 229.

48 The 12 essay-answers were published as a two-full-pages ad in the *New York Times Week in Review* of 7 October, 2007, paid for by the John Templeton Foundation.

49 Davies: *The Goldilocks Enigma*, pp. 295–303.

50 Adi Setia, 'Taskhir, fine-tuning, intelligent design and the scientific appreciation of nature', *Islam & Science*, Summer 2004.

51 M. Basil Altaie, 'The anthropic principle', http://www.cosmokalam.com/general/articles/anthropic.doc

52 Geoffroy: *Averroes – Islam and Reason*, p. 131, footnote 103 (my translation).

53 Adi Setia: *Taskhir*.

54 Ibn Rushd: al-Kashf, cited by Geoffroy: *Averroes – Islam and Reason*, pp. 139–140.

55 For instance, Jaafar Sheikh Idrees, *al-Fizya' wa wujud al-khaliq* (*Physics and the Existence of the Creator*) (al-Muntada al-Islamiy, 2nd edition, 2001).

56 Idrees: *Physics and the existence of the Creator*, p. 55.

57 Idrees: *Physics and the existence of the Creator*, p. 55.

58 Faheem Ashraf, 'Islamic concept of creation of universe: big bang and science-religion interaction', *Science-Religion Dialogue*, Spring 2003, http://www.hssrd.org/journal/spring2003/creationunverse.htm

59 Ashraf: Islamic concept of creation of universe.

60 Taneli Kukkonen, 'Averroes and the teleological argument', *Religious Studies* 38:4 (2002), pp. 503–528.

61 Barrow and Tipler: *The Anthropic Cosmological Principle*, pp. 46–7.

62 Kukkonen: 'Averroes and the teleological argument'.

nine

1   Polkinghorne, John, 'The Inbuilt Potentiality of Creation', in W. A. Dembski & M. Ruse (eds), *Debating Design* (Cambridge: Cambridge University Press, 2004, p. 256).

2   Ruse, Michael, *The Evolution-Creation Struggle* (Cambridge, Massachussetts & London: Harvard University Press, 2005, p. 70).

3   Ikhwan as-Safa, *Rasa'il (Epistles)* (Beirut: Dar Sadir, 1975, vol. 2, p. 112).

4   al-Qaradawi, Yusuf, 'Bidayat al-khalq wa naDhariyyat at-Tatawwar ('The beginning of creation and the theory of evolution'), al-Shari'ah wal Hayat TV program, 03 March 2009, http://www.qaradawi.net/site/topics/article.asp?cu_no=2&item_no=6842&version=1&template_id=105&parent_id=16

5   For a brief review and summary of the main results of polls since 1991, see: 'Public beliefs about evolution and creation' (http://www.religioustolerance.org/ev_publi.htm) and 'Reading the polls on evolution and creationism – Pew Research Center pollwatch' (http://people-press.org/commentary/display.php3?AnalysisID=118).

6   Recent surveys include Zogby International – results from nationwide poll, 6 March 2006 (http://www.discovery.org/scripts/viewDB/filesDB-download.php?command=download&id=719); Pew Forum on Religion and Public Life: 'Public divided on origins of life', 30 August 2005 (http://pewforum.org/surveys/origins/).

7   Poll conducted from 21–24 November 1991 by the Gallup Organization (http://library.thinkquest.org/29178/gallup.htm).

8   'Public beliefs about evolution and creation' (http://www.religioustolerance.org/ev_publi.htm).

9   Gallup poll, 'Evolution beliefs', 11 June 2007 (http://www.gallup.com/video/27838/Evolution-Beliefs.aspx).

10 See http://www.hcdi.net/News/PressRelease.cfm?ID=93 for a press release on the poll, and http://www.hcdi.net/polls/J5776/ for the results by religious affiliation of the respondents.

11 There have been a few very recent surveys of attitudes towards evolution among the Muslims showing very similar and consistent results. Hamid, Salman 'Bracing for Islamic creationism', *Science* 322 (2008), pp. 1637–8; Hokayem, Hayat and BouJaoude, Saouma, '[Lebanese] college students' perceptions of the theory of evolution', *Journal of Research in Science Teaching* 45/4 (2008), pp. 395–419; Chanet, Bruno and Lusignan, François, 'Teaching evolution in primary schools: an example in French classrooms', *Evo Edu Outreach* 2 (2009), pp. 136–140.

12 http://www.islamonline.net/askaboutislam/display.asp?hquestionID=4881

13 http://www.islamonline.net/fatwa/english/FatwaDisplay.asp?hFatwaID=104342

14 http://www.readingislam.com/servlet/Satellite?cid=1123996016602&pagename=IslamOnline-English-AAbout_Islam/AskAboutIslamE/AskAboutIslamE

15 http://www.islamonline.net/fatwa/english/FatwaDisplay.asp?hFatwaID=104342

16 http://www.readingislam.com/servlet/Satellite?cid=1123996016602&pagename=IslamOnline-English-AAbout_Islam/AskAboutIslamE/AskAboutIslamE

17 http://www.islamonline.net/fatwa/english/FatwaDisplay.asp?hFatwaID=34982

18 http://www.islamonline.net/fatwa/english/FatwaDisplay.asp?hFatwaID=34982

19 All quotes from Zaghloul al-Naggar's lecture on evolution in this section are taken from http://www.isoc-unsw.org.au/main/index.php?option=com_content&task=view&id=22&Itemid=73

20 http://www.kirjasto.sci.fi/bkurten.htm

21 All quotes of Nuh Ha Min Keller here are taken from his article: 'Islam and Evolution', dated 21 November 2005, available at http://www.islamonline.net/english/Contemporary/2005/11/article01.shtml

22 Charlesworth, Brian & Charlesworth, Deborah, *Evolution – A Very Short Introduction* (Oxford and New York: Oxford University Press, 2003, p. 46).

23 Ayala, Francisco, *Darwin's Gift to Science and Religion* (Washington, DC: Joseph Henry Press, 2007, p. 8).

24 Ayala: *Darwin's Gift to Science and Religion*, p. 9.

25 Ayala: *Darwin's Gift to Science and Religion*, p. 53.

26 See Conway-Morris, Simon's seminal book *Life's Solution: Inevitable Humans in a Lonely Universe* (Cambridge: Cambridge University Press, 2003).

27 Charlesworth & Charlesworth: *Evolution – A Very Short Introduction*, p. 29.

28 Charlesworth & Charlesworth: *Evolution – A Very Short Introduction*, p. 103.

29 Miller, Kenneth R., *Only a Theory – Evolution and the Battle for America's Soul* (New York: Viking, 2008).

30 Miller: *Only a Theory*, pp. 97–107.

31 Miller: *Only a Theory*, p. 80.

32 Charlesworth & Charlesworth: *Evolution*, p. 6.

33 Ayala: *Darwin's Gift to Science and Religion*, p. 58.

34 Charlesworth & Charlesworth: *Evolution*, p. 75.

35 Staune, Jean, *Notre existence a-t-elle un sens? Une enquête scientifique et philosophique (Does Our Existence Have a Meaning? A Scientific and Philosophical Investigation)* (Paris: Presses de la Renaissance, 2007).

36 Staune: *Does Our Existence Have a Meaning?*, p. 2

37 Conway-Morris: *Life's Solution*.

38 Ward, 'Theistic evolution', Dembski and Ruse (eds), *Debating Design*, p. 262.

39 Haught, 'Darwin, Design, and Divine Providence', Dembski and Ruse (eds), *Debating Design*, p. 243.

40 Ruse: *The Evolution-Creation Struggle*, p. 231.

41 Haught, 'Darwin, design, and divine providence', p. 240.

42 Haught, 'Darwin, design, and divine providence', p. 241.

43 Haught, 'Darwin, design, and divine providence', p. 242.

44 Polkinghorne, John, 'The inbuilt potentiality of creation', *Debating Design*, op. cit., pp. 246–260.

45 Peacocke, Arthur, *Paths from Science towards God* (Oxford, UK: OneWorld, 2001, p. 77).

46 Peacocke: *Paths from Science towards God*, p. 256.

47 Peacocke: *Paths from Science towards God*, p. 258.

48 Ward, 'Theistic evolution', p. 263.

49 Ward, 'Theistic evolution', p. 262.

50 Ward, 'Theistic evolution', p. 269.

51 Ward, 'Theistic evolution', p. 273.

52 Ayala: *Darwin's Gift to Science and Religion*, p. 103.

53 Ayala: *Darwin's Gift to Science and Religion*, p. 104.

54 Charlesworth & Charlesworth: *Evolution*, p. 103; Ayala: *Darwin's Gift to Science and Religion*, p. 108.

55 Ayala: *Darwin's Gift to Science and Religion*, pp. 105–6.

56 *Science et Vie Hors Series*, no. 235, June 2006, p. 18, referring to a paper in Nature but without full citation.

57 'Hominid' refers to the common lineage of humans, apes, chimpanzees, gorillas etc.

58 http://news.bbc.co.uk/2/hi/science/nature/5363328.stm

59 http://www.aljazeera.net/NR/exeres/54C0A6AA-246F-45DE-A79D-
    2042487364BB.htm

60 al-Bouti is a former dean of the college of Islamic Law in Syria; he is the author of more
    than 60 books.

61 Ramadan, Mohammed Said, *al-Bouti Kubra al-Yaqiniyyat al-Kawniyyah: Wujud al-Khaliq wa
    Wadhifat al-Makhluq* (*The Greatest Cosmic Certainties: the Existence of the Creator and the Role
    of the Created*) (Damascus, Syria: Dar al-Fikr & Beirut, Lebanon: Dar al-Fikr al-Mu'asir,
    2004, reprint of the 8th edition (1982); originally published in 1969).

62 The expression 'Peace be upon him' is spoken when one has mentioned a prophet or
    messenger of God.

63 al-Bouti: *The Greatest Cosmic Certainties*, p. 245.

64 al-Bouti: *The Greatest Cosmic Certainties*, p. 250.

65 al-Bouti: *The Greatest Cosmic Certainties*, pp. 260–72.

66 Ziadat, Adel A., *Western Science in the Arab World: The Impact of Darwinism, 1860–1930*
    (London: MacMillan, 1986, p. 25).

67 Ziadat: *Western Science in the Arab World*, p. 25.

68 Ommaya, Ayub Khan, 'Rise and decline of science in the Islamic world', *The World and I*
    16/7 (July 2001), p. 148.

69 Ommaya: 'Rise and decline of science in the Islamic world'.

70 Azzam, Mahfuz Ali, *Mabda' at-Tatawwur al-Hayawiyy lada Falasifat al-Islam* (*The Principle
    of Biological Evolution in the Works of the Classical Muslim Philosophers*) (Beirut, Lebanon:
    al-Mu'assassah al-Jami'iyyah li-d-Dirassat wa-l-Nashr wa-t-Tawzi', 1996).

71 al-Farabi, in *Ara' Ahl al-Madina al-Fadila* (*The Virtuous/Model City/State*), as cited by
    Azzam, pp. 12–16.

72 Azzam: *The Principle of Biological Evolution in the Works of the Classical Muslim Philosophers*,
    p. 90.

73 Azzam: *The Principle of Biological Evolution in the Works of the Classical Muslim Philosophers*,
    p. 91.

74 Azzam: *The Principle of Biological Evolution in the Works of the Classical Muslim Philosophers*,
    p. 120.

75 Ikhwan as-Safa, 'Rasa'il' ('Epistles'), cited in Azzam: *The Principle of Biological Evolution in
    the Works of the Classical Muslim Philosophers*, p. 172.

76 Azzam: *The Principle of Biological Evolution in the Works of the Classical Muslim Philosophers*,
    p. 47.

77 Azzam: *The Principle of Biological Evolution in the Works of the Classical Muslim Philosophers*,
    p. 13, citing Ibn Khaldun from his famous Muqaddima (Prolegomena).

78 Azzam: *The Principle of Biological Evolution in the Works of the Classical Muslim Philosophers*,
    pp. 117–8.

79 My own translation from Arabic.

80 For example, Ziadat: *Western Science in the Arab World*, p. 86.

81 al-Afghani, Jamal Eddine, 'Refutation of the Materialists', as quoted by Keddie, Nikki,
    Sayyid Jamal ad-Din al-Afghani: A Political Biography (Berkeley, 1972, p. 135).

82 Keddie: *Sayyid Jamal ad-Din al-Afghani*, p. 173.

83 Ziadat: *Western Science in the Arab World*, p. 90.

84 Related by Muhammad al-Mukhzumi, Khatirat *Jamal al-Din al-Afghani* (*The Thought of
    al-Afghani*) (Damascus, 1965, p. 115).

85 al-Jisr, Hussein, *al-Risala al-Hamidiyya fi Haqiqat al-Diana al-Islamiyya wa Haqiqat al-Shari'a
    al-Muhamadiyya* (*A Hamedian Essay on the Truthfulness of the Islamic Religion and the Truthfulness
    of the Islamic Canon Law*) (Beirut, 1887).

86 Majid, Abdul, 'The Muslim responses to evolution', Metanexus Views, 2002.

87 See Abdul Majid's article for a description of these thinkers' views on evolution and Islam.

88 Ziadat: *Western Science in the Arab World*, p. 121.

89 Ziadat: *Western Science in the Arab World*, p. 127.

90 Ziadat: *Western Science in the Arab World*, pp. 127–128.

91 For instance, Mansour al-'Abbadi, 'Qul Siru fil Ard fanDhuru kayfa bada-l-Khalq' ('Say walk on Earth and see how creation was begun'), available at http://www.science-islam.net/article.php3?id_article=804&lang=ar

92 Shahrour, Mohamad, *al-Kitab wal Qur'an: qira'a mu'asirah* (*The Book and the Koran: A Modern Reading*) (Damascus, 1990, pp. 281–5).

93 Shahrour: *al-Kitab wal Qur'an*, pp. 299–300.

94 Ayala: *Darwin's Gift to Science and Religion*, p. 105.

95 Yahya, Harun, *Allah Is Known Through Reason* (New Delhi, India: Goodword Books, 2003 (first published in 2000)).

96 Yahya: *Allah Is Known Through Reason*, p. 101.

97 Yahya: *Allah Is Known Through Reason*, p. 105.

98 Yahya: *Allah Is Known Through Reason*, p. 108.

99 Arnould, Jacques, *Dieu versus Darwin: Les créationnistes vont-ils triompher de la science?* (*God vs. Darwin: Will Creationists Defeat Science?*) (Paris: Albin Michel, 2007; Taner Edis, op. cit.).

100 Arnould: *God vs. Darwin*, p. 140.

101 Yahya: *Allah Is Known Through Reason*, p. 66.

102 Brunet, Michel et al., 'A new hominid from the upper miocene of Chad, Central Africa', *Nature* 418 (11 July 2002), p. 145.

103 Yahya, Harun, 'New fossil discovery sinks evolutionary theories' available ahttp://www.harunyahya.com/articles/70New_Fossil_Discovery_sci32.php

104 Gee, Henry, 'Face of yesterday – on the dramatic discovery of a seven-million-year-old hominid', The Guardian (11 July 2002).

105 Fortey, Richard, 'The Cambrian explosion exploded?', Science 293/5529 (20 July 2001).

106 Yahya, Harun, 'The collapse of Darwinism – part one: the fossil record refutes evolution', 17 August 2002, available at http://www.islamonline.net/english/Science/2002/08/article03.shtml; also available at http://www.harunyahya.com/articles/20evolution01.php.

107 Yahya: 'The Collapse of Darwinism'.

108 Yahya, Harun, 'Thermodynamics falsifies evolution' (17 December 2002), available at http://www.islamonline.net/english/Science/2002/12/article04.shtml

109 Yahya, Harun, 'Proteins challenge chance' (10 September 2002), available at http://198.65.147.194/English/Science/2002/09/article06.shtml; see also in Chapter 10 of 'The Evolution Deceit', available at: http://www.harunyahya.com/evolutiondeceit10.php

110 See Dessibourg, Olivier, 'Le créationnisme islamiste débarque en Suisse', Le Temps (29 March 2007); also, Dean, Cornelia, 'Islamic creationist and a book sent round the world', *New York Times* (17 July 2007).

111 al-Najami, Siham, 'Debating the origin of life', *Gulf News* (17 February 2007), p. 4.

112 Nasr, Seyyed Hossein, 'On the question of biological origins', *Islam & Science*, 4/2 (Winter 2006).

113 Most notably, Osman Bakar has published *Critique of Evolutionary Theory: A Collection of Essays* (Kuala Lumpur: Islamic Academy of Science and Nurin Enterprise, 1987).

114 Chahine, Abdes-Sabour, *Abi Adam* (*My Father, Adam*) first published in Cairo in 1998; I have used a version published in 2003.

115 Chahine: *My Father, Adam*, p. 11.

116 Chahine: *My Father, Adam*, p.

117 Hammad, Mohammad, 'Takhareef Abdes-Sabour Chahine' ('The ramblings of Chahine') *al-Araby magazine* (Cairo), 17 October 2004.

118 Golshani, Mehdi, 'Does science offer evidence of a transcendent reality and purpose?', *Science & Islam* 1/1 (June 2003), pp. 45–58.

ten

1 Robert Griffiths, quoted by Hugh Ross, in *The Creator and the Cosmos* 2nd ed. (Colorado Springs, CO: NavPress, 1995, p. 123).

2 The Université Interdisciplinaire de Paris, http://www.uip.edu/

3 The project 'Science and Religion in Islam' (http://www.science-islam.net/) was funded by the John Templeton Foundation and managed by the Université Interdisciplinaire de Paris.

4 For example, Nichols, Terence L., 'Miracles in science and theology', *Zygon* 37/3 (Sept. 2002), pp. 703–15; Ward, Keith, 'Believing in miracles', *Zygon* 37/3 (Sept. 2002), pp. 741–50.

5 Testard-Vaillant, Philippe, '67 miracles en 150 ans', *Science et Vie Hors Serie* 236 (Sept. 2006), pp. 28–35.

6 Poll conducted by HCD Research and the Louis Finkelstein Institute for Religious and Social Studies of the Jewish Theological Seminary in New York, December 2004.

7 Vaudaine, Sarah, 'Des guérisons bientot explicables?' ('Healings soon explainable?'), *Science et Vie Hors Serie* 236 (Sept. 2006), pp. 75–81.

8 'La force de guérir', interview given to Gwen-Haël Denigot, *Science et Vie Hors Serie* 236 (Sept. 2006), pp. 97–101.

9 Vennetier, Perrine, 'Effet Placebo: le côté obscure de la force thérapeutique', *Science et Vie Hors Serie* 236 (Sept. 2006), pp. 102–7.

10 Nichols: 'Miracles in science and theology'.

11 Ward: 'Believing in miracles'.

12 Ahmed, Absar, 'Miracles – a philosophical analysis', *Science-Religion Dialogue* (Fall 2003/Spring 2004), available at: http://www.hssrd.org/journal/fallsummer2003-2004/english/miracles.htm

13 Khan, Syed Ahmad, *Tafsir-ul-Qur'an*, vol. III, p. 28, cited by Absar Ahmed.

14 Golshani, Mehdi, *The Holy Qur'an and the Sciences of Nature: A Theological Reflection* (New York: Global Scholarly Publications, 2003, pp. 309–10).

15 Vaudaine: 'Healings soon explainable?'

16 Benson, Herbert, 'How large is faith', in John Marks Templeton (ed) *How Large is God* (Philadelphia & London: Templeton Foundation Press, 1997).

17 All the Benson quotes are taken from his essay 'How Large is Faith'.

18 The description of the experiment and its results have been taken from the *New York Times* report, 'Long-awaited medical study questions the power of prayer' by Benedict Carey, 31 March 2006.

19 *Zygon* 37/3 (Sept. 2002), p. 717.

20 *Theology and Science* 4/1 (March 2006), p. 51.

21 *Ratio* 19/4 (Dec. 2006), p. 495.

22 *International Journal for Philosophy of Religion* 57/1 (Feb 2005), p. 67.

23 *The Journal of Religion* 84/4 (Oct 2004), p. 648.

24 *Modern Theology* 20/4 (Oct 2004), p. 613.

25 Peacocke, Arthur, *Paths from Science towards God* (Oxford, UK: OneWorld, 2001, p. 45).

26 Peacocke: *Paths from Science towards God*, p. 59. Peacocke justifies his 'self-limited omnipotence' thesis by: 'in the Christian understanding, divine omnipotence has always been regarded as limited [. . .] by God's nature as Love.'

27 Peacocke: *Paths from Science towards God*, p. 45.

28 Peacocke: *Paths from Science towards God*, p. 102.

29 Peacocke: *Paths from Science towards God*, p. 106.

30 Berry, R. J., 'Divine action, expected and unexpected', *Zygon* 37/3 (Sept. 2002), pp. 717–27.

31 A search through the aforementioned electronic library database for 'divine action' and 'Islam' or 'Islamic' in the title turned up no results.

32 *The Muslim World* 94/1 (Jan. 2004), p. 65.

33 This paper was delivered at the Third International Conference on Islamic Philosophy at Cairo University in 1998; only its abstract is available through BiblioIslam.net (http://biblioislam.net/Elibrary/arabic/library/card.asp?tblID=3&id=648).

34 See, for example, Polkinghorne, John C., *The Faith of a Physicist: Reflections of a Bottom-up Thinker* (Princeton, NJ: Princeton University Press, 1994).

35 Russell, Robert J., 'An appreciative response to Niels Henrik Gregersen's JKR Research Conference Lecture', *Theology and Science* 4/2 (July 2006), p. 129.

36 Aspect, Alain, PhD thesis, *Three Experimental Tests of Bell Inequalities by the Measurement of Polarization Correlations between Photons* (Orsay, 1983).

37 Gisin, Nicholas, Brendel, J., Tittel W. and Zbinden, H., Quantum correlation over more than 10 km, *Optics and Photonics News* (highlights in Optics 1998) (9 December 1998), p. 41.

38 Boutamina, N. E., *L'Islam fondateur de la Science* (Paris: al-Bouraq, 2006, pp. 383–5).

39 This is Taner Edis's thesis, which he boldly presents in his recent book *An Illusion of Harmony: Science and Religion in Islam* (Amherst, NY: Prometheus Books, 2007).

40 Ashraf, Faheem, 'Islamic concept of creation of universe: Big Bang and science-religion interaction', *Science-Religion Dialogue* (Spring 2003), available at http://www.hssrd.org/journal/spring2003/creationunverse.htm

41 Sardar, Ziauddin, 'How to take Islam back to reason', *New Statesman* 17/801 (5 April 2004), p. 28.

42 Sardar: 'How to take Islam back to reason'.

43 Clayton, Philip, 'The state of the international religion-science discussion today: "science and the spiritual quest": a model for the renewal of the religion-science discourse', *Islam & Science* (Summer 2004).

44 Moore, James F., 'Interfaith Dialogue and the Science-and-Religion Discussion', *Zygon* 37/1 (March 2002), pp. 37–43.

index